Introduction to Mechatronics and Measurement Systems

Fourth Edition

David G. Alciatore
Department of Mechanical Engineering
Colorado State University

Michael B. Histand
Professor Emeritus
Department of Mechanical Engineering
Colorado State University

Mc Graw Hill

Connect
Learn
Succeed™

The McGraw-Hill Companies

INTRODUCTION TO MECHATRONICS AND MEASUREMENT SYSTEMS, FOURTH EDITION

Published by McGraw-Hill, a business unit of The McGraw-Hill Companies, Inc., 1221 Avenue of the Americas, New York, NY 10020. Copyright © 2012 by The McGraw-Hill Companies, Inc. All rights reserved. Previous editions © 2007, 2003 and 1999. No part of this publication may be reproduced or distributed in any form or by any means, or stored in a database or retrieval system, without the prior written consent of The McGraw-Hill Companies, Inc., including, but not limited to, in any network or other electronic storage or transmission, or broadcast for distance learning.

Some ancillaries, including electronic and print components, may not be available to customers outside the United States.

This book is printed on acid-free paper.

2 3 4 5 6 7 8 9 0 DOC/DOC 1 0 9 8 7 6 5 4 3 2

ISBN 978-0-07-338023-0
MHID 0-07-338023-7

Vice President & Editor-in-Chief: *Marty Lange*
Vice President EDP/Central Publishing Services: *Kimberly Meriwether David*
Publisher: *Raghothaman Srinivasan*
Executive Editor: *Bill Stenquist*
Development Editor: *Lorraine Buczek*
Marketing Manager: *Curt Reynolds*
Project Manager: *Melissa M. Leick*
Design Coordinator: *Margarite Reynolds*
Cover Designer: *Studio Montage, St. Louis, Missouri*
Cover Images: *Burke/Triolo/Brand X Pictures/Jupiterimages; © Chuck Eckert/Alamy; Royalty-Free/CORBIS; Imagestate Media (John Foxx); Chad Baker/Getty Images (clockwise, left to right)*
Buyer: *Nicole Baumgartner*
Media Project Manager: *Balaji Sundararaman*
Compositor: *Laserwords Private Limited*
Typeface: *10/12 Times Roman*
Printer: *R. R. Donnelley*

All credits appearing on page or at the end of the book are considered to be an extension of the copyright page.

Library of Congress Cataloging-in-Publication Data

Alciatore, David G.
  Introduction to mechatronics and measurement systems / David G. Alciatore.—4th ed.
      p. cm.
  Includes index.
  ISBN 978-0-07-338023-0
  1. Mechatronics. 2. Measurement. I. Title.
  TJ163.12.H57 2011
  621—dc22
                                                                    2010052867

www.mhhe.com

CONTENTS

CLASS DISCUSSION ITEMS

EXAMPLES

DESIGN EXAMPLES

THREADED DESIGN EXAMPLES

MCGRAW-HILL DIGITAL OFFERINGS INCLUDE:

McGraw-Hill Create™

Craft your teaching resources to match the way you teach! With McGraw-Hill Create™, www.mcgrawhill-create.com, you can easily rearrange chapters, combine material from other content sources, and quickly upload content you have written like your course syllabus or teaching notes. Find the content you need in Create by searching through thousands of leading McGraw-Hill textbooks. Arrange your book to fit your teaching style. Create even allows you to personalize your book's appearance by selecting the cover and adding your name, school, and course information. Order a Create book and you'll receive a complimentary print review copy in 3–5 business days or a complimentary electronic review copy (eComp) via email in minutes. Go to www.mcgrawhillcreate.com today and register to experience how McGraw-Hill Create™ empowers you to teach your students your way.

McGraw-Hill Higher Education and Blackboard have teamed up.

Blackboard, the Web-based course-management system, has partnered with McGraw-Hill to better allow students and faculty to use online materials and activities to complement face-to-face teaching. Blackboard features exciting social learning and teaching tools that foster more logical, visually impactful and active learning opportunities for students. You'll transform your closed-door classrooms into communities where students remain connected to their educational experience 24 hours a day.

This partnership allows you and your students access to McGraw-Hill's Create™ right from within your Blackboard course–all with one single sign-on. McGraw-Hill and Blackboard can now offer you easy access to industry leading technology and content, whether your campus hosts it, or we do. Be sure to ask your local McGraw-Hill representative for details.

Electronic Textbook Options

This text is offered through CourseSmart for both instructors and students. CourseSmart is an online resource where students can purchase the complete text online at almost half the cost of a traditional text. Purchasing the eTextbook allows students to take advantage of CourseSmart's web tools for learning, which include full text search, notes and highlighting, and email tools for sharing notes between classmates. To learn more about CourseSmart options, contact your sales representative or visit www.CourseSmart.com.

APPROACH

The formal boundaries of traditional engineering disciplines have become fuzzy following the advent of integrated circuits and computers. Nowhere is this more evident than in mechanical and electrical engineering, where products today include an assembly of interdependent electrical and mechanical components. The field of mechatronics has broadened the scope of the traditional field of electromechanics. *Mechatronics* is defined as the field of study involving the analysis, design, synthesis, and selection of systems that combine electronic and mechanical components with modern controls and microprocessors.

This book is designed to serve as a text for (1) a modern instrumentation and measurements course, (2) a hybrid electrical and mechanical engineering course replacing traditional circuits and instrumentation courses, (3) a stand-alone mechatronics course, or (4) the first course in a mechatronics sequence. The second option, the hybrid course, provides an opportunity to reduce the number of credit hours in a typical mechanical engineering curriculum. Options 3 and 4 could involve the development of new interdisciplinary courses and curricula.

Currently, many curricula do not include a mechatronics course but include some of the elements in other, more traditional courses. The purpose of a course in mechatronics is to provide a focused interdisciplinary experience for undergraduates that encompasses important elements from traditional courses as well as contemporary developments in electronics and computer control. These elements include measurement theory, electronic circuits, computer interfacing, sensors, actuators, and the design, analysis, and synthesis of mechatronic systems. This interdisciplinary approach is valuable to students because virtually every newly designed engineering product is a mechatronic system.

NEW TO THE FOURTH EDITION

The fourth edition of *Introduction of Mechatronics and Measurement Systems* has been improved, updated, and expanded beyond the previous edition. Additions and new features include:

- New sections throughout the book dealing with the "practical considerations" of mechatronic system design and implementation, including circuit construction, electrical measurements, power supply options, general integrated circuit design, and PIC microcontroller circuit design.
- Expanded section on LabVIEW data acquisition, including a complete music sampling example with Web resources.

- More website resources, including Internet links and online video demonstrations, cited and described throughout the book.
- Expanded section on Programmable Logic Controllers (PLCs) including the basics of ladder logic with examples.
- Interesting new clipart images next to each Class Discussion Item to help provoke thought, inspire student interest, and improve the visual look of the book.
- Additional end-of-chapter questions throughout the book provide more homework and practice options for professors and students.
- Corrections and many small improvements throughout the entire book.

CONTENT

Chapter 1 introduces mechatronic and measurement system terminology. Chapter 2 provides a review of basic electrical relations, circuit elements, and circuit analysis. Chapter 3 deals with semiconductor electronics. Chapter 4 presents approaches to analyzing and characterizing the response of mechatronic and measurement systems. Chapter 5 covers the basics of analog signal processing and the design and analysis of operational amplifier circuits. Chapter 6 presents the basics of digital devices and the use of integrated circuits. Chapter 7 provides an introduction to microcontroller programming and interfacing, and specifically covers the PIC microcontroller and PicBasic Pro programming. Chapter 8 deals with data acquisition and how to couple computers to measurement systems. Chapter 9 provides an overview of the many sensors common in mechatronic systems. Chapter 10 introduces a number of devices used for actuating mechatronic systems. Finally, Chapter 11 provides an overview of mechatronic system control architectures and presents some case studies. Chapter 11 also provides an introduction to control theory and its role in mechatronic system design. The appendices review the fundamentals of unit systems, statistics, error analysis, and mechanics of materials to support and supplement measurement systems topics in the book.

It is practically impossible to write and revise a large textbook without introducing errors by mistake, despite the amount of care exercised by authors, editors, and typesetters. When errors are found, they will be published on the book website at: **www.mechatronics.colostate.edu/book/corrections_4th_edition.html.** You should visit this page now to see if there are any corrections to record in your copy of the book. If you find any additional errors, please report them to *David.Alciatore@ colostate.edu* so they can be posted for the benefit of others. Also, please let me know if you have suggestions or requests concerning improvements for future editions of the book. Thank you.

LEARNING TOOLS

Class discussion items (CDIs) are included throughout the book to serve as thought-provoking exercises for the students and instructor-led cooperative learning activities in the classroom. They can also be used as out-of-class homework assignments to supplement the questions and exercises at the end of each chapter. Hints and partial answers for many of the CDIs are available on the book website at **www. mechatronics.colostate.edu.** Analysis and design examples are also provided throughout the book to improve a student's ability to apply the material. To enhance student learning, carefully designed laboratory exercises coordinated with the lectures should accompany a course using this text. A supplemental Laboratory Exercises Manual is available for this purpose (see **www.mechatronics.colostate.edu/ lab_book.html** for more information). The combination of class discussion items, design examples, and laboratory exercises exposes a student to a real-world practical approach and provides a useful framework for future design work.

In addition to the analysis Examples and design-oriented Design Examples that appear throughout the book, Threaded Design Examples are also included. The examples are mechatronic systems that include microcontrollers, input and output devices, sensors, actuators, support electronics, and software. The designs are presented incrementally as the pertinent material is covered throughout the chapters. This allows the student to see and appreciate how a complex design can be created with a divide-and-conquer approach. Also, the threaded designs help the student relate to and value the circuit fundamentals and system response topics presented early in the book. The examples help the students see the "big picture" through interesting applications beginning in Chapter 1.

ACKNOWLEDGMENTS

To ensure the accuracy of this text, it has been class-tested at Colorado State University and the University of Wyoming. We'd like to thank all of the students at both institutions who provided us valuable feedback throughout this process. In addition, we'd like to thank our many reviewers for their valuable input.

YangQuan Chen   *Utah State University*
Meng-Sang Chew   *Lehigh University*
Mo-Yuen Chow   *North Carolina State University*
Burford Furman   *San José State University*
Venkat N. Krovi   *State University of New York- Buffalo*
Satish Nair   *University of Missouri*
Ramendra P. Roy   *Arizona State University*
Ahmad Smaili   *Hariri Canadian University, Lebanon*
David Walrath   *University of Wyoming*

SUPPLEMENTAL MATERIALS ARE AVAILABLE ONLINE AT:
www.mechatronics.colostate.edu

Cross-referenced visual icons appear throughout the book to indicate where additional information is available on the book website at **www.mechatronics.colostate.edu.**

Shown below are the icons used, along with a description of the resources to which they point:

Video Demo

Indicates where an online video demonstration is available for viewing. The online videos are Windows Media (WMV) files viewable in an Internet browser. The clips show and describe electronic components, mechatronic device and system examples, and laboratory exercise demonstrations.

Internet Link

Indicates where a link to additional Internet resources is available on the book website. These links provide students and instructors with reliable sources of information for expanding their knowledge of certain concepts.

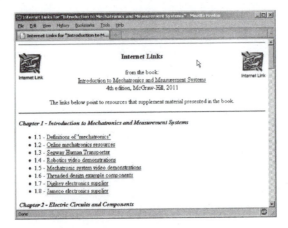

Indicates where MathCAD files are available for performing analysis calculations. The files can be edited to perform similar and expanded analyses. PDF versions are also posted for those who don't have access to MathCAD software.

MathCAD Example

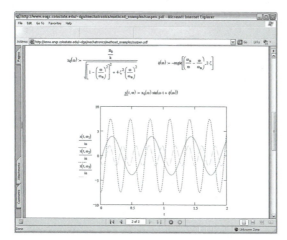

Indicates where a laboratory exercise is available in the supplemental Laboratory Exercises Manual that parallels the book. The manual provides useful hands-on laboratory exercises that help reinforce the material in the book and that allow students to apply what they learn. Resources and short video demonstrations of most of the exercises are available on the book website. For information about the Laboratory Exercises Manual, visit **www.mechatronics.colostate.edu/lab_book.html.**

Lab Exercise

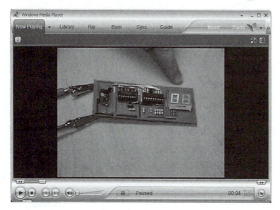

ADDITIONAL SUPPLEMENTS

More information, including a recommended course outline, a typical laboratory syllabus, Class Discussion Item hints, and other supplemental material, is available on the book website.

In addition, a complete password-protected Solutions Manual containing solutions to all end-of-chapter problems is available at the McGraw-Hill book website at **www.mhhe.com/alciatore.**

These supplemental materials help students and instructors apply concepts in the text to laboratory or real-world exercises, enhancing the learning experience.

CHAPTER 1

Introduction

After you read, discuss, study, and apply ideas in this chapter, you will be able to:

1. Define mechatronics and appreciate its relevance to contemporary engineering design

2. Identify a mechatronic system and its primary elements

3. Define the elements of a general measurement system

1.1 MECHATRONICS

Mechanical engineering, as a widespread professional practice, experienced a surge of growth during the early 19th century because it provided a necessary foundation for the rapid and successful development of the industrial revolution. At that time, mines needed large pumps never before seen to keep their shafts dry, iron and steel mills required pressures and temperatures beyond levels used commercially until then, transportation systems needed more than real horse power to move goods; structures began to stretch across ever wider abysses and to climb to dizzying heights, manufacturing moved from the shop bench to large factories; and to support these technical feats, people began to specialize and build bodies of knowledge that formed the beginnings of the engineering disciplines.

The primary engineering disciplines of the 20th century—mechanical, electrical, civil, and chemical—retained their individual bodies of knowledge, textbooks, and professional journals because the disciplines were viewed as having mutually exclusive intellectual and professional territory. Entering students could assess their individual intellectual talents and choose one of the fields as a profession. We are now witnessing a new scientific and social revolution known as the information revolution, where engineering specialization ironically seems to be simultaneously focusing and diversifying. This contemporary revolution was spawned by the engineering development of semiconductor electronics, which has driven an information and communications explosion that is transforming human life. To practice engineering today, we must understand new ways to process information and be able to utilize semiconductor electronics within our products, no matter what label we put on ourselves as practitioners. Mechatronics is one of the new and exciting fields on the engineering landscape, subsuming parts of traditional engineering fields and requiring a broader approach to the design of systems that we can formally call mechatronic systems.

Then what precisely is mechatronics? The term **mechatronics** is used to denote a rapidly developing, interdisciplinary field of engineering dealing with the design of products whose function relies on the integration of mechanical and electronic components coordinated by a control architecture. Other definitions of the term "mechatronics" can be found online at Internet Link 1.1. The word mechatronics was coined in Japan in the late 1960s, spread through Europe, and is now commonly used in the United States. The primary disciplines important in the design of mechatronic systems include mechanics, electronics, controls, and computer engineering. A mechatronic system engineer must be able to design and select analog and digital circuits, microprocessor-based components, mechanical devices, sensors and actuators, and controls so that the final product achieves a desired goal.

Mechatronic systems are sometimes referred to as smart devices. While the term smart is elusive in precise definition, in the engineering sense we mean the inclusion of elements such as logic, feedback, and computation that in a complex design may appear to simulate human thinking processes. It is not easy to compartmentalize mechatronic system design within a traditional field of engineering because such design draws from knowledge across many fields. The mechatronic system designer must be a generalist, willing to seek and apply knowledge from a broad range of sources. This may intimidate the student at first, but it offers great benefits for individuality and continued learning during one's career.

Today, practically all mechanical devices include electronic components and some type of computer monitoring or control. Therefore, the term mechatronic system encompasses a myriad of devices and systems. Increasingly, microcontrollers are embedded in electromechanical devices, creating much more flexibility and control possibilities in system design. Examples of mechatronic systems include an aircraft flight control and navigation system, automobile air bag safety system and antilock brake systems, automated manufacturing equipment such as robots and numerically controlled (NC) machine tools, smart kitchen and home appliances such as bread machines and clothes washing machines, and even toys.

Figure 1.1 illustrates all the components in a typical mechatronic system. The actuators produce motion or cause some action; the sensors detect the state of the system parameters, inputs, and outputs; digital devices control the system; conditioning and interfacing circuits provide connections between the control circuits and the input/output devices; and graphical displays provide visual feedback to users. The subsequent chapters provide an introduction to the elements listed in this block diagram and describe aspects of their analysis and design. At the beginning of each chapter, the elements presented are emphasized in a copy of Figure 1.1. This will help you maintain a perspective on the importance of each element as you gradually build your capability to design a mechatronic system. Internet Link 1.2 provides links to various vendors and sources of information for researching and purchasing different types of mechatronics components.

Internet Link

1.1 Definitions of "mechatronics"

Internet Link

1.2 Online mechatronics resources

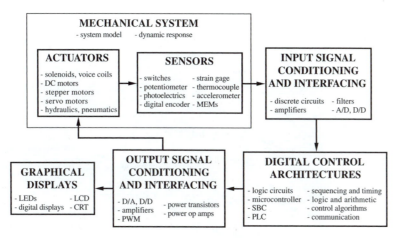

Figure 1.1 Mechatronic system components.

Internet Link

1.3 Segway human transporter

Video Demo

1.1 Adept One robot demonstration

1.2 Adept One robot internal design and construction

1.3 Honda Asimo Raleigh, NC, demonstration

1.4 Sony "Qrio" Japanese dance demo

1.5 Inkjet printer components

Example 1.1 describes a good example of a mechatronic system—an office copy machine. All of the components in Figure 1.1 can be found in this common piece of office equipment. Other mechatronic system examples can be found on the book website. See the Segway Human Transporter at Internet Link 1.3, the Adept pick-and-place industrial robot in Video Demos 1.1 and 1.2, the Honda Asimo and Sony Qrio humanoid-like robots in Video Demos 1.3 and 1.4, and the inkjet printer in Video Demo 1.5. As with the copy machine in Example 1.1, these robots and printer contain all of the mechatronic system components shown in Figure 1.1. Figure 1.2 labels the specific components mentioned in Video Demo 1.5. Video demonstrations of many more robotics-related devices can be found

Mechatronic System—Copy Machine | EXAMPLE 1.1

An office copy machine is a good example of a contemporary mechatronic system. It includes analog and digital circuits, sensors, actuators, and microprocessors. The copying process works as follows: The user places an original in a loading bin and pushes a button to start the process; the original is transported to the platen glass; and a high intensity light source scans the original and transfers the corresponding image as a charge distribution to a drum. Next, a blank piece of paper is retrieved from a loading cartridge, and the image is transferred onto the paper with an electrostatic deposition of ink toner powder that is heated to bond to the paper. A sorting mechanism then optionally delivers the copy to an appropriate bin.

Analog circuits control the lamp, heater, and other power circuits in the machine. Digital circuits control the digital displays, indicator lights, buttons, and switches forming the user interface. Other digital circuits include logic circuits and microprocessors that coordinate all of the functions in the machine. Optical sensors and microswitches detect the presence or absence of paper, its proper positioning, and whether or not doors and latches are in their correct positions. Other sensors include encoders used to track motor rotation. Actuators include servo and stepper motors that load and transport the paper, turn the drum, and index the sorter.

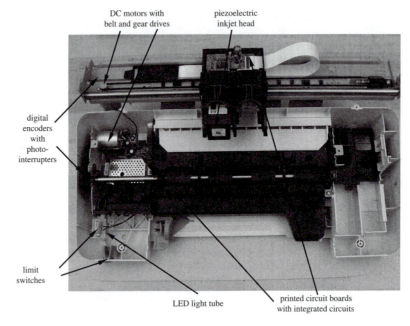

Figure 1.2 Inkjet printer components.

Internet Link

1.4 Robotics
video
demonstrations

1.5 Mechatronic
system video
demonstrations

at Internet Link 1.4, and demonstrations of other mechatronic system examples can be found at Internet Link 1.5.

■ **CLASS DISCUSSION ITEM 1.1**
Household Mechatronic Systems

What typical household items can be characterized as mechatronic systems? What components do they contain that help you identify them as mechatronic systems? If an item contains a microprocessor, describe the functions performed by the microprocessor.

1.2 MEASUREMENT SYSTEMS

A fundamental part of many mechatronic systems is a **measurement system** composed of the three basic parts illustrated in Figure 1.3. The **transducer** is a sensing device that converts a physical input into an output, usually a voltage. The **signal processor** performs filtering, amplification, or other signal conditioning on the transducer output. The term **sensor** is often used to refer to the transducer or to the combination of transducer and signal processor. Finally, the **recorder** is an instrument, a computer, a hard-copy device, or simply a display that maintains the sensor data for online monitoring or subsequent processing.

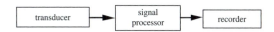

Figure 1.3 Elements of a measurement system.

These three building blocks of measurement systems come in many types with wide variations in cost and performance. It is important for designers and users of measurement systems to develop confidence in their use, to know their important characteristics and limitations, and to be able to select the best elements for the measurement task at hand. In addition to being an integral part of most mechatronic systems, a measurement system is often used as a stand-alone device to acquire data in a laboratory or field environment.

Measurement System—Digital Thermometer | EXAMPLE 1.2

The following figure shows an example of a measurement system. The thermocouple is a transducer that converts temperature to a small voltage; the amplifier increases the magnitude of the voltage; the A/D (analog-to-digital) converter is a device that changes the analog signal to a coded digital signal; and the LEDs (light emitting diodes) display the value of the temperature.

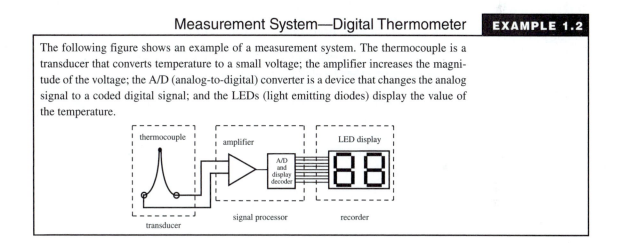

Supplemental information important to measurement systems and analysis is provided in Appendix A. Included are sections on systems of units, numerical precision, and statistics. You should review this material on an as-needed basis.

1.3 THREADED DESIGN EXAMPLES

Throughout the book, there are Examples, which show basic analysis calculations, and Design Examples, which show how to select and synthesize components and subsystems. There are also three more complicated Threaded Design Examples, which build upon new topics as they are covered, culminating in complete mechatronic systems by the end. These designs involve systems for controlling the position and speed of different types of motors in various ways. Threaded Design Examples A.1, B.1, and C.1 introduce each thread. All three designs incorporate components important in mechatronic systems: microcontrollers, input devices, output devices, sensors, actuators, and support electronics and software. Please read through the following information and watch the introductory videos. It will also be helpful to watch the videos again when follow-on pieces are presented so that you can see how everything fits in the "big picture." The list of Threaded Design Example citations at the beginning of the book, with the page numbers, can be useful for looking ahead or reflecting back as new portions are presented.

All of the components used to build the systems in all three threaded designs are listed at Internet Link 1.6, along with descriptions and price information. Most of the parts were purchased through Digikey Corporation (see Internet Link 1.7) and Jameco Electronics Corporation (see Internet Link 1.8), two of the better online suppliers of electronic parts. By entering part numbers from Internet Link 1.6 at the supplier websites, you can access technical datasheets for each product.

THREADED DESIGN EXAMPLE

A.1   *DC motor power-op-amp speed controller—Introduction*

Internet Link

1.6 Threaded design example components

1.7 Digikey electronics supplier

1.8 Jameco electronics supplier

Video Demo

1.6 DC motor power-op-amp speed controller

This design example deals with controlling the rotational speed of a direct current (DC) permanent magnet motor. Figure 1.4 illustrates the major components and interconnections in the system. The light-emitting diode (LED) provides a visual cue to the user that the microcontroller is running properly. The speed input device is a potentiometer (or pot), which is a variable resistor. The resistance changes as the user turns the knob on top of the pot. The pot can be wired to produce a voltage input. The voltage signal is applied to a microcontroller (basically a small computer on a single integrated circuit) to control a DC motor to rotate at a speed proportional to the voltage. Voltage signals are "analog" but microcontrollers are "digital," so we need analog-to-digital (A/D) and digital-to-analog (D/A) converters in the system to allow us to communicate between the analog and digital components. Finally, because a motor can require significant current, we need a power amplifier to boost the voltage and source the necessary current. Video Demo 1.6 shows a demonstration of the complete working system shown in Figure 1.5.

With all three Threaded Design Examples (A, B, and C), as you progress sequentially through the chapters in the book you will gain fuller understanding of the components in the design.

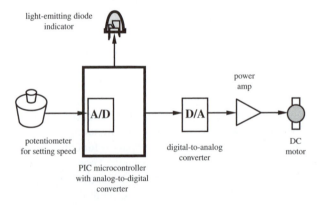

Figure 1.4 Functional diagram of the DC motor speed controller.

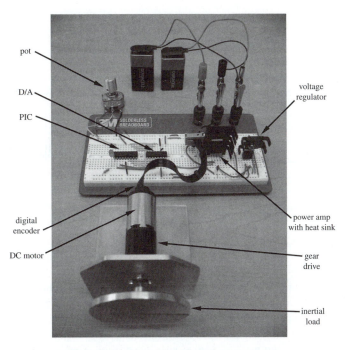

Figure 1.5 Photograph of the power-amp speed controller.

Note that the PIC microcontroller (with the A/D) and the external D/A converter are not actually required in this design, in its current form. The potentiometer voltage output could be attached directly to the power amp instead, producing the same functionality. The reason for including the PIC (with A/D) and the D/A components is to show how these components can be interfaced within an analog system (this is useful to know in many applications). Also, the design serves as a platform for further development, where the PIC can be used to implement feedback control and a user interface, in a more complex design. An example where you might need the microcontroller in the loop is in robotics or numerically controlled mills and lathes, where motors are often required to follow fairly complex motion profiles in response to inputs from sensors and user programming, or from manual inputs.

THREADED DESIGN EXAMPLE

Stepper motor position and speed controller—Introduction   B.1

This design example deals with controlling the position and speed of a stepper motor, which can be commanded to move in discrete angular increments. Stepper motors are useful in position indexing applications, where you might need to move parts or tools to and from various fixed positions (e.g., in an automated assembly or manufacturing line). Stepper motors are also useful in accurate speed control applications (e.g., controlling the spindle speed of a computer hard-drive or DVD player), where the motor speed is directly proportional to the step rate.

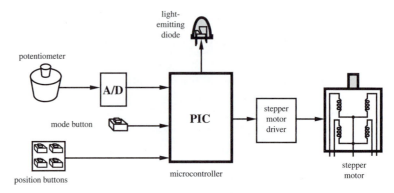

Figure 1.6 Functional diagram of the stepper motor position and speed controller.

Figure 1.6 shows the major components and interconnections in the system. The input devices include a pot to control the speed manually, four buttons to select predefined positions, and a mode button to toggle between speed and position control. In position control mode, each of the four position buttons indexes the motor to specific angular positions relative to the starting point (0°, 45°, 90°, 180°). In speed control mode, turning the pot clockwise (counterclockwise) increases (decreases) the speed. The LED provides a visual cue to the user to indicate that the PIC is cycling properly. As with Threaded Design Example A, an A/D converter is used to convert the pot's voltage to a digital value. A microcontroller uses that value to generate signals for a stepper motor driver circuit to make the motor rotate.

Video Demo 1.7 shows a demonstration of the complete working system shown in Figure 1.7. As you progress through the book, you will learn about the different elements in this design.

Video Demo

1.7 Stepper motor position and speed controller

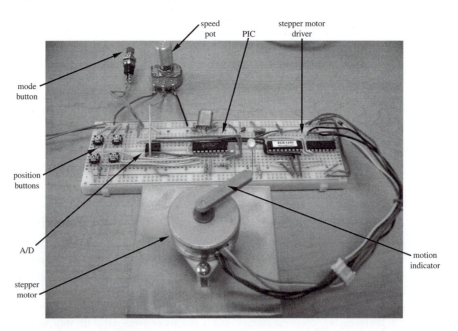

Figure 1.7 Photograph of the stepper motor position and speed controller.

DC motor position and speed controller—Introduction `C.1`

This design example illustrates control of position and speed of a permanent magnet DC motor. Figure 1.8 shows the major components and interconnections in the system. A numerical keypad enables user input, and a liquid crystal display (LCD) is used to display messages and a menu-driven user interface. The motor is driven by an H-bridge, which allows the voltage applied to the motor (and therefore the direction of rotation) to be reversed. The H-bridge also allows the speed of the motor to be easily controlled by pulse-width modulation (PWM), where the power to the motor is quickly switched on and off at different duty cycles to change the average effective voltage applied.

A digital encoder attached to the motor shaft provides a position feedback signal. This signal is used to adjust the voltage signal to the motor to control its position or speed. This is called a servomotor system because we use feedback from a sensor to control the motor. Servomotors are very important in automation, robotics, consumer electronic devices, flow-control valves, and office equipment, where mechanisms or parts need to be accurately positioned or moved at certain speeds. Servomotors are different from stepper motors (see Threaded Design Example B.1) in that they move smoothly instead of in small incremental steps.

Two PIC microcontrollers are used in this design because there are a limited number of input/output pins available on a single chip. The main (master) PIC gets input from the user, drives the LCD, and sends the PWM signal to the motor. The secondary (slave) PIC monitors the digital encoder and transmits the position signal back to the master PIC upon command via a serial interface.

Video Demo 1.8 shows a demonstration of the complete working system shown in Figure 1.9. You will learn about each element of the design as you proceed sequentially through the book.

Video Demo

1.8 DC motor position and speed controller

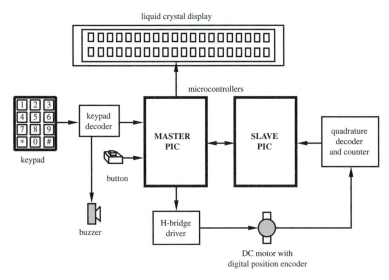

Figure 1.8 Functional diagram for the DC motor position and speed controller.

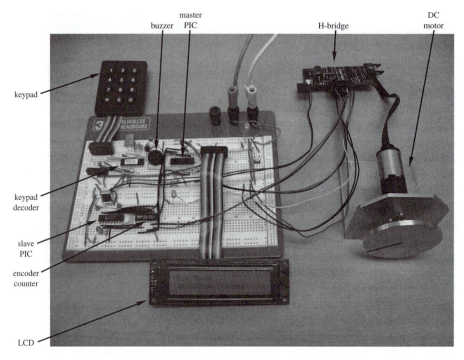

Figure 1.9 Photograph of the DC motor position and speed controller.

BIBLIOGRAPHY

Alciatore, D. and Histand, M., "Mechatronics at Colorado State University," *Journal of Mechatronics,* Mechatronics Education in the United States issue, Pergamon Press, May, 1995.

Alciatore, D. and Histand, M., "Mechatronics and Measurement Systems Course at Colorado State University," *Proceedings of the Workshop on Mechatronics Education,* pp. 7–11, Stanford, CA, July, 1994.

Ashley, S., "Getting a Hold on Mechatronics," *Mechanical Engineering,* pp. 60–63, ASME, New York, May, 1997.

Beckwith, T., Marangoni, R., and Lienhard, J., *Mechanical Measurements,* Addison-Wesley, Reading, MA, 1993.

Craig, K., "Mechatronics System Design at Rensselaer," *Proceedings of the Workshop on Mechatronics Education,* pp. 24–27, Stanford, CA, July, 1994.

Doeblin, E., *Measurement Systems Applications and Design,* 4th edition, McGraw-Hill, New York, 1990.

Morley, D., "Mechatronics Explained," *Manufacturing Systems,* p. 104, November, 1996.

Shoureshi, R. and Meckl, P., "Teaching MEs to Use Microprocessors," *Mechanical Engineering,* v. 166, n. 4, pp. 71–74, April, 1994.

CHAPTER 2

Electric Circuits and Components

T his chapter reviews the fundamentals of basic electrical components and discrete circuit analysis techniques. These topics are important in understanding and designing all elements in a mechatronic system, especially discrete circuits for signal conditioning and interfacing. ■

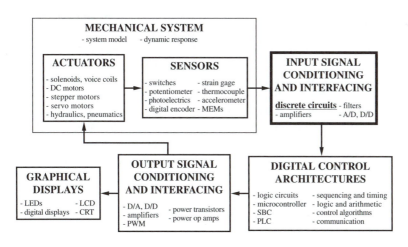

CHAPTER OBJECTIVES

After you read, discuss, study, and apply ideas in this chapter, you will:

1. Understand differences among resistance, capacitance, and inductance

2. Be able to define Kirchhoff's voltage and current laws and apply them to passive circuits that include resistors, capacitors, inductors, voltage sources, and current sources

3. Know how to apply models for ideal voltage and current sources

4. Be able to predict the steady-state behavior of circuits with sinusoidal inputs

5. Be able to characterize the power dissipated or generated by a circuit

6. Be able to predict the effects of mismatched impedances

7. Understand how to reduce noise and interference in electrical circuits

8. Appreciate the need to pay attention to electrical safety and to ground components properly

9. Be aware of several practical considerations that will help you assemble actual circuits and make them function properly and reliably

10. Know how to make reliable voltage and current measurements

2.1 INTRODUCTION

Practically all mechatronic and measurement systems contain electrical circuits and components. To understand how to design and analyze these systems, a firm grasp of the fundamentals of basic electrical components and circuit analysis techniques is a necessity. These topics are fundamental to understanding everything else that follows in this book.

When electrons move, they produce an electrical current, and we can do useful things with the energized electrons. The reason they move is that we impose an electrical field that imparts energy by doing work on the electrons. A measure of the electric field's potential is called **voltage.** It is analogous to potential energy in a gravitational field. We can think of voltage as an "across variable" between two points in the field. The resulting movement of electrons is the current, a "through variable," that moves through the field. When we measure current through a circuit, we place a meter in the circuit and let the current flow through it. When we measure a voltage, we place two conducting probes on the points across which we want to measure the voltage. Voltage is sometimes referred to as **electromotive force,** or **emf.**

Current is defined as the time rate of flow of charge:

$$I(t) = \frac{dq}{dt} \tag{2.1}$$

where I denotes current and q denotes quantity of charge. The charge is provided by the negatively charged electrons. The SI unit for current is the **ampere** (A), and charge is measured in **coulombs** ($C = A \cdot s$). When voltage and current in a circuit are constant (i.e., independent of time), their values and the circuit are referred to as **direct current,** or DC. When the voltage and current vary with time, usually sinusoidally, we refer to their values and the circuit as **alternating current,** or AC.

An electrical circuit is a closed loop consisting of several conductors connecting electrical components. Conductors may be interrupted by components called switches. Some simple examples of valid circuits are shown in Figure 2.1.

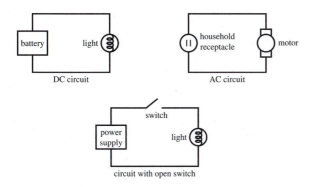

Figure 2.1 Electrical circuits.

The terminology and current flow convention used in the analysis of an electrical circuit are illustrated in Figure 2.2a. The voltage source, which provides energy to the circuit, can be a power supply, battery, or generator. The voltage source adds electrical energy to electrons, which flow from the negative terminal to the positive terminal, through the circuit. The positive side of the source attracts electrons, and the negative side releases electrons. The negative side is usually not labeled in a circuit schematic (e.g., with a minus sign) because it is implied by the positive side, which is labeled with a plus sign. Standard convention assumes that positive charge flows in a direction opposite from the electrons. **Current** describes the flow of this positive charge (not electrons). We owe this convention to Benjamin Franklin, who thought current was the result of the motion of positively charged particles. A **load** consists of a network of circuit elements that may dissipate or store electrical energy. Figure 2.2b shows two alternative ways to draw a circuit schematic. The **ground** indicates a reference point in the circuit where the voltage is assumed to be zero. Even though we do not show a connection between the ground symbols in the top circuit, it is implied that both ground symbols represent a single reference voltage (i.e., there is a "common ground"). This technique can be applied when drawing complicated circuits to reduce the number of lines. The bottom circuit is an equivalent representation.

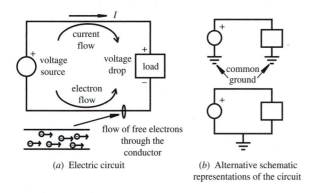

(a) Electric circuit

(b) Alternative schematic representations of the circuit

Figure 2.2 Electric circuit terminology.

2.2 BASIC ELECTRICAL ELEMENTS

There are three basic passive electrical elements: the resistor (R) , capacitor (C), and inductor (L). Passive elements require no additional power supply, unlike active devices such as integrated circuits. The passive elements are defined by their voltage-current relationships, as summarized below, and the symbols used to represent them in circuit schematics are shown in Figure 2.3.

There are two types of ideal energy sources: a **voltage source** (V) and a **current source** (I). These ideal sources contain no internal resistance, inductance, or capacitance. Figure 2.3 also illustrates the schematic symbols for ideal sources. Figure 2.4 shows some examples of actual components that correspond to the symbols in Figure 2.3.

2.2.1 Resistor

A **resistor** is a dissipative element that converts electrical energy into heat. **Ohm's law** defines the voltage-current characteristic of an ideal resistor:

$$V = IR \tag{2.2}$$

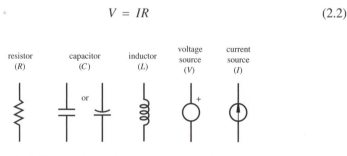

Figure 2.3 Schematic symbols for basic electrical elements.

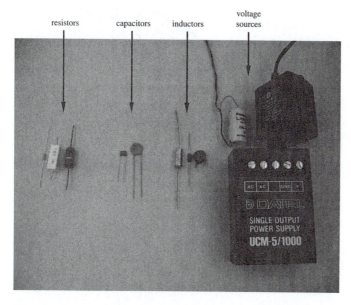

Figure 2.4 Examples of basic circuit elements.

The unit of resistance is the **ohm** (Ω). Resistance is a material property whose value is the slope of the resistor's voltage-current curve (see Figure 2.5). For an ideal resistor, the voltage-current relationship is linear, and the resistance is constant. However, real resistors are typically nonlinear due to temperature effects. As the current increases, temperature increases resulting in higher resistance. Also a real resistor has a limited power dissipation capability designated in watts, and it may fail when this limit is exceeded.

If a resistor's material is homogeneous and has a constant cross-sectional area, such as the cylindrical wire illustrated in Figure 2.6, then the resistance is given by

$$R = \frac{\rho L}{A} \tag{2.3}$$

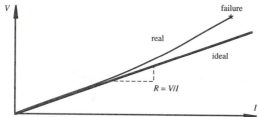

Figure 2.5 Voltage-current relation for an ideal resistor.

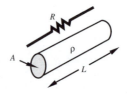

Figure 2.6 Wire resistance.

Table 2.1 Resistivities of common conductors

Material	Resistivity ($10^{-8}\Omega$m)
Aluminum	2.8
Carbon	4000
Constantan	44
Copper	1.7
Gold	2.4
Iron	10
Silver	1.6
Tungsten	5.5

Internet Link

2.1 Conductor sizes

2.2 Conductor current ratings

where ρ is the **resistivity,** or specific resistance of the material; L is the wire length; and A is the cross-sectional area. Resistivities for common conductors are given in Table 2.1. Example 2.1 demonstrates how to determine the resistance of a wire of given diameter and length. Internet Links 2.1 and 2.2 list the standard conductor diameters and current ratings.

EXAMPLE 2.1   Resistance of a Wire

As an example of the use of Equation 2.3, we will determine the resistance of a copper wire 1.0 mm in diameter and 10 m long.

From Table 2.1, the resistivity of copper is

$$\rho = 1.7 \times 10^{-8} \ \Omega m$$

Because the wire diameter, area, and length are

$$D = 0.0010 \text{ m}$$
$$A = \pi D^2/4 = 7.8 \times 10^{-7} \text{ m}^2$$
$$L = 10 \text{ m}$$

the total wire resistance is

$$R = \rho L/A = 0.22 \ \Omega$$

Actual resistors used in assembling circuits are packaged in various forms including axial-lead components, surface mount components, and the **dual in-line package (DIP)** and the **single in-line package (SIP),** which contain multiple resistors in a package that conveniently fits into circuit boards. These four types are illustrated in Figures 2.7 and 2.8. Video Demo 2.1 also shows several examples of resistor types and packages.

An axial-lead resistor's value and tolerance are usually coded with four colored bands (*a, b, c, tol*) as illustrated in Figure 2.9. The colors used for the bands are listed with their respective values in Table 2.2 and at Internet Link 2.3 (for easy reference). A resistor's value and tolerance are expressed as

$$R = ab \times 10^c \pm \text{tolerance} (\%) \qquad (2.4)$$

where the *a* band represents the tens digit, the *b* band represents the ones digit, the *c* band represents the power of 10, and the *tol* band represents the tolerance or uncertainty as a percentage of the coded resistance value. Here is a popular (and politically correct) mnemonic you can use to remember the resistor color codes when you don't have a table handy: "Bob BROWN Ran Over YELLOW Grass, But VIOLET Got Wet." The capitalized letters identify the colors: black, brown, red, orange, yellow, green, blue, violet, gray, and white. The set of **standard values** for the first two

Video Demo

2.1 Resistors

Internet Link

2.3 Resistor color codes

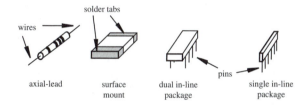

Figure 2.7 Resistor packaging.

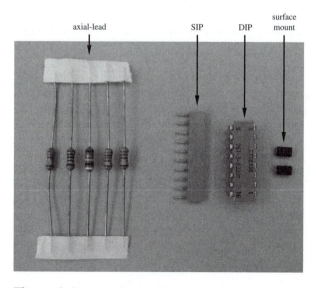

Figure 2.8 Examples of resistor packaging.

Figure 2.9 Axial-lead resistor color bands.

Table 2.2 Resistor color band codes

| *a*, *b*, and *c* **Bands** | | *tol* **Band** | |
Color	Value	Color	Value
Black	0	Gold	±5%
Brown	1	Silver	±10%
Red	2	Nothing	±20%
Orange	3		
Yellow	4		
Green	5		
Blue	6		
Violet	7		
Gray	8		
White	9		

digits (*ab*) are 10, 11, 12, 13, 14, 15, 16, 18, 20, 22, 24, 27, 30, 33, 36, 39, 43, 47, 51, 56, 62, 68, 75, 82, and 91. Often, resistance values are in the kΩ range and sometimes the unit is abbreviated as k instead of kΩ. For example, 10 k next to a resistor on an electrical schematic implies 10 kΩ.

The most common resistors you will use in ordinary electronic circuitry are 1/4 watt, 5% tolerance carbon or metal-film resistors. Resistor values of this type range in value between 1 Ω and 24 MΩ. Resistors with higher power ratings are also available. The 1/4 watt rating means the resistor can fail if it is required to dissipate more power than this.

Precision metal-film resistors have 1% or smaller uncertainties and are available in a wider range of values than the lower tolerance resistors. They usually have a numerical four-digit code printed directly on the body of the resistor. The first three digits denote the value of the resistor, and the last digit indicates the power of 10 by which to multiply.

EXAMPLE 2.2 | **Resistance Color Codes**

An axial-lead resistor has the following color bands:

$$a = \text{green}, \ b = \text{brown}, \ c = \text{red}, \ \text{and} \ tol = \text{gold}$$

From Equation 2.4 and Table 2.2, the range of possible resistance values is

$$R = 51 \times 10^2 \, \Omega \pm 5\% = 5100 \pm (0.05 \times 5100) \ \Omega$$

or

$$4800 \, \Omega < R < 5300 \, \Omega$$

Resistors come in a variety of shapes and sizes. As with many electrical components, the size of the device often has little to do with the characteristic value (e.g., resistance) of the device. Capacitors are one exception, where a larger device usually implies a higher capacitance value. With most devices that carry continuous current, the physical size is usually related to the maximum current or power rating, both of which are related to the power dissipation capabilities.

Video Demo 2.2 shows various types of components of various sizes to illustrate this principle. The best place to find detailed information on various components is online from vendor websites. Internet Link 2.4 points to a collection of links to the largest and most popular suppliers.

Variable resistors are available that provide a range of resistance values controlled by a mechanical screw, knob, or linear slide. The most common type is called a **potentiometer,** or **pot.** The various schematic symbols for a potentiometer are shown in Figure 2.10. A potentiometer that is included in a circuit to adjust or fine-tune the resistance in the circuit is called a **trim pot.** A trim pot is shown with a little symbol to denote the screw used to adjust ("trim") its value. The direction to rotate the potentiometer for increasing resistance is usually indicated on the component. Potentiometers are discussed further in Sections 4.8 and 9.2.2.

Conductance is defined as the reciprocal of resistance. It is sometimes used as an alternative to resistance to characterize a dissipative circuit element. It is a measure of how easily an element conducts current as opposed to how much it resists it. The unit of conductance is the **siemen** ($S = 1/\Omega$ = mho).

2.2.2 Capacitor

A **capacitor** is a passive element that stores energy in the form of an electric field. This field is the result of a separation of electric charge. The simplest capacitor consists of a pair of parallel conducting plates separated by a dielectric material as illustrated in Figure 2.11. The **dielectric material** is an insulator that increases the capacitance as a result of permanent or induced electric dipoles in the material.

Video Demo

2.2 Electronics components of various types and sizes

Internet Link

2.4 Electronic component online resources and vendors

Figure 2.10 Potentiometer schematic symbols.

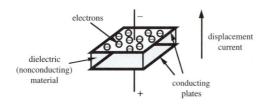

Figure 2.11 Parallel plate capacitor.

Strictly, direct current (DC) does not flow through a capacitor; rather, charges are displaced from one side of the capacitor through the conducting circuit to the other side, establishing the electric field. The displacement of charge is called a **displacement current** because current appears to flow through the device as it charges or discharges. The capacitor's voltage-current relationship is defined as

$$V(t) = \frac{1}{C} \int_0^t I(\tau)\,d\tau = \frac{q(t)}{C} \tag{2.5}$$

where $q(t)$ is the amount of accumulated charge measured in coulombs and C is the capacitance measured in farads (F = coulombs/volts). By differentiating this equation, we can relate the displacement current to the rate of change of voltage:

$$I(t) = C \frac{dV}{dt} \tag{2.6}$$

Capacitance is a property of the dielectric material and the plate geometry and separation. Values for typical capacitors range from 1 pF to 1000 μF, but they are also available with much larger values. Because the voltage across a capacitor is the integral of the displacement current (see Equation 2.5), the voltage cannot change instantaneously. As we will see several times throughout the book, this characteristic can be used for timing purposes in electrical circuits using a simple RC circuit, which is a resistor and capacitor in series.

The primary types of commercial capacitors are electrolytic capacitors, tantalum capacitors, ceramic disk capacitors, and mylar capacitors. Electrolytic capacitors are polarized, meaning they have a positive end and a negative end. The positive lead of a polarized capacitor must be held at a higher voltage than the negative side; otherwise, the device will usually be damaged (e.g., it will short and/or explode with a popping sound). Capacitors come in many sizes and shapes (see Video Demo 2.3). Often the capacitance is printed directly on the component, typically in μF or pF, but sometimes a three-digit code is used. The first two digits are the value and the third is the power of 10 multiplied times picofarads (e.g., 102 implies 10×10^2 pF = 1 nF). If there are only two digits, the value reported is in picofarads (e.g., 22 implies 22 pF). For more information, see Section 2.10.1.

Video Demo

2.3 Capacitors

2.2.3 Inductor

An **inductor** is a passive energy storage element that stores energy in the form of a magnetic field. The simplest form of an inductor is a wire coil, which has a tendency to maintain a magnetic field once established. The inductor's characteristics are a direct result of Faraday's law of induction, which states

$$V(t) = \frac{d\lambda}{dt} \tag{2.7}$$

where λ is the total **magnetic flux** through the coil windings due to the current. Magnetic flux is measured in webers (Wb). The magnetic field lines surrounding an inductor are illustrated in Figure 2.12. The south-to-north direction of the magnetic

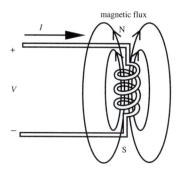

Figure 2.12 Inductor flux linkage.

field lines, shown with arrowheads in the figure, is found using the **right-hand rule** for a coil. The rule states that, if you curl the fingers of your right hand in the direction of current flow through the coil, your thumb will point in the direction of magnetic north. For an ideal coil, the flux is proportional to the current:

$$\lambda = LI \tag{2.8}$$

where L is the inductance of the coil, which is assumed to be constant. The unit of measure of inductance is the **henry** (H = Wb/A). Using Equations 2.7 and 2.8, an inductor's voltage-current relationship can be expressed as

$$V(t) = L\frac{dI}{dt} \tag{2.9}$$

The magnitude of the voltage across an inductor is proportional to the rate of change of the current through the inductor. If the current through the inductor is increasing ($dI/dt > 0$), the voltage polarity is as shown in Figure 2.12. If the current through the inductor is decreasing ($dI/dt < 0$), the voltage polarity is opposite to that shown.

Integrating Equation 2.9 results in an expression for current through an inductor given the voltage:

$$I(t) = \frac{1}{L}\int_0^t V(\tau)\,d\tau \tag{2.10}$$

where τ is a dummy variable of integration. From this we can infer that the current through an inductor cannot change instantaneously because it is the integral of the voltage. This is important in understanding the function or consequences of inductors in circuits. It takes time to increase or decrease the current flowing through an inductor. An important mechatronic system component, the electric motor, has large inductance due to its internal coils, so it is difficult to start or stop the motor very quickly. This is true of electromagnetic relays and solenoids as well.

Typical inductor components range in value from 1 μH to 100 mH. Inductance is important to consider in motors, relays, solenoids, some power supplies, and high-frequency circuits. Although some manufacturers have coding systems for inductors, there is no standard method. Often, the value is printed on the device directly, typically in μH or mH.

2.3 KIRCHHOFF'S LAWS

Now we are ready to put circuit elements and sources together in circuits and calculate voltages and currents anywhere in the circuit. Kirchhoff's laws are essential for the analysis and understanding of circuits, regardless of how simple or complex the circuit may be. In fact, these laws are the basis for even the most complex circuit analysis such as that involved with transistor circuits, operational amplifiers, or integrated circuits with hundreds of elements. **Kirchhoff's voltage law (KVL)** states that the sum of voltages around a closed loop or path is 0 (see Figure 2.13):

$$\sum_{i=1}^{N} V_i = 0 \tag{2.11}$$

Note that the loop must be closed, but the conductors themselves need not be closed (i.e., the loops can go through open circuits).

To apply KVL to a circuit, as illustrated in Figure 2.13, you first assume a current direction on each branch of the circuit. Next assign the appropriate polarity to the voltage across each passive element assuming that the voltage drops across each element in the direction of the current. Where assumed current enters a passive element, a plus is shown, and where the assumed current leaves the element, a minus is shown. The polarity of voltage across a voltage source and the direction of current through a current source must always be maintained as given. Now, starting at any point in the circuit (such as node A in Figure 2.13) and following either a clockwise or counterclockwise loop direction (clockwise in Figure 2.13), form the sum of the voltages across each element, assigning to each voltage the first algebraic sign encountered at each element in the loop. For Figure 2.13, the result would be

$$-V_1 - V_2 + V_3 + \cdots - V_N = 0 \tag{2.12a}$$

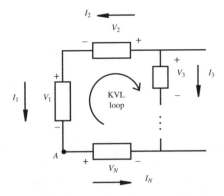

Figure 2.13 Kirchhoff's voltage law.

Alternatively, you can assign the signs based on whether the voltage increases (from − to +, assigning +) or drops (from + to −, assigning −) across the element. Using this convention, the equation would be:

$$V_1 + V_2 - V_3 + \cdots + V_N = 0 \qquad (2.12b)$$

Equations 2.12a and 2.12b are equivalent, but the second convention is more intuitive because it represents what actually occurs in the circuit. However, the first convention is more common, probably because it involves less thought.

Kirchhoff's Voltage Law **EXAMPLE 2.3**

KVL will be used to find the current I_R in the following circuit.

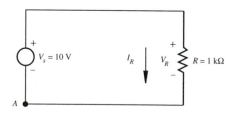

The first step is to assume the current direction for I_R. The chosen direction is shown in the figure. With a circuit this simple, the current direction is obvious based on the polarity of the source; but in more complex circuits, current directions might not be so obvious. Then we use the current direction through the resistor to assign the voltage-drop polarity. If the current were assumed to flow in the opposite direction instead, the voltage polarity across the resistor would also have to be reversed. The polarity for the voltage source is fixed regardless of current direction. Starting at point A and progressing clockwise around the loop, we assign the first voltage sign we come to on each element yielding

$$-V_s + V_R = 0$$

Applying Ohm's law,

$$-V_s + I_R R = 0$$

Therefore,

$$I_R = V_s / R = 10/1000 \text{ A} = 10 \text{ mA}$$

Kirchhoff's current law (KCL) states that the sum of the currents flowing into a closed surface or node is 0. Referring to Figure 2.14a,

$$I_1 + I_2 - I_3 = 0 \qquad (2.13)$$

More generally, referring to Figure 2.14b,

$$\sum_{i=1}^{N} I_i = 0 \qquad (2.14)$$

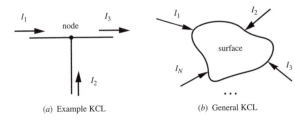

(*a*) Example KCL   (*b*) General KCL

Figure 2.14 Kirchhoff's current law.

Lab Exercise

Lab 1
Introduction—
Resistor codes,
breadboard,
and basic
measurements

Video Demo

2.4 Breadboard
construction

2.5 Instrumentation
for powering
and making
measurements in
circuits

Note that currents entering a node or surface are assigned a positive value, and currents leaving are assigned a negative value.

It is important to note that, when analyzing a circuit, you arbitrarily assume current directions and denote the directions with arrows on the schematic. If the calculated result for a current is negative, the current actually flows in the opposite direction. Also, assumed voltage drops must be consistent with the assumed current directions. If a calculated voltage is negative, its actual polarity is opposite to that shown.

Lab Exercise 1 introduces many of the basic concepts presented so far in this chapter. The following practical skills are developed:

■ Assembling basic circuits using a breadboard (see Video Demo 2.4)

■ Making voltage and current measurements (see Video Demo 2.5)

■ Reading resistor and capacitor values

More information and resources dealing with all of these topics can also be found in Section 2.10.

2.3.1 Series Resistance Circuit

Applying KVL to the simple series resistor circuit illustrated in Figure 2.15 yields some useful results. Assuming a current direction I, starting at node A, and following a clockwise direction yields

$$-V_s + V_{R_1} + V_{R_2} = 0 \tag{2.15}$$

From Ohm's law,

$$V_{R_1} = IR_1 \tag{2.16}$$

and

$$V_{R_2} = IR_2 \tag{2.17}$$

Substituting these two equations into Equation 2.15 gives

$$-V_s + IR_1 + IR_2 = 0 \tag{2.18}$$

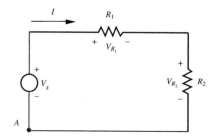

Figure 2.15 Series resistance circuit.

and solving for I yields

$$I = \frac{V_s}{(R_1 + R_2)} \tag{2.19}$$

Note that, if we had a single resistor of value $R_1 + R_2$, we would have the same result. Therefore resistors in series add, and the equivalent resistance of a series resistance circuit is

$$R_{eq} = R_1 + R_2 \tag{2.20}$$

In general, N resistors connected in series can be replaced by a single equivalent resistance given by

$$R_{eq} = \sum_{i=1}^{N} R_i \tag{2.21}$$

By applying KVL to capacitor and inductor circuits, it can be shown (Questions 2.11 and 2.13) that two capacitors in series combine as

$$C_{eq} = \frac{C_1 C_2}{C_1 + C_2} \tag{2.22}$$

and two inductors in series add:

$$L_{eq} = L_1 + L_2 \tag{2.23}$$

A circuit containing two resistors in series is referred to as a **voltage divider** because the source voltage V_s divides between each resistor. Expressions for the resistor voltages can be obtained by substituting Equation 2.19 into Equations 2.16 and 2.17 giving

$$V_{R_1} = \frac{R_1}{R_1 + R_2} V_s, \quad V_{R_2} = \frac{R_2}{R_1 + R_2} V_s \tag{2.24}$$

In general, for N resistors connected in series with a total applied voltage of V_s, the voltage V_{R_i} across any resistor R_i is

$$V_{R_i} = \frac{R_i}{R_{eq}} V_s = \frac{R_i}{\sum\limits_{j=1}^{N} R_j} V_s \tag{2.25}$$

Voltage dividers are useful because they allow us to create different reference voltages in a circuit even if the circuit is energized only by a single output supply. However, care must be exercised that attached loads do not drain significant current and affect the voltage references produced with the dividers (see Class Discussion Item 2.2).

■ **CLASS DISCUSSION ITEM 2.2**
Improper Application of a Voltage Divider

Your car has a 12 V battery that powers some circuits in the car at lower voltage levels. Why is it inappropriate to use a simple voltage divider to create a lower voltage level for circuits that might draw variable current?

2.3.2 Parallel Resistance Circuit

Applying KCL to the simple parallel resistor circuit illustrated in Figure 2.16 also yields some useful results. Because each resistor experiences the same voltage V_s, as they are both in parallel with the source, Ohm's law gives

$$I_1 = \frac{V_s}{R_1} \tag{2.26}$$

and

$$I_2 = \frac{V_s}{R_2} \tag{2.27}$$

Applying KCL at node A gives

$$I - I_1 - I_2 = 0 \tag{2.28}$$

Substituting the currents from Equations 2.26 and 2.27 yields

$$I = \frac{V_s}{R_1} + \frac{V_s}{R_2} = V_s\left(\frac{1}{R_1} + \frac{1}{R_2}\right) \tag{2.29}$$

Replacing the resistance values R_1 and R_2 with their conductance equivalents $1/G_1$ and $1/G_2$ gives

$$I = V_s(G_1 + G_2) \tag{2.30}$$

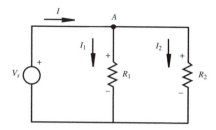

Figure 2.16 Parallel resistance circuit.

A single resistor with a conductance of value $(G_1 + G_2)$ would have given the same result; therefore, conductances in parallel add. We can write Equation 2.30 as

$$I = V_s G_{eq} = \frac{V_s}{R_{eq}} \tag{2.31}$$

where G_{eq} is the equivalent conductance and R_{eq} is the equivalent resistance. By comparing the right-hand side of this equation to Equation 2.29, we get

$$\frac{1}{R_{eq}} = \frac{1}{R_1} + \frac{1}{R_2} \tag{2.32}$$

or

$$R_{eq} = \frac{R_1 R_2}{R_1 + R_2} \tag{2.33}$$

In general, N resistors connected in parallel can be replaced by a single equivalent resistance given by

$$\frac{1}{R_{eq}} = \sum_{i=1}^{N} \frac{1}{R_i} \tag{2.34}$$

or

$$R_{eq} = 1 \Big/ \sum_{i=1}^{N} \frac{1}{R_i} \tag{2.35}$$

By applying KCL to capacitor and inductor circuits, it can be shown (Questions 2.12 and 2.14) that two capacitors in parallel add:

$$C_{eq} = C_1 + C_2 \tag{2.36}$$

and two inductors in parallel combine as

$$L_{eq} = \frac{L_1 L_2}{L_1 + L_2} \tag{2.37}$$

A circuit containing two resistors connected in parallel is called a **current divider** because the source current I divides between each resistor. Expressions for the divided currents can be obtained by solving Equation 2.29 for V_s and substituting into Equation 2.26 and 2.27 giving

$$I_1 = \frac{R_2}{R_1 + R_2} I, \quad I_2 = \frac{R_1}{R_1 + R_2} I \tag{2.38}$$

Video Demo

2.6 Light bulb series and parallel circuit comparison

Video Demo 2.6 illustrates the differences between parallel and series wiring of lighting. The demonstration illustrates voltage and current division and the effects on power output.

When drawing circuit schematics, by hand or with software tools, it is important to be consistent with how you show connections (or the lack thereof)

EXAMPLE 2.4   ## Circuit Analysis

As an example of how the tools presented in the previous sections apply to a nontrivial circuit, consider the following network, where the goal is to find I_{out} and V_{out}. At any node in the circuit, such as the one labeled by V_{out}, the voltage is defined with respect to the ground reference denoted by the ground symbol ⏚. Voltage differences between any two points can be obtained by taking the difference between the ground-referenced values at the points.

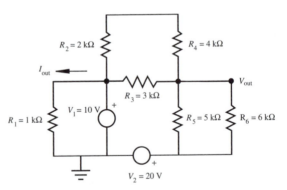

The first step is to combine resistor clusters between and around the sources (V_1 and V_2) and the branches of interest (those dealing with I_{out} and V_{out}) using the series and parallel resistance formulas (Equations 2.20 and 2.33). Resistors R_2 and R_4 are in series, with an equivalent resistance of ($R_2 + R_4$), and this is in parallel with resistor R_3. Resistors R_5 and R_6 are also in parallel. Therefore, the resultant resistances for the equivalent circuit that follows are

$$R_{234} = \frac{(R_2 + R_4)R_3}{(R_2 + R_4) + R_3} = 2.00 \text{ k}\Omega$$

$$R_{56} = \frac{R_5 R_6}{R_5 + R_6} = 2.73 \text{ k}\Omega$$

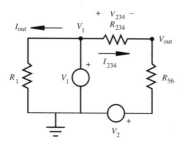

Applying KVL to the left loop gives

$$V_1 = I_{out} R_1$$

so

$$I_{out} = V_1/R_1 = 10 \text{ V}/1 \text{ k}\Omega = 10 \text{ mA}$$

Applying KVL to the right loop tells us that the total voltage across R_{234} and R_{56} in the assumed direction of I_{234} is $(V_1 - V_2)$. Voltage division (Equation 2.24) can then be used to determine the voltage drop across R_{234} in the assumed direction of I_{234}:

$$V_{234} = \frac{R_{234}}{R_{234} + R_{56}}(V_1 - V_2) = -4.23 \text{ V}$$

Because V_1 is referenced to ground, the voltage on the left side of resistor R_{234} is V_1; and because the voltage drops by V_{234} across the resistor, the desired output voltage is

$$V_{out} = V_1 - V_{234} = 14.2 \text{ V}$$

Note that because V_{234} was found to be negative, the actual flow of current through R_{234} would be in the opposite direction from that assumed in this solution.

A myriad of methods may be used to solve this problem (e.g., see Question 2.24), and the one presented here is just an example solution, not necessarily the best method.

at intersecting lines on the drawing. Figure 2.17 illustrates two conventions for doing this. The first convention (Figure 2.17a) is the most common and is what was used in Example 2.4. With this convention, a dot implies a connection, and the absence of a dot (at crossing lines only) implies no connection. Figure 2.17b shows an alternative convention where a dot is not required to indicate a connection as long as a crossing arc is used to indicate a nonconnection. Because the circuit diagrams in this chapter have been very simple, we really didn't need a convention—any crossing lines have been assumed to be connected. Even with the circuit in Example 2.4, the connection dots are not really required. People will assume there are connections at all intersecting lines in simple diagrams unless dot or arc features appear at one or more intersections. However, with more complicated circuits (e.g., those in Chapter 7 dealing with complicated microcontroller-based solutions),

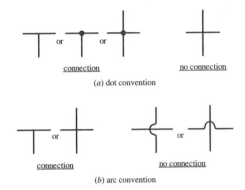

(a) dot convention

(b) arc convention

Figure 2.17 Circuit schematic connection conventions.

Lab Exercise

Lab 2 Instrument familiarization and basic electrical relations

Video Demo

2.5 Instrumentation for powering and making measurements in circuits

2.7 Connectors (BNC, banana plugs, alligator clips)

Internet Link

2.5 All about circuits - Vol. I - DC

a clear and consistent convention is very important to present and interpret the intent of the designer.

Lab Exercise 2 provides experience with using various instruments including an oscilloscope, multimeter, power supply, and function generator (see Video Demo 2.5). The Lab also covers practical application of Ohm's law, KVL, and KCL, as applied to making voltage and current measurements in circuits. Video Demo 2.7 shows the various types of cables and connectors that are used to connect instruments to each other and to circuits. Internet Link 2.5 is an excellent resource reviewing many topics related to electricity and DC circuit analysis.

2.4 VOLTAGE AND CURRENT SOURCES AND METERS

When we analyze electrical networks on paper, we usually assume that sources and meters are ideal. However, actual physical devices are not ideal, and it is sometimes necessary to account for their limitations when circuits contain these devices. The following ideal behavior is usually assumed:

- An **ideal voltage source** has zero output resistance and can supply infinite current.
- An **ideal current source** has infinite output resistance and can supply infinite voltage.
- An **ideal voltmeter** has infinite input resistance and draws no current.
- An **ideal ammeter** has zero input resistance and no voltage drop across it.

Unfortunately, real sources and meters have terminal characteristics that are somewhat different from the ideal cases. However, the terminal characteristics of the real sources and meters can be modeled using ideal sources and meters with their associated input and output resistances.

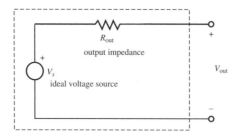

Figure 2.18 Real voltage source with output impedance.

As shown in Figure 2.18, a "real" voltage source can be modeled as an ideal voltage source in series with a resistance called the **output impedance** of the device. When a load is attached to the source and current flows, the output voltage V_{out} will be different from the ideal source voltage V_s due to voltage division. The output impedance of most commercially available voltage sources (e.g., a power supply) is very small, usually a fraction of an ohm. For most applications, this impedance is small enough to be neglected. However, the output impedance can be important when driving a circuit with small resistance because the impedance adds to the resistance of the circuit. Figure 2.19 shows examples of two commercially available voltage sources. The top unit is a triple-output power supply that can provide three different voltages relative to ground, adjustable from 0 V to 9 V, 20 V, and -20 V. The bottom unit is a programmable power supply that provides digitally controlled voltage sources.

As shown in Figure 2.20, a "real" current source can be modeled as an ideal current source in parallel with a resistance called the output impedance. When a load

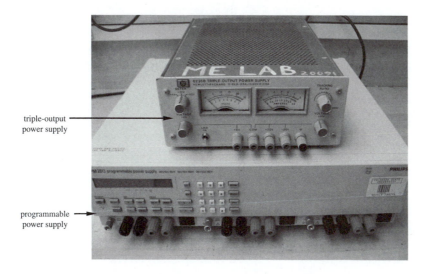

Figure 2.19 Example of commercially available voltage sources.

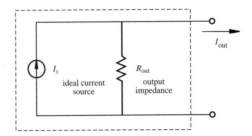

Figure 2.20 Real current source with output impedance.

is attached to the source, the source current I_s divides between the output impedance and the load. The output impedance of most commercially available current sources is very large, minimizing the current division effect. However, this impedance can be important when driving a circuit with a large resistance.

As shown in Figure 2.21, a "real" ammeter can be modeled as an ideal amme-ter in series with a resistance called the **input impedance** of the device. The input impedance of most commercially available ammeters is very small, minimizing the voltage drop V_R added in the circuit. However, this resistance can be important when making a current measurement through a circuit branch with small resistance because the output impedance adds to the resistance of the branch.

As shown in Figure 2.22, a "real" voltmeter can be modeled as an ideal volt-meter in parallel with an input impedance. The input impedance of most commer-cially available voltmeters (e.g., an oscilloscope or multimeter) is very large, usually on the order of 1 to 10 MΩ. However, this resistance must be considered when mak-ing a voltage measurement across a circuit branch with large resistance because the

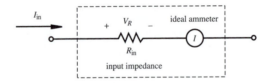

Figure 2.21 Real ammeter with input impedance.

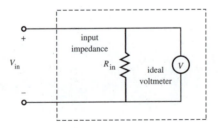

Figure 2.22 Real voltmeter with input impedance.

parallel combination of the meter input impedance and the circuit branch would result in significant error in the measured value.

Figure 2.23 shows examples of commercially available **digital multimeters** (DMMs) that contain, among other things, ammeters and voltmeters. Figure 2.24 shows an example of a commercially available oscilloscope that contains a voltmeter capable of digitizing, displaying, and recording dynamic measurements. Internet Link 2.6 provides links to various online resources and vendors that offer an assortment of instrumentation (power supplies, function generators, multimeters, oscilloscopes, data acquisition equipment, and more).

Lab Exercise 2 provides experience with the effects of input and output impedance of various instruments. It is important to know how these instrument characteristics can affect voltage and current measurements. Section 2.10.3 has more information and resources on these topics. Lab Exercise 3 provides a complete overview of how to use an oscilloscope. Features and concepts covered include how to connect signals, grounding, coupling, and triggering. Video Demo 2.8 demonstrates how to use a typical analog oscilloscope. Many of the concepts involved with using an analog scope are also relevant with other scopes, even more sophisticated digital scopes. More information and resources dealing with how to use an oscilloscope properly can be found in Section 2.10.5.

Internet Link

2.6 Instrumentation online resources and vendors

Lab Exercise

Lab 2 Instrument familiarization and basic electrical relations

Lab 3 The oscilloscope

Video Demo

2.8 Oscilloscope demonstrations using the Tektronix 2215 analog scope

Figure 2.23 Examples of commercially available digital multimeters. *(Courtesy of Hewlett Packard, Santa Clara, CA)*

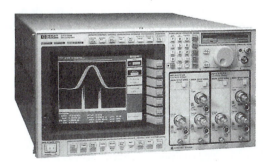

Figure 2.24 Example of a commercially available oscilloscope. *(Courtesy of Hewlett Packard, Santa Clara, CA)*

| **EXAMPLE 2.5** | Input and Output Impedance |

This example illustrates the effects of source and meter output and input impedance on making measurements in a circuit. Consider the following circuit with voltage source V_s and voltage meter V_m.

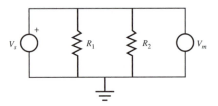

The equivalent resistance for this circuit is

$$R_{eq} = \frac{R_1 R_2}{R_1 + R_2}$$

If the source and meter were both ideal, the measured voltage V_m would be equal to V_s, and the equivalent circuit would look like this:

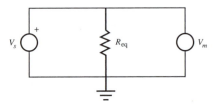

However, if the source has output impedance Z_{out} and the meter has input impedance Z_{in}, the "real" circuit actually looks like this:

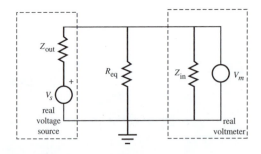

The parallel combination of R_{eq} and Z_{in} yields the following circuit (a). Z_{out} and the parallel combination of R_{eq} and Z_{in} are now effectively in series because no current flows into the ideal meter V_m. Thus, the total equivalent resistance shown in circuit (b) is

$$R'_{eq} = \frac{R_{eq} Z_{in}}{R_{eq} + Z_{in}} + Z_{out}$$

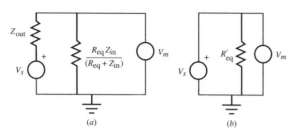

(a)   (b)

Note that R'_{eq} defined in the previous equation approaches R_{eq} as Z_{in} approaches infinity and as Z_{out} approaches 0. From voltage division in circuit (a), the voltage measured by the actual meter would be

$$V_m = \frac{\dfrac{R_{eq}Z_{in}}{(R_{eq} + Z_{in})}}{\dfrac{R_{eq}Z_{in}}{(R_{eq} + Z_{in})} + Z_{out}} V_s = \frac{R'_{eq} - Z_{out}}{R'_{eq}} V_s$$

The measured voltage V_m equals V_s for $Z_{in} = \infty$ and $Z_{out} = 0$, but with a real source and real meter, the measured voltage could differ appreciably from the expected ideal result. For example, if $R_1 = R_2 = 1$ kΩ,

$$R_{eq} = \frac{1 \cdot 1}{1 + 1} \text{ k}\Omega = 0.5 \text{ k}\Omega$$

and if $Z_{in} = 1$ MΩ and $Z_{out} = 50$ Ω,

$$R'_{eq} = \frac{0.5 \cdot 1000}{0.5 + 1000} + 0.05 \text{ k}\Omega = 0.550 \text{ k}\Omega$$

Therefore, if $V_s = 10$ V,

$$V_m = \left(\frac{0.550 - 0.05}{0.550}\right) 10 \text{ V} = 9.09 \text{ V}$$

This differs substantially from the result that would be expected (10 V) with an ideal source and meter.

2.5 THEVENIN AND NORTON EQUIVALENT CIRCUITS

Sometimes, to simplify the analysis of more complex circuits, we wish to replace voltage sources and resistor networks with an equivalent voltage source and series resistor. This is called a **Thevenin equivalent** of the circuit. Thevenin's theorem states that, given a pair of terminals in a linear network, the network may be replaced by an ideal voltage source V_{OC} in series with a resistance R_{TH}. V_{OC} is equal to the open circuit voltage across the terminals, and R_{TH} is the equivalent resistance across the terminals when independent voltage sources are shorted and independent current sources are replaced with open circuits.

We will illustrate Thevenin's theorem with the circuit shown in Figure 2.25. The part of the circuit in the dashed box will be replaced by its Thevenin equivalent. The open circuit voltage V_{OC} is found by disconnecting the rest of the circuit and determining the voltage across the terminals of the remaining open circuit. For this example, the voltage divider rule gives

$$V_{OC} = \frac{R_2}{R_1 + R_2} V_s \tag{2.39}$$

To find R_{TH}, the supply V_s is shorted (i.e., $V_s = 0$), grounding the left end of R_1. If there were current sources in the circuit, they would be replaced with open circuits. Because R_1 and R_2 are in parallel relative to the open terminals, the equivalent resistance is

$$R_{TH} = \frac{R_1 R_2}{R_1 + R_2} \tag{2.40}$$

The Thevenin equivalent circuit is shown in Figure 2.26.

Another equivalent circuit representation is the **Norton equivalent,** shown in Figure 2.27. Here the linear network is replaced by an ideal current source I_{SC} and the Thevenin resistance R_{TH} in parallel with this source. I_{SC} is found by calculating

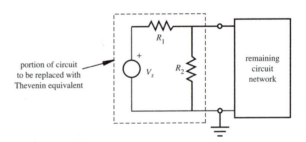

Figure 2.25 Example illustrating Thevenin's theorem.

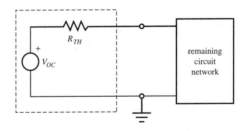

Figure 2.26 Thevenin equivalent circuit.

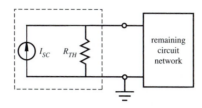

Figure 2.27 Norton equivalent circuit.

the current that would flow through the terminals if they were shorted together, having removed the remaining load circuit. It can be shown that the current I_{SC} flowing through R_{TH} produces the Thevenin voltage V_{OC} just discussed.

The Thevenin and Norton equivalents are independent of the remaining circuit network representing a load. This is useful because it is possible to make changes in the load without reanalyzing the Thevenin or Norton equivalent.

2.6 ALTERNATING CURRENT CIRCUIT ANALYSIS

When linear circuits are excited by alternating current (AC) signals of a given frequency, the current through and voltage across every element in the circuit are AC signals of the same frequency. A sinusoidal AC voltage $V(t)$ is illustrated in Figure 2.28 and can be expressed mathematically as

$$V(t) = V_m \sin(\omega t + \phi) \tag{2.41}$$

where V_m is the signal **amplitude,** ω is the **radian frequency** measured in radians per second, and ϕ is the **phase angle** relative to the reference sinusoid $V_m \sin(\omega t)$ measured in radians. The phase angle is related to the **time shift** (Δt) between the signal and reference:

$$\phi = \omega \Delta t \tag{2.42}$$

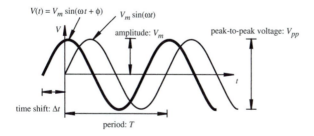

Figure 2.28 Sinusoidal waveform.

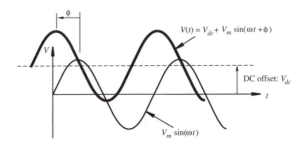

Figure 2.29 Sinusoidal signal DC offset.

A positive phase angle ϕ implies a **leading** waveform (i.e., it occurs earlier on the time axis), and a negative angle implies a **lagging** waveform (i.e., it occurs later on the time axis). The **period** T of the waveform is the time required for a full cycle. The frequency of the signal, measured in hertz (Hz = cycles/sec), is related to the period and radian frequency as

$$f = \frac{1}{T} = \frac{\omega}{2\pi} \tag{2.43}$$

Figure 2.29 illustrates another important sinusoidal waveform parameter called the **DC offset.** It represents the vertical shift of the signal from the reference sinusoid. Mathematically, the DC offset is represented by the term V_{dc} in the equation:

$$V(t) = V_{dc} + V_m \sin(\omega t + \phi) \tag{2.44}$$

Figures 2.28 and 2.29 illustrate a positive phase angle (ϕ), where the voltage signal $V(t)$ leads (i.e., occurs earlier in time relative to) the reference sine wave.

EXAMPLE 2.6	**AC Signal Parameters**

As an example of how the AC signal parameters are discerned in a signal equation, consider the following AC voltage:

$$V(t) = 5.00 \sin (t - 1) \text{ V}$$

The signal amplitude is

$$V_m = 5.00 \text{ V}$$

The signal radian frequency is

$$\omega = 1.00 \text{ rad/sec}$$

ω is the coefficient of the time variable t in the argument of the sinusoid. Likewise, the frequency in hertz is

$$f = \frac{\omega}{2\pi} = \frac{1}{2\pi} \text{ Hz} = 0.159 \text{ Hz}$$

and the phase angle is

$$\phi = -1 \text{ rad} = -57.3°$$

The negative phase indicates the signal lags (i.e., occurs later in time relative to) the reference ($\sin(t)$). The arguments of the sinusoids are always assumed to be specified in radians for computational purposes.

Alternating current power is used in many applications where direct current (DC) power is impractical or infeasible. Principal reasons for using AC power include

- AC power is more efficient to transmit over long distances because it is easily transformed to a high-voltage, low-current form, minimizing power losses (see Section 2.7) during transmission. In residential areas, it is easily transformed back to required levels. Note that the voltage drop in the transmission line is small compared to the voltage level at the source.
- AC power is easy to generate with rotating machinery (e.g., an electric generator).
- AC power is easy to use to drive rotating machinery (e.g., an AC electric motor).
- AC power provides a fixed frequency signal (60 Hz in the United States, 50 Hz in Europe) that can be used for timing purposes and synchronization.

■ CLASS DISCUSSION ITEM 2.3
Reasons for AC

Justify and fully explain the reasons AC power is used in virtually all commercial and public utility systems. Refer to the reasons just listed.

The steady state analysis of AC circuits is simplified by the use of **phasor** analysis, which uses complex numbers to represent sinusoidal signals. **Euler's formula** forms the basis for this analysis:

$$e^{j(\omega t + \phi)} = \cos(\omega t + \phi) + j \sin(\omega t + \phi) \qquad (2.45)$$

where $j = \sqrt{-1}$. This implies that sinusoidal signals can be expressed as real and imaginary components of **complex exponentials.** Because of the mathematical ease of manipulating exponential expressions vs. trigonometric expressions, this form of analysis is convenient for making and interpreting calculations.

Once all transients have dissipated in an AC circuit after power is applied, the voltage across and current through each element will oscillate with the same frequency ω as the input. The amplitude of the voltage and current for each element will be constant but may differ in phase from the input. This fact lets us treat circuit variables V and I as complex exponentials with magnitudes V_m and I_m and phase ϕ for a "steady state" analysis. A phasor (e.g., voltage V) is a vector representation of the complex exponential:

$$V = V_m e^{j(\omega t + \phi)} = V_m \langle \phi \rangle = V_m[\cos(\omega t + \phi) + j\sin(\omega t + \phi)] \quad (2.46)$$

where $V_m e^{j(\omega t + \phi)}$ is the complex exponential form, $V_m \langle \phi \rangle$ is the **polar form,** and $V_m[\cos(\omega t + \phi) + j\sin(\omega t + \phi)]$ is the complex **rectangular form** of the phasor. A graphical interpretation of these quantities is shown on the complex plane in Figure 2.30. Note that the phase angle ϕ is measured from the ωt reference.

Useful mathematical relations for manipulating complex numbers and phasors include

$$r = \sqrt{x^2 + y^2} \quad (2.47)$$

$$\phi = \tan^{-1}\left(\frac{y}{x}\right) \quad (2.48)$$

$$x = r\cos(\phi) \quad (2.49)$$

$$y = r\sin(\phi) \quad (2.50)$$

$$(x_1 + y_1 j) + (x_2 + y_2 j) = (x_1 + x_2) + (y_1 + y_2)j \quad (2.51)$$

$$r_1 \langle \phi_1 \rangle \cdot r_2 \langle \phi_2 \rangle = r_1 \cdot r_2 \langle \phi_1 + \phi_2 \rangle \quad (2.52)$$

$$r_1 \langle \phi_1 \rangle / r_2 \langle \phi_2 \rangle = r_1 / r_2 \langle \phi_1 - \phi_2 \rangle \quad (2.53)$$

where r is the phasor magnitude, ϕ is the phasor angle, x is the real component, and y is the imaginary component. Note that the quadrant determined by the arguments (x, y) of the arctangent function must be carefully considered when converting from rectangular to polar form. For example, if $x = y = -1$, $\phi = -135°$, not $45°$ that you

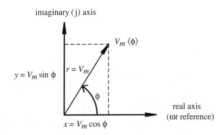

imaginary (j) axis

$V_m \langle \phi \rangle$

$y = V_m \sin \phi$ $r = V_m$

ϕ

real axis
(ωt reference)

$x = V_m \cos \phi$

Figure 2.30 Phasor representation of a sinusoidal signal.

would get if you carelessly used a single argument $\tan^{-1}$ function on a calculator or in a computer program.

Ohm's law can be extended to the AC circuit analysis of resistor, capacitor, and inductor elements as

$$V = ZI \tag{2.54}$$

where Z is called the **impedance** of the element. This is a complex number, and you can imagine Z as a complex, frequency-dependent resistance. Impedances can be derived from the fundamental constitutive equations for the elements using complex exponentials. The unit of impedance is the ohm (Ω).

For the resistor, because $V = IR$,

$$Z_R = R \tag{2.55}$$

For the inductor, because $V = L\dfrac{dI}{dt}$, if $I = I_m e^{j(\omega t + \phi)}$, then

$$V = Lj\omega I_m e^{j(\omega t + \phi)} = (Lj\omega)I \tag{2.56}$$

Therefore, the impedance of an inductor is given by

$$Z_L = j\omega L = \omega L \langle 90° \rangle \tag{2.57}$$

which implies that the voltage will lead the current by 90°. Note that because a DC signal can be considered an AC signal with zero frequency ($\omega = 0$), the impedance of an inductor in a DC circuit is 0. Therefore, it acts as a short in a DC circuit. At very high AC frequencies ($\omega = \infty$), the inductor has infinite impedance, so it behaves as an open circuit.

For the capacitor, because $I = C\dfrac{dV}{dt}$, if $V = V_m e^{j(\omega t + \phi)}$, then

$$I = Cj\omega V_m e^{j(\omega t + \phi)} = (Cj\omega)V \tag{2.58}$$

giving

$$V = \left(\frac{1}{Cj\omega}\right)I \tag{2.59}$$

Therefore, the impedance of a capacitor is given by

$$Z_C = \frac{1}{j\omega C} = \frac{-j}{\omega C} = \frac{1}{\omega C}\langle -90° \rangle \tag{2.60}$$

which implies the voltage will lag the current by 90°. The impedance of a capacitor in a DC circuit ($\omega = 0$) is infinite, so it acts as an open circuit. At very high AC frequencies ($\omega = \infty$), the capacitor has zero impedance, so it acts as a short circuit.

As illustrated in Example 2.7, every result presented in previous sections for analyzing simple DC circuits, including Ohm's law, series and parallel resistance combinations, voltage division, and current division, applies to the AC signals and impedances just presented! Internet Link 2.7 is an excellent resource that reviews AC electricity, circuit analysis, and devices.

In circuits with multiple sources, it is important to express them all in either their sine or cosine form consistently so that the phase relationships are relative to a consistent reference. The following trigonometric identities are useful in accomplishing this:

$$\sin(\omega t + \phi) = \cos(\omega t + \phi - \pi/2) \tag{2.61}$$

$$\cos(\omega t + \phi) = \sin(\omega t + \phi + \pi/2) \tag{2.62}$$

EXAMPLE 2.7	**AC Circuit Analysis**

Internet Link

2.7 All about circuits - Vol. II - AC

The following is an illustrative example of AC circuit analysis. The goal is to find the steady state current I through the capacitor in the following circuit:

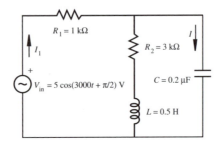

Because the input voltage source is

$$V_{in} = 5\cos\left(3000\,t + \frac{\pi}{2}\right) \text{ V}$$

each element in the circuit will respond at the radian frequency:

$$\omega = 3000 \text{ rad/sec}$$

Because the voltage source has a magnitude of 5 V and a phase of $\pi/2$, relative to $\cos(3000t)$, the phasor and complex form of the source is

$$V_{in} = 5\langle 90° \rangle \ V = (0 + 5j) \ V$$

The complex and phasor form of the capacitor impedance is

$$Z_C = -j/\omega C = -1666.67j \ \Omega = 1666.67\langle -90° \rangle \ \Omega$$

The complex and phasor form of the inductor impedance is

$$Z_L = j\omega L = 1500j \ \Omega = 1500\langle 90° \rangle \ \Omega$$

To find the current (I) through the capacitor, we will first find the current through the entire circuit (I_1) and then use current division. Therefore, we need the impedance of the middle branch of the circuit, along with the equivalent impedance of the entire circuit.

Resistor R_2 and inductor L are in series, so their combined impedance, in both rectangular and phasor form, using Equations 2.47 and 2.48, is:

$$R_2 + Z_L = (3000 + 1500\text{j}) \ \Omega = 3354.1 \ \langle 26.57° \rangle \ \Omega$$

This impedance is in parallel with capacitor C, and the combined impedance of this parallel combination, using Equation 2.33, is:

$$\frac{(R_2 + Z_L)Z_C}{(R_2 + Z_L) + Z_C}$$

The numerator of this expression can be calculated using Equation 2.52:

$$(R_2 + Z_L)Z_C = 3354.1 \ \langle 26.57° \rangle \cdot 1666.67 \ \langle -90° \rangle \ \Omega = 5{,}590{,}180 \ \langle -63.43° \rangle \ \Omega$$

The denominator can be found using Equation 2.51:

$$(R_2 + Z_L) + Z_C = ((3000 + 1500\text{j}) - 1666.67\text{j}) \ \Omega = (3000 - 166.67\text{j}) \ \Omega$$

Using Equations 2.47 and 2.48, the phasor form of this impedance, which is required to perform the division with the numerator, is:

$$(R_2 + Z_L) + Z_C = (3000 - 166.67\text{j}) \ \Omega = (3004.63 \ \langle -3.18° \rangle \ \Omega$$

Therefore, the parallel combination of $(R_2 + Z_L)$ and Z_C, using Equation 2.53, is:

$$\frac{(R_2 + Z_L)Z_C}{(R_2 + Z_L) + Z_C} = \frac{5{,}590{,}180 \langle -63.43° \rangle}{3004.63 \langle -3.18° \rangle} \ \Omega = 1860.52 \ \langle -60.25° \rangle \ \Omega$$

The rectangular form of this impedance, using Equations 2.49 and 2.50, is:

$$\frac{(R_2 + Z_L)Z_C}{(R_2 + Z_L) + Z_C} = 1860.52 \ \langle -60.25° \rangle \ \Omega = (923.22 - 1615.30\text{j}) \ \Omega$$

This impedance is in series with resistor R_1, so the equivalent impedance of the entire circuit is:

$$Z_{eq} = R_1 + \frac{(R_2 + Z_L)Z_C}{(R_2 + Z_L) + Z_C} = 1000 + (923.22 - 1615.30\text{j})) \ \Omega = 1923.22 - 1615.30\text{j} \ \Omega$$

From Equations 2.47 and 2.48, the phasor form of this impedance is:

$$Z_{eq} = 1923.22 - 1615.30\text{j} \ \Omega = 2511.57 \ \langle -40.03° \rangle \ \Omega$$

We can now find I_1 from Ohm's law:

$$I_1 = \frac{V_{in}}{Z_{eq}} = \frac{5 \langle 90° \rangle}{2511.57 \langle -40.03° \rangle} = 1.991 \ \langle 130.03° \rangle \ \text{mA}$$

Current division is used to find I

$$I = \frac{(R_2 + Z_L)}{(R_2 + Z_L) + Z_C} I_1 = \frac{3354.1 \langle 26.57° \rangle}{3004.63 \langle -3.18° \rangle} 1.991 \ \langle 130.03° \rangle \ \text{mA}$$

(continued)

(concluded)

which, using Equations 2.52 and 2.53, gives

$$I = 2.22 \langle 159.8° \rangle \text{ mA}$$

so the capacitor current leads the input reference by 159.8° or 2.789 rad, and the resulting current is

$$I(t) = 2.22 \cos(3000t + 2.789) \text{ mA}$$

Note that if the input voltage were $V_{in} = 5 \sin(3000t + \pi/2)$ V instead, the resulting current would be $I(t) = 2.22 \sin(3000t + 2.789)$ mA. But in this example, the reference was $\cos(3000t)$.

MathCAD Example 2.1 executes all of the analyses above in software. Phasors can be entered or displayed in polar or rectangular form, and all calculations are performed with ease. If you are not familiar with MathCAD, you might want to watch Video Demo 2.9, which describes and demonstrates the software and its capabilities.

MathCAD Example

2.1 AC circuit analysis

2.7 POWER IN ELECTRICAL CIRCUITS

Video Demo

2.9 MathCAD analysis software demo

All circuit elements dissipate, store, or deliver power through the physical interaction between charges and electromagnetic fields. An expression for power can be derived by first looking at the infinitesimal work (dW) done when an infinitesimal charge (dq) moves through an electric field resulting in a change in potential represented by a voltage V. This infinitesimal work is given by

$$dW = V dq \tag{2.63}$$

Because **power** is the rate of work done,

$$P = \frac{dW}{dt} = V\frac{dq}{dt} = VI \tag{2.64}$$

Therefore, the power consumed or generated by an element is simply the product of the voltage across and the current through the element. If the current flows in the direction of decreasing voltage as shown in Figure 2.31, P is negative, implying that the element is dissipating or storing energy. If the current flows in the direction of increasing voltage, P is positive, implying that the element is generating or releasing energy. The instantaneous power in a resistive circuit can be expressed as

$$P = VI = I^2R = V^2/R \tag{2.65}$$

For AC signals, because $V = V_m \sin(\omega t + \phi_V)$ and $I = I_m \sin(\omega t + \phi_I)$, the power changes continuously over a period of the AC waveform. Instantaneous power is not a useful quantity by itself, but if we look at the average power delivered over a period, we get a good measure of the circuit's or component's overall

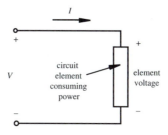

Figure 2.31 Power in a circuit element.

power characteristics. It can be shown (Question 2.42) that the average power over a period is

$$P_{avg} = \frac{V_m I_m}{2} \cos(\theta) \tag{2.66}$$

where θ is the difference between the voltage and current phase angles ($\phi_V - \phi_I$), which is the phase angle of the complex impedance $Z = V/I$.

If we use the rms, or **root-mean-square** values of the voltage and current defined by

$$I_{rms} = \sqrt{\frac{1}{T} \int_0^T I^2 \, dt} = \frac{I_m}{\sqrt{2}} \quad \text{and} \quad V_{rms} = \sqrt{\frac{1}{T} \int_0^T V^2 \, dt} = \frac{V_m}{\sqrt{2}} \tag{2.67}$$

the average AC power consumed by a resistor can be expressed in the same form as with DC circuits (see Question 2.43):

$$P_{avg} = V_{rms} I_{rms} = R I_{rms}^2 = V_{rms}^2 / R \tag{2.68}$$

■ **CLASS DISCUSSION ITEM 2.4**
Transmission Line Losses

When power is transmitted from power plants over large distances, high-voltage lines are used. Transformers (see Section 2.8) are used to change voltage levels both before and after transmission. Because current is lower at the higher voltage, less power is lost during the transmission, based on the middle expression for power in Equation 2.65. But doesn't the last expression imply that more power is lost at a higher voltage? How do you explain this apparent discrepancy?

■ CLASS DISCUSSION ITEM 2.5
International AC

In European countries, the household AC signal is 220 V_{rms} at 50 Hz. What effect does this have on electrical devices such as an electric razor purchased in the United States but used in these countries?

For AC networks with inductance and capacitance in addition to resistance, the average power consumed by the network can be expressed, using Equations 2.66 and 2.67, as

$$P_{avg} = I_{rms}V_{rms} \cos\theta = I_{rms}^2|Z| \cos\theta = (V_{rms}^2/|Z|) \cos\theta \tag{2.69}$$

where $|Z|$ is the magnitude of the complex impedance. $\cos\theta$ is called the **power factor,** because the average power dissipated by the network is dependent on this term.

■ CLASS DISCUSSION ITEM 2.6
AC Line Waveform

Draw a figure that represents one cycle of the AC voltage signal present at a typical household wall receptacle. What is the amplitude, frequency, period, and rms value for the voltage? Also, what is a typical rms current capacity for a household circuit?

2.8 TRANSFORMER

A transformer is a device used to change the relative amplitudes of voltage and current in an AC circuit. As illustrated in Figure 2.32, it consists of primary and secondary windings whose magnetic fluxes are linked by a ferromagnetic core.

Video Demo 2.10 shows an example of an actual transformer, in this case a laminated core, shell-type power transformer.

Using Faraday's law of induction and neglecting magnetic losses, the voltage per turn of wire is the same for both the primary and secondary windings, because the windings experience the same alternating magnetic flux. Therefore, the primary and secondary voltages (V_P and V_S) are related by

$$\frac{V_P}{N_P} = \frac{V_S}{N_S} = -\frac{d\phi}{dt} \tag{2.70}$$

where N_P is the number of turns in the primary winding, N_S is the number of turns in the secondary winding, and ϕ is the magnetic flux linked between the two coils. Thus, the secondary voltage is related to the primary voltage by

$$V_S = \frac{N_S}{N_P}V_P \tag{2.71}$$

Video Demo

2.10 Power transformer with laminated core

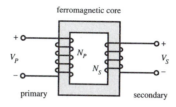

Figure 2.32 Transformer.

where N_S/N_P is the turns ratio of the transformer. If $N_S > N_P$, the transformer is called a **step-up transformer** because the voltage increases. If $N_S < N_P$, it is called a **step-down transformer** because the voltage decreases. If $N_S = N_P$, it is called an **isolation transformer,** and the output voltage is the same as the input voltage. All transformers electrically isolate the output circuit from the input circuit.

If we neglect losses in the transformer due to winding resistance and magnetic effects, the power in the primary and secondary circuits is equal:

$$I_P V_P = I_S V_S \qquad (2.72)$$

Substituting Equation 2.71 results in the following relation between the secondary and primary currents:

$$I_S = \frac{N_P}{N_S} I_P \qquad (2.73)$$

Thus, a step-up transformer results in lower current in the secondary and a step-down transformer results in higher current. An isolation transformer has equal alternating currents in both the primary and secondary. Note that any DC component of voltage or current in a transformer primary will not appear in the secondary. Only alternating currents are transformed.

■ **CLASS DISCUSSION ITEM 2.7**
DC Transformer

Can a transformer be used to increase voltage in a DC circuit? Why or why not?

2.9 IMPEDANCE MATCHING

Often we must be careful when connecting different devices and circuits together. For example, when using certain function generators to drive a circuit, proper **signal termination,** or loading, may be required as illustrated in Figure 2.33. Placing the 50 Ω termination resistance in parallel with a higher impedance network helps match the receiving network input impedance to the function generator output

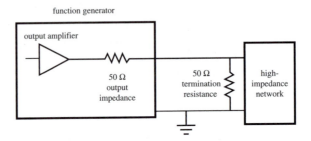

Figure 2.33 Signal termination.

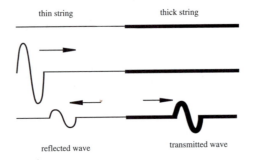

Figure 2.34 Impedance matching—string analogy.

impedance. This is called **impedance matching.** If we do not match impedances, a high-impedance network will reflect frequency components of the driving circuit (e.g., the function generator), especially the high-frequency components. A good analogy to this effect is a thin string attached to a thicker string. As illustrated in Figure 2.34, if we propagate transverse vibrations along the thin string, there will be partial transmission to the thick string and partial reflection back to the source. This is a result of the mismatch of the properties at the interface between the two strings.

In addition to signal termination concerns, impedance matching is important in applications where it is desired to transmit maximum power to a load from a source. This concept is easily illustrated with the simple resistive circuit shown in Figure 2.35 with source voltage V_s, source output impedance R_s, and load resistance R_L. The voltage across the load is given by voltage division:

$$V_L = \frac{R_L}{R_L + R_s} V_s \tag{2.74}$$

Therefore, the power transmitted to the load is

$$P_L = \frac{V_L^2}{R_L} = \frac{R_L}{(R_L + R_s)^2} V_s^2 \tag{2.75}$$

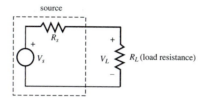

Figure 2.35 Impedance matching.

To find the load resistance that maximizes this power, we set the derivative of the power equal to 0 and solve for the load resistance:

$$\frac{dP_L}{dR_L} = V_s^2 \frac{(R_L + R_s)^2 - 2R_L(R_L + R_s)}{(R_L + R_s)^4} = 0 \qquad (2.76)$$

The derivative is 0 only when the numerator is 0, so

$$(R_L + R_s)^2 = 2R_L(R_L + R_s) \qquad (2.77)$$

Solving for R_L gives

$$R_L = R_s \qquad (2.78)$$

The second derivative of power can be checked to verify that this solution results in a maximum and not a minimum. The result of this analysis is as follows: To maximize power transmission to a load, the load's impedance should match the source's impedance.

■ **CLASS DISCUSSION ITEM 2.8**
Audio Stereo Amplifier Impedances

Why are audio stereo amplifier output impedances important specifications when selecting speakers?

■ **CLASS DISCUSSION ITEM 2.9**
Common Usage of Electrical Components

Cite specific examples in your experience where and how each of the following electrical components is used:

■ Battery
■ Resistor
■ Capacitor
■ Inductor
■ Voltage divider
■ Transformer

2.10 PRACTICAL CONSIDERATIONS

This chapter has presented all of the fundamentals and theory of basic electrical circuits. This final section presents various practical considerations that come up when trying to assemble actual circuits that function properly and reliably. The Laboratory Exercises book (see *mechatronics.colostate.edu/lab_book.html*) that accompanies this textbook provides some useful experiences to help you develop prototyping and measurement skills, and the sections below provide some additional supporting information.

2.10.1 Capacitor Information

As we saw in Section 2.2.1, determining resistance values from a discrete resistor component is very easy—a simple matter of looking up color values in a table. Unfortunately, capacitor labeling is not nearly as straightforward.

A capacitor is sometimes referred to as a "cap." Large caps are usually the electrolytic type that must be attached to a circuit with an indicated polarity. Because large capacitors have a large package size, the manufacturer usually prints the value clearly on the package, including the unit prefix. The only thing you need to be careful with is the capital letter M, which is often used to indicate micro, not mega. For example, an electrolytic capacitor labeled "+500MF" indicates a 500 μF capacitor.

It is very important to be careful with electrolytic-capacitor polarity. The capacitor's internal construction is not symmetrical, and you can destroy the cap if you apply the wrong polarity to the terminals: the terminal marked + must be at a higher voltage than the other terminal. Sometimes, violating this rule will result in gas formation internally that can cause the cap to explode. Improper polarity can also cause the cap to become shorted.

As the caps get smaller, determining the value becomes more difficult. Tantalum caps are silver-colored cylinders. They are polarized: a + mark and/or a metal nipple mark the positive end. An example label is +4R7m. This is fairly clear as long as you know that the "R" marks the decimal place: A +4R7m is a 4.7 mF (millifarad) cap. The same cap could also be labeled +475K, which you might think is 475 kilofarads, but you would be wrong. Here, the "K" is a tolerance indicator, not a unit prefix. "K" means $\pm 10\%$ (see more below). Capacitance values are usually quite small on the Farad scale. The values are usually in the microfarad (μF = 10^{-6} F) to picofarad (pF = 10^{-12} F) range. Labeling on tantalum caps mimics the resistor code system: 475 indicates 47 times ten to the fifth power, and the unit prefix pF is assumed. In general, if a cap's numerical value is indicated as a fraction (e.g., 0.01), the unit prefix will almost always be micro (μ); and if the value is a large integer (e.g., 47×10^5), pF will apply. The prefix nano (n = 10^{-9}) is usually not used for capacitance values. Returning to the example, a tantalum cap labeled "475" must be 47×10^5 pF, which is 4.7×10^6 pF, which is 4.7×10^{-6} F or 4.7 μF.

Mylar capacitors are usually yellow cylinders that are rather clearly marked. For example, ".01M" is just 0.01 μF. Mylar caps are not polarized, so you can orient them at random in your circuits. Because they are fabricated as long coils of metal foil (separated by a thin dielectric—the "mylar" that gives them their name), mylar caps betray their function at very high frequencies where the inductance of the coil become significant, blocking the very high frequencies you would expect a cap to pass. Ceramic caps, described next, are better in this respect, although they are very poor in other characteristics.

Ceramic caps have a flat, round shape (like pancakes) and are usually orange-colored. Because of their shape and construction (in contrast to the coiled mylars), they act like capacitors even at high frequencies. The trick in reading these is to ignore the markings that might accidentally be interpreted as units. For example, a ceramic cap labeled "Z5U .02M 1kV" is a 0.02 μF cap with a maximum voltage rating of 1kV. The M is a tolerance marking, in this case ± 20%.

CK05 caps have a small box shape, with their leads 0.2 inches apart so they can be easily inserted in prototyping or printed circuit boards. Therefore, they are common and useful. An example marking is 101K, which is 100 pF (10×10^1 pF), as described above.

The tolerance codes that often appear on capacitors are listed in Table 2.3. These codes apply to both capacitors and resistors with printed labels. Note that the Z tolerance code indicates a very tight tolerance if on a resistor, but a very large tolerance if on a capacitor!

Table 2.3 Capacitor and resistor tolerance codes

Code Letter	Meaning
Z	+80%, −20% for caps, ± 0.025 (precision resistors)
M	± 20%
K	± 10%
J	± 5%
G	± 2%
F	± 1%
D	± 0.5%
C	± 0.25%
B	± 0.1%
N	± 0.02%
A	± 0.005

More information dealing with practical considerations concerning capacitors can be found at Internet Link 2.8.

Internet Link

2.8 Capacitor practical considerations

2.10.2 Breadboard and Prototyping Advice

A **breadboard** is a convenient device for prototyping circuits in a form that can be easily tested and modified. Figure 2.36 illustrates a typical breadboard layout consisting of a rectangular matrix of insertion points spaced 0.1 inches apart. As

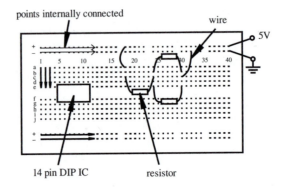

Figure 2.36 Breadboard.

shown in the figure, each column a through e, and f through j, is internally connected, as illustrated by the arrows in the diagram. The + and − rows that lie along the top and bottom edges of the breadboard are also internally connected to provide convenient DC voltage and ground busses. As illustrated in the figure, integrated circuits (ICs, or "chips") are usually inserted across the gap between columns a through e, and f through j. A 14-pin dual in-line package (DIP) IC is shown here. When the IC is placed across the gap, each pin of the IC is connected to a separate numbered column, making it easy to connect wires to and from the IC pins. The figure also shows an example of how to construct a simple resistor circuit. The schematic for this circuit is shown in Figure 2.37. The techniques for measuring voltage V_1 and current I_3 are described in Section 2.10.3. Figure 2.38 shows an example of a wired breadboard including resistors, an integrated circuit, and a push-button switch. When constructing such circuits, care should be exercised in trimming leads so the components lie on top of the breadboard in an organized geometric pattern. This will make it easier to see connections and find potential problems and errors later.

Video Demo 2.11 shows how a breadboard is constructed, and Video Demo 2.12 provides "rules of thumb" for how to properly assemble circuits on the board. Internet Link 2.9 is an excellent resource, providing useful tips when prototyping with breadboards (solderless protoboards), perf boards (soldered protoboards), and printed circuit boards (PCBs) for when a design is finalized.

Video Demo

2.11 Breadboard construction
2.12 Breadboard advice and rules of thumb

Internet Link

2.9 Prototyping tips

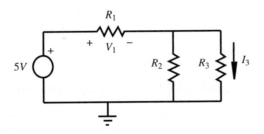

Figure 2.37 Example resistor circuit schematic.

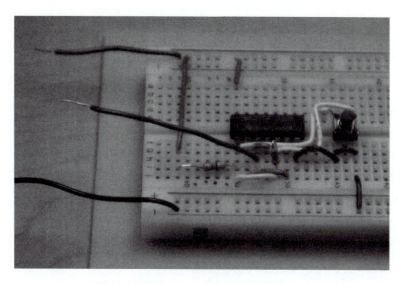

Figure 2.38 Example Breadboard Circuit.

Below is a set of basic guidelines you should follow when using breadboards to prototype circuits involving integrated circuits. Generally, if you carefully follow this protocol, you will save a lot of time and avoid a lot of frustration:

a. Start with a clearly drawn schematic illustrating all components, inputs, outputs, and connections.

b. Draw a detailed wiring diagram, using the information from datasheets regarding device pin-outs. Label and number each pin used on each IC and fully specify each component. This will be your wiring guide.

c. Double-check the functions you want to perform with each device and test them individually.

d. Insert the ICs into your breadboard.

e. Wire all connections carefully, checking off or highlighting each line on your schematic as you insert each wire. Select wire colors in a consistent and meaningful way (e.g., red for +5V, black for ground, other colors for signals), and use appropriate lengths (~ 1/4 in) for exposed wire ends. If the ends are too short, you might not establish good connections; and if too long, you might damage the breadboard or risk shorts. Also be careful to not insert component (e.g., resistor and capacitor) leads too far into the breadboard holes. This can also result in breadboard damage or shorting problems.

f. Be very gentle with the breadboards. Don't force wires into or out of the holes. If you do this, the breadboard might be damaged, and you will no longer be able to create reliable connections in the damaged holes or rows. Use a "chip puller" (small tool) to remove ICs from the breadboard to prevent bent or broken pins.

g. Make sure your wiring is very neat (i.e., not a "rat's nest"), and keep all of your wires as short as possible to minimize electrical magnetic interference (EMI) and added resistance, inductance, and capacitance.

h. Make sure all components and wires are firmly seated in the breadboard, establishing good connections. This is especially important with large ICs like PIC microcontrollers.

i. Double check the +5V and ground connections to each IC.

j. Before connecting the power supply, set the output to +5V and turn it off.

k. Connect the power supply to your breadboard and then turn it on.

l. Measure signals at inputs and outputs to verify proper functionality.

m. If your circuit is not functioning properly, go back through the above steps in reverse order checking everything carefully. If you are still having difficulty, use the beep continuity-check feature on a multimeter to verify all connections and to check for shorts (see Video Demo 2.13).

n. When prototyping with a soldered protoboard or PCB, use IC sockets to allow easy installation and removal of ICs.

Video Demo

2.13 Current measurement and checking continuity

2.10.3 Voltage and Current Measurement

It is very important that you know how to measure voltage and current, especially when prototyping a circuit. Figure 2.39 illustrates how you measure voltage across an element in a circuit, in this case a resistor. To measure voltage, the leads of the voltmeter are simply placed across the element. However, as shown in Figure 2.40, when measuring current through an element, the ammeter must be connected in series with the element. This requires physically altering the circuit to insert the ammeter in series. For the example in the figure, the top lead of resistor R_3 is removed from the breadboard to make the necessary connections to the ammeter. A demonstration of these techniques can be found in Video Demo 2.13. It is also important to be aware of input impedance effects, especially when measuring voltage across a large resistance or measuring current through a circuit branch with low resistance (see Section 2.4 for more information).

2.10.4 Soldering

Once a prototype circuit has been tested on a breadboard, a permanent prototype can be created by soldering components and connections using a **protoboard** (also called a perf board, perforated board, or vector board). These boards are manufactured with a regular square matrix of holes spaced 0.1 inch apart as with the insertion points in a breadboard. Unlike with the breadboard, there are no prewired connections between the holes. All connections must be completed with external wire and solder joints. The result is a prototype that is more robust and reliable.

For multiple versions of a prototype or production version of a circuit, a **printed circuit board (PCB)** is usually manufactured. Here, components are inserted and soldered to perforations in the board and all connections between the components

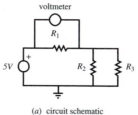

(a) circuit schematic

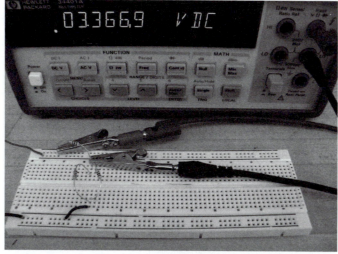

(b) photograph

Figure 2.39 Measuring voltage.

are "printed" on the board with a conducting medium. For more information about PCBs and how to make them, see Internet Links 2.10 and 2.11.

Solder is a metallic alloy of tin, lead, and other elements that has a low melting point (approximately 375°F). The solder usually is supplied in wire form, often with a core of flux, which facilitates melting, helps enhance wetting of the metal surfaces, and helps prevent oxidation. Solder is applied using a soldering iron consisting of a heated tip and support handle (see Figure 2.41). Some soldering irons also include a rheostat to control tip temperature. When using a soldering iron, be sure the tip is securely installed. Also, after heating, make sure the tip is clean and shiny, wiping it on a wet sponge or steel wool if necessary.

Here is a helpful list of steps you should follow to create a good solder connection:

(1) Before soldering, make sure you have everything you need: hot soldering iron, solder, components, wire, protoboard or PCB, tip-cleaning pad, and magnifying glass.

(2) Clean any surfaces that are to be joined. You can use fine emery paper, steel wool, or a metal brush to remove oxide layers and dirt so that the solder can easily wet the surface. Rosin core (flux) solder will enhance the wetting process.

Internet Link

2.10 ExpressPCB free PCB layout software and inexpensive manufacturing

2.11 Printed circuit board information resource

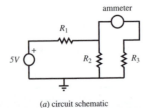

(a) circuit schematic

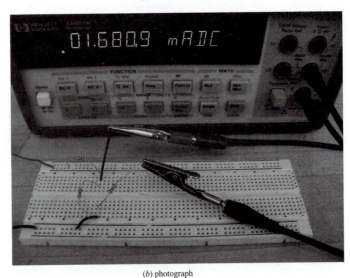

(b) photograph

Figure 2.40 Measuring current.

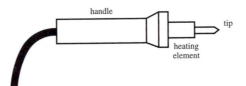

Figure 2.41 Soldering iron.

(3) Make a mechanical connection between the wires to be joined, either by bending or twisting, and ensure the components are secure so that they will not move when you apply the iron. Figure 2.42 illustrates two wires twisted together and a component inserted into a protoboard in preparation for soldering.

(4) Heating the wires and metal surfaces to be joined is necessary so that the solder properly wets the metal for a strong bond to result. When soldering electronic components, practice in heating is necessary so that the process is swift enough to not damage components. Soldering irons with sharper tips are convenient

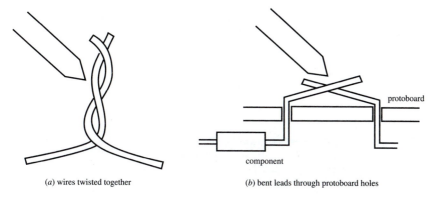

(*a*) wires twisted together   (*b*) bent leads through protoboard holes

Figure 2.42 Preparing a soldered joint.

for joining small electronic components, because they can deliver the heat very locally.

(5) When the surfaces have been heated momentarily, apply the solder to the work (not the soldering iron) and it should flow fluidly over the surfaces. Feed enough solder to provide a robust but not blobby joint. (If the solder balls up on the work, the iron is not hot enough.) Smoothly remove the iron and allow the joint to solidify momentarily. You should see a slight change in surface texture of the solder when it solidifies. If the joint is ragged or dull, you may have a cold joint, one where the solder has not properly wetted the surfaces. Such a joint will not have adequate or reliable conductivity and must be repaired by resoldering. Figure 2.43 illustrates a successful solder joint where the solder has wet both surfaces, in this case a component lead in a metal-rim-hole perforated board.

(6) If flux solvent is available, wipe the joint clean.

(7) Inspect your work with a magnifying glass to make sure the joint looks good.

Often you may have a small component or IC that you do not want to heat excessively. To avoid excessive heat, you can use a heat sink. A heat sink is a piece of metal like an alligator clip connected to the wire between the component and the connection to help absorb some of the heat that would be conducted to the component. However, if the heat sink is too close to the connection, it will be

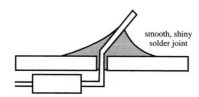

smooth, shiny
solder joint

Figure 2.43 Successful solder joint.

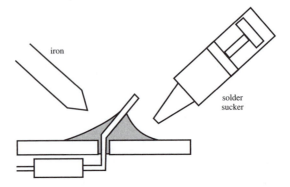

iron

solder
sucker

Figure 2.44 Removing a soldered joint.

Internet Link

2.12 Electronics Club "Guide to Soldering"

2.13 Curious Inventor–"How to Solder"

Video Demo

2.14 How and why to solder correctly

Lab Exercise

Lab 3 The oscilloscope

Video Demo

2.8 Oscilloscope demonstrations using the Tektronix 2215 analog scope

difficult to heat the wires. When using an IC, a socket can be soldered into the protoboard first, and then the IC inserted, thereby avoiding any thermal stress on the IC.

When using hook-up wire, be sure to use solid (nonbraided) wire on a protoboard, because it will be easy to manipulate and join. Wire must be stripped of its insulating cover before soldering. When using hook-up wire in a circuit, tinning the wire first (covering the end with a thin layer of solder) facilitates the joining process.

Often you may make mistakes in attaching components and need to remove one or more soldered joints. A solder sucker makes this a lot easier. To use a solder sucker (see Figure 2.44), cock it first, heat the joint with the soldering iron, then trigger the solder sucker to absorb the molten solder. Then the component can easily be removed, because very little solder will be left to hold it.

For more information and advice on how to solder properly, see Internet Links 2.12 and 2.13. Video Demo 2.14 is also an excellent resource.

2.10.5   The Oscilloscope

The oscilloscope, or o-scope for short, is probably one of the most widely used electrical instruments and is one of the most misunderstood. Lab Exercise 3 provides experiences to become familiar with the proper methods for connecting inputs, grounding, coupling, and triggering the oscilloscope. Video Demo 2.8 provides a demonstration of typical oscilloscope functionality, and this section provides information on some important oscilloscope concepts.

Most oscilloscopes are provided with a switch to select between AC or DC coupling of a signal to the oscilloscope input amplifier. When AC coupling is selected, the DC component of the signal is blocked by a capacitor inside the oscilloscope that is connected between the input terminal and the amplifier stage. Both AC and DC coupling configurations are illustrated in Figure 2.45. R_{in} is the input resistance (impedance) and C_{in} is the input capacitance. C_c is the coupling capacitor, which is present only when AC coupling is selected.

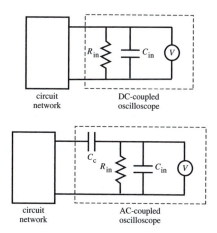

Figure 2.45 Oscilloscope coupling.

AC coupling must be selected when the intent is to block any DC component of a signal. This is important, for example, when measuring small AC spikes and transients on a 5 V TTL (transistor-transistor logic) supply voltage. However, it must be kept in mind that with AC coupling:

- One is not aware of the presence of any DC level with respect to ground.
- The lower-frequency components of a signal are attenuated.
- When the oscilloscope is switched from DC to AC coupling, it takes a little time before the display stabilizes. This is due to the time required to charge the coupling capacitor C_c to the value of the DC component (average value) of the signal.
- Sometimes the input time constant ($\tau = R_{in}C_c$) is quoted among the oscilloscope specifications. This number is useful, because after about five time constants (5τ), the displayed signal is stable.

AC coupling can be explained by considering the impedance of the coupling capacitor as a function of frequency: $Z_C = 1/(j\omega C)$. With a DC voltage ($\omega = 0$) the impedance of the capacitor is infinite, and all of the DC voltage at the input is blocked by the capacitor. For AC signals, the impedance is less than infinite, resulting in attenuation of the input signal dependent upon the frequency. As the input frequency increases, the attenuation decreases to zero.

Triggering is another important concept when attempting to display a signal on an oscilloscope. A trigger is an event that causes the signal to sweep across the display. If the signal being measured is periodic, and the trigger is consistent with each sweep, the signal will appear static on the display, which is desirable. The oscilloscope may be level triggered, where the sweep starts when the signal reaches a selectable voltage level, or slope triggered, where a certain rate of signal change triggers the sweep. Another triggering option available is line triggering, where the

AC power input is used to synchronize the sweep. Thus, any terminal voltage synchronized with the line frequency of 60Hz or multiples of 60Hz can be triggered in this mode. This is useful to detect if 60 Hz noise from various line-related sources is superimposed on a signal.

Normally all measurement instruments, power sources, and signal sources in a circuit must be referenced to a **common ground,** as shown in Figure 2.46. However, to measure a differential voltage ΔV, it is correct to connect the scope as shown in Figure 2.47. Note that the oscilloscope signal ground and external network ground are not common, allowing measurement of a potential difference anywhere in a circuit. However, in some oscilloscopes, each channel's "−" signal reference is attached to chassis ground, which is attached to the AC line ground. Therefore, to make a differential voltage measurement, you must use a two-channel signal difference feature, using the "+" leads of each channel for the measurement. An alternative for DC circuits is to measure the voltage at each node separately, relative to ground, and then manually subtract the voltage readings.

The **input impedance** of an oscilloscope is typically in the 1 MΩ range, which is fairly large. However, as described in Section 2.4, when measuring the voltage drop across an element whose impedance is of similar or greater magnitude than the input impedance, serious errors can result. One approach to avoid this problem is to increase the input impedance of the oscilloscope using an attenuator probe, which increases the input impedance by some known factor, at the same time decreasing

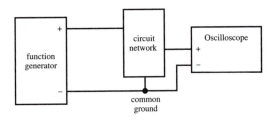

Figure 2.46 Common ground connection.

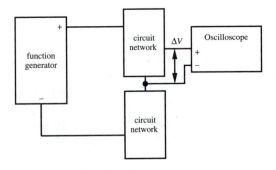

Figure 2.47 Relative ground connection.

the amplitude of the input signal by the same factor. Thus a 10X probe will increase the magnitude of the input impedance of the oscilloscope by a factor of 10, but the displayed voltage will be only 1/10 of the amplitude of the actual terminal voltage. Most oscilloscopes offer an alternative scale to be used with a 10X probe.

2.10.6 Grounding and Electrical Interference

It is important to provide a common ground defining a common voltage reference among all instruments and power sources used in a circuit or system. As illustrated in Figure 2.48, many power supplies have both a positive DC output (+ output) and a negative DC output (− output). These outputs produce both positive and negative voltages referenced to a common ground, usually labeled COM. On other instruments and circuits that may be connected to the power supply, all input and output voltages must be referenced to the same common ground. It is wise to double-check the integrity of each signal ground connection when assembling a group of devices.

It is important not to confuse the signal ground with the chassis ground. The **chassis ground** is internally connected to the ground wire on the power cord and may not be connected to the signal ground (COM). The chassis ground is attached to the metal case enclosing an instrument to provide user safety if there is an internal fault in the instrument (see Section 2.10.7).

Figure 2.49 illustrates an interference problem where high-frequency electrical **noise** can be induced in a signal by magnetic induction in the measurement leads. The area circumscribed by the leads encloses external magnetic fields from any AC magnetic sources in the environment, such as AC power lines or electric machinery. This would result in an undesirable magnetically induced AC voltage, as a result of Faraday's law of induction, given by

$$V_{\text{noise}} = A \cdot \frac{\mathrm{d}B}{\mathrm{d}t} \tag{2.79}$$

where A is the area enclosed by the leads and B is the external magnetic field. The measured voltage differs from the actual value according to

$$V_{\text{measured}} = V_{\text{actual}} + V_{\text{noise}} \tag{2.80}$$

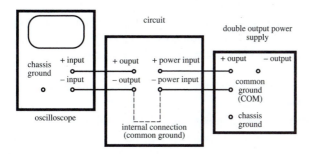

Figure 2.48 Common ground.

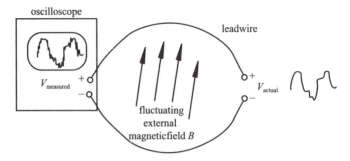

Figure 2.49 Inductive coupling.

■ CLASS DISCUSSION ITEM 2.10
Automotive Circuits

Often, electrical components in an automobile such as the alternator or starter motor are grounded to the frame. Explain how this results in an electrical circuit.

Many types of **electromagnetic interference** (EMI) can reduce the effectiveness and reliability of a circuit or system. Also, poorly designed connections within a circuit can cause noise and unwanted signals. These effects can be mitigated using a number of standard methods. The first approach is to eliminate or move the source of the interference, if possible. The source may be a switch, motor, or AC power line in close proximity to the circuit. It may be possible to remove, relocate, shield, or improve grounding of the interference source. However, this is not usually possible, and standard methods to reduce external EMI or internal coupling may be applied. Some standard methods are

■ Eliminate potential differences caused by **multiple point grounding.** A common ground bus (large conductor, plate, or solder plane) should have a resistance small enough that voltage drops between grounding points are negligibly small. Also, make the multiple point connections close to ensure that each ground point is at approximately the same potential.

■ Isolate sensitive signal circuits from high-power circuits using **optoisolators** or transformer couplings. Optoisolators are LED-phototransistor pairs (described in Chapter 3) that electrically decouple two sides of a circuit by transmitting a signal as light rather than through a solid electrical connection. One advantage is that the sensitive signal circuits are isolated from current spikes in the high-power circuit.

■ Eliminate inductive coupling caused by **ground loops.** When the distance between multiple ground points is large, noise can be inductively coupled to the circuit through the conducting loops created by the multiple ground

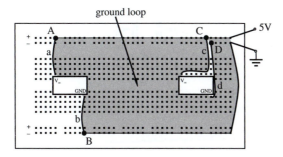

Figure 2.50 Ground loop.

points. Figure 2.50 illustrates how you should be careful to not create large ground loops when wiring a circuit on a breadboard. Both circuits shown are providing power (5V and ground) to an integrated circuit. The wiring on the left, connected at points *A* and *B* via wires *a* and *b* creates a large ground loop area which can pick up induced voltages in the presence of fluctuating magnetic fields, as described above. The wiring on the right, connected to points *C* and *D* via wires *c* and *d* creates very little area in the resulting circuit; therefore, very little induced voltage will occur even in the presence of external fields.

- Shield sensitive circuits with earth-grounded metal covers to block external electric and magnetic fields.

- Use short leads in connecting all circuits to reduce capacitive and inductive coupling between leads.

- Use "**decoupling**" or "**bypass**" **capacitors** (e.g., 0.1 µF) between the power and ground pins of integrated circuits to provide a short circuit for high-frequency noise.

- Use coaxial cable or twisted pair cable for high-frequency signal lines to minimize the effects of external magnetic fields.

- Use multiple-conductor shielded cable instead of ribbon cable for signal lines in the presence of power circuits (where large currents produce large magnetic fields) to help maintain signal integrity.

- If printed circuit boards are being designed, ensure that adequate **ground planes** are provided. A ground plane is a large surface conductor that minimizes potential differences among ground points.

Much more information and advice concerning how to prototype circuits properly and carefully can be found in Sections 2.10.2 and 7.10.4.

2.10.7 Electrical Safety

When using and designing electrical systems, safety should always be a concern. In the United States, electrical codes require outlets with three terminals: hot,

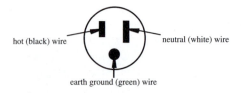

hot (black) wire   neutral (white) wire

earth ground (green) wire

Figure 2.51 Three-prong AC power plug.

neutral, and ground. Figure 2.51 illustrates the prongs on a plug that is inserted into an outlet. The wires in the plug cable include a black wire connected to the hot prong, a white wire connected to the neutral prong, and a bare or green wire connected to the ground prong. The two flat prongs (hot and neutral) of a plug complete the active circuit, allowing alternating current to flow from the wall outlet through an electrical device. The round ground prong is connected only to the chassis of the device and not to the power circuit ground in the device. The chassis ground provides an alternative path to earth ground, reducing the danger to a person who may contact the chassis when there is a fault in the power circuit. Without a separation between chassis and power ground, a high voltage can exist on the chassis, creating a safety hazard for the user because he or she can complete a path to ground. Removing the ground prong or using a three-prong-to-two-prong adapter carelessly creates a hazard (see v Discussion Items 2.11 and 2.12).

■ **CLASS DISCUSSION ITEM 2.11**
Safe Grounding

Consider the following oscilloscope whose power cord ground prong has been broken so that the chassis is not connected to ground. If you use this instrument, describe the possible danger you face.

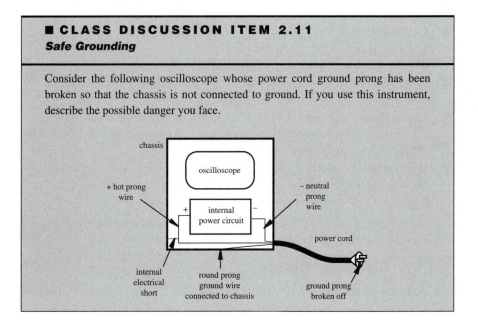

■ **CLASS DISCUSSION ITEM 2.12**
Electric Drill Bathtub Experience

The following electric drill runs on household power and has a metal housing. You use a three-prong-to-two-prong adapter to plug the drill into the wall socket. You are standing in a wet bathtub drilling a hole in the wall. You are unaware that the black wire's insulation has worn thin and the bare copper black wire is contacting the metal housing of the drill. How have you created a lethal situation for yourself? How could it have been prevented or mitigated?

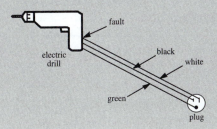

■ **CLASS DISCUSSION ITEM 2.13**
Dangerous EKG

A cardiac patient is lying in his hospital bed with electrocardiograph (EKG) leads attached to his chest to monitor his cardiac rhythm. An electrical short occurs in the next room, and our patient experiences a cardiac arrest. You and the hospital facilities engineer have determined that there were multiple grounding points in the patient's room (see the illustration), and a fault in electrical equipment in the next room caused current to flow in the ground wire from the piece of the equipment. You are on the scene to determine if there could have been a lethal current through the patient. Consider the fact that ground lines have finite resistance per unit length and that a few microamps through the heart can cause ventricular fibrillation (a fatal malfunction).

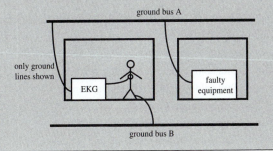

Video Demo

2.15 Human circuit toy ball

2.16 Squirrel zapped by power lines

2.17 Stupid man zapped by power lines

Electricity passing through a person can cause discomfort, injury, and even death. The human body, electrically speaking, is roughly composed of a low-resistance core (on the order of 500 Ω across the abdomen) surrounded by high-resistance skin (on the order of 10 kΩ through the skin when dry). When the skin is wet, its resistance drops dramatically. Currents through the body below 1 mA are usually not perceived. Currents as low as 10 mA can cause tingling and muscle contractions. Currents through the thorax as low as 100 mA can affect normal heart rhythm. Currents above 5 A can cause tissue burning. Video Demo 2.15 shows an electronic toy that illustrates how current can flow through human skin. In this case a person's hand is being used to complete a circuit to control a blinking LED. Video Demos 2.16 and 2.17 graphically illustrate what can happen to animals and humans when they are not careful around high-voltage lines.

QUESTIONS AND EXERCISES

Section 2.2   Basic Electrical Elements

2.1. What is the resistance of a kilometer-long piece of 14-gage (0.06408 inch diameter) copper wire?

2.2. Determine the possible range of resistance values for each of the following:
  a. Resistor R_1 with color bands: red, brown, yellow.
  b. Resistor R_2 with color bands: black, violet, orange.
  c. The series combination of R_1 and R_2.
  d. The parallel combination of R_1 and R_2.

  Note: the color bands are listed in order, starting with the first.

2.3. What colors should bands *a, b, c,* and *d* be for the following circuit B to have the equivalent resistance of circuit A?

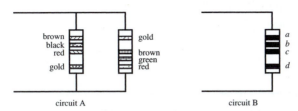

circuit A   circuit B

2.4. When using a trim pot in a circuit, it is usually placed in series with another fixed value resistor. Why is it not placed in parallel instead? Support your conclusions with analysis.

2.5. Document a complete and thorough answer to Class Discussion Item 2.1.

Section 2.3 Kirchhoff's Laws

2.6. Does it matter in which direction you assume the current flows when applying Kirchhoff's laws to a circuit? Why?

2.7. You quickly need a 50 Ω resistor but have a store of only 100 Ω resistors. What can you do?

2.8. Using Ohm's law, KVL, and KCL, derive an expression for the equivalent resistance of three parallel resistors (R_1, R_2, and R_3).

2.9. Derive current division formulas, similar to Equation 2.38, for three resistors in parallel.

2.10. Given two resistors R_1 and R_2, where R_1 is much greater than R_2, prove that the parallel combination is approximately equal to R_2.

2.11. Derive an expression for the equivalent capacitance of two capacitors attached in series.

2.12. Derive an expression for the equivalent capacitance of two capacitors attached in parallel.

2.13. Derive an expression for the equivalent inductance of two inductors attached in series.

2.14. Derive an expression for the equivalent inductance of two inductors attached in parallel.

2.15. Find I_{out} and V_{out} in the following circuit:

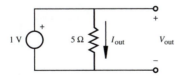

2.16. Find V_{out} in the following circuit:

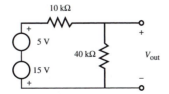

2.17. For the circuit in Question 2.27, with $R_1 = 1\ \text{k}\Omega$, $R_2 = 2\ \text{k}\Omega$, $R_3 = 3\ \text{k}\Omega$, and $V_{\text{in}} = 5$ V, find

    a. the current through R_1

    b. the current through R_3

    c. the voltage across R_2

2.18. For the circuit in Example 2.4, find

    a. the current through R_4

    b. the voltage across R_5

    You can use results from the example to help with your calculations.

2.19. Given the following circuit with $R_1 = 1\ \text{k}\Omega$, $R_2 = 2\ \text{k}\Omega$, $R_3 = 3\ \text{k}\Omega$, $R_4 = 4\ \text{k}\Omega$, $R_5 = 1\ \text{k}\Omega$, and $V_s = 10$ V, determine

    a. the total equivalent resistance seen by V_s

    b. the voltage at node A

    c. the current through resistor R_5

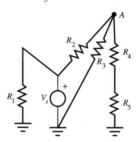

2.20. Given the following circuit with $R_1 = 2\ \text{k}\Omega$, $R_2 = 4\ \text{k}\Omega$, $R_3 = 5\ \text{k}\Omega$, $R_4 = 3\ \text{k}\Omega$, $R_5 = 1\ \text{k}\Omega$, and $V_s = 10$ V, determine

    a. the total equivalent resistance seen by V_s

    b. the voltage at node A

    c. the current through resistor R_5

    Also, how is this circuit different from the circuit in Question 2.19? If the resistance values were the same, would the circuits be identical? If not, what parts are different?

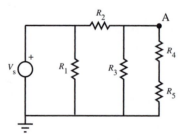

2.21. For the following circuit with $V_1 = 1$ V, $I_1 = 1$ A, $R_1 = 10\ \Omega$, and $R_2 = 100\ \Omega$, what is V_{R_2}?

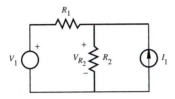

2.22. For the following circuit with $R_1 = 1$ kΩ, $R_2 = 9$ kΩ, $R_3 = 10$ kΩ, $R_4 = 1$ kΩ, $R_5 = 1$ kΩ, $V_1 = 5$ V, and $V_2 = 10$ V, find I and the voltage at node A.

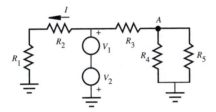

2.23. Find the equivalent resistance of the circuit below, as seen by voltage source V. Use the following values for the resistors: $R_1 = 1$ kΩ, $R_2 = 2$ kΩ, $R_3 = 3$ kΩ, $R_4 = 4$ kΩ, and $R_5 = 5$ kΩ.

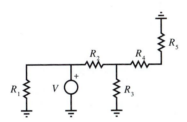

2.24. Solve for I_{out} and V_{out} in Example 2.4 by writing and solving KVL and KCL equations for all loops and nodes in the original circuit.

Section 2.4   Voltage and Current Sources and Meters

2.25. What is the output impedance of your laboratory DC power supply? What is the input impedance of your laboratory oscilloscope when DC coupled?

2.26. Explain why measuring voltages with an oscilloscope across impedances on the order of 1 MΩ may result in significant errors. Document your conclusions with analysis.

2.27. For the following circuit, what is V_{out} in terms of V_{in} for

    a. $R_1 = 50\ \Omega, R_2 = 10\ k\Omega, R_3 = 1.0\ M\Omega$?

    b. $R_1 = 50\ \Omega, R_2 = 500\ k\Omega, R_3 = 1.0\ M\Omega$?

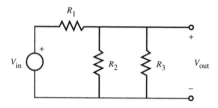

If R_3 represents the input impedance associated with a device measuring the voltage across R_2, what conclusions can you make about the two voltage measurements?

2.28. For the circuit in Question 2.27, if R_1 represents the output impedance of a voltage source and R_3 is assumed to be infinite (representing an ideal voltmeter), what effect does R_1 have on the voltage measurement being made? Also, what would the effect be for each of the R_2 values in Question 2.27? Please comment on the results.

Section 2.5 Thevenin and Norton Equivalent Circuits

2.29. What is the Thevenin equivalent of your laboratory DC power supply?

2.30. What is the Thevenin equivalent of the circuit in Question 2.27 for the resistance values in part "a?"

Section 2.6 Alternating Current Circuit Analysis

2.31. For the circuit in Example 2.7, find the steady state voltage across the capacitor as a function of time. You can use results from the example to help with your calculations.

2.32. For the following circuit, what are the steady state voltages across R_1, R_2, and C, if $V_s = 10\ V\ DC, R_1 = 1\ k\Omega, R_2 = 1\ k\Omega$, and $C = 0.01\ \mu F$?

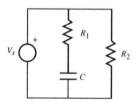

2.33. Find the steady state current $I(t)$ in the following circuit, where $R_1 = R_2 = 100\ k\Omega$, $C = 1\ \mu F$, and $L = 20\ H$ for

    a. $V_s = 5\ V\ DC$

    b. $V_s = 5\ \cos(\pi t)\ V$

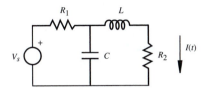

2.34. For each of these waveforms, what is the frequency in Hz and in rad/sec, the peak-to-peak amplitude, and the DC offset?

    a. $2.0 \sin(\pi t)$

    b. $10.0 + \cos(2\pi t)$

    c. $3.0 \sin(2\pi t + \pi)$

    d. $\sin(\pi) + \cos(\pi)$

Section 2.7   Power in Electrical Circuits

2.35. If 100 volts rms is applied across a 100 Ω power resistor, what is the power dissipated in watts?

2.36. If 100 volts peak-to-peak is applied across a 100 Ω power resistor, what is the power dissipated in watts?

2.37. If standard U.S. household voltage is 120 volts rms, what is the peak voltage that would be observed on an oscilloscope?

2.38. Write a function to represent a typical household voltage signal.

2.39. A circuit designer needs to choose an appropriate size resistor to be used in series with a light emitting diode (LED). The LED manufacturer claims that the LED requires 2 V to keep it on and 10 mA to generate bright light. Also, the current should not exceed 100 mA. Assuming that a 5 V source is being used to drive the LED circuit, what range of resistance values would be appropriate for the job? Also, what resistor power rating would be required?

2.40. For the following circuit with $R_1 = 1$ kΩ, $R_2 = 2$ kΩ, $R_3 = 3$ kΩ, $R_4 = 4$ kΩ, $V_1 = 10$ V, $V_2 = 5$ V, and $V_3 = 10$ V, find

    a. V_{out}

    b. the power produced by each voltage source

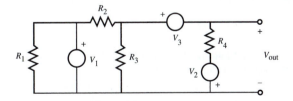

2.41. Solve the previous question with $R_3 = 2$ kΩ and $R_4 = 1$ kΩ, keeping everything else the same.

2.42. Prove Equation 2.66.

2.43. Derive the rms expressions in Equation 2.67 and show that Equation 2.68 is correct.

2.44. Sketch the output waveform for V_{out} in the following circuit on the axes as shown:

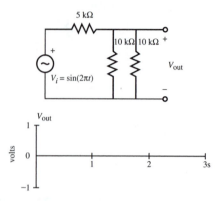

2.45. Document a complete and thorough answer to Class Discussion Item 2.6.

Section 2.8   Transformer

2.46. If you were to design a transformer for 24 V_{rms} low-voltage lighting in a new kitchen, what should the turns ratio of the primary to secondary windings be to provide a satisfactory voltage source?

Section 2.9   Impedance Matching

2.47. If your audio stereo amplifier has an output impedance of 8 Ω, what resistance should your speaker coils have to maximize the generated sound power?

Section 2.10   Grounding and Electrical Interference

2.48. When making high-frequency voltage measurements with an oscilloscope, why is it good practice to use BNC (coaxial) cable rather than two separate wires to the probe?

BIBLIOGRAPHY

Horowitz, P. and Hill, W., *The Art of Electronics,* 2nd Edition, Cambridge University Press, New York, 1989.

Johnson, D., Hilburn, J., and Johnson, J., *Basic Electric Circuit Analysis,* 2nd Edition, Prentice-Hall, Englewood Cliffs, NJ, 1984.

Lerner, R. and Trigg, G., *Encyclopedia of Physics,* VCH Publishers, New York, 1991.

McWhorter, G. and Evans, A., *Basic Electronics,* Master Publishing, Richardson, TX, 1994.

Mims, F., *Getting Started in Electronics,* Radio Shack Archer Catalog No. 276-5003A, 1991.

CHAPTER 3

Semiconductor Electronics

This chapter presents semiconductor diodes and transistors, important for sensing, interfacing, and display in mechatronic systems. ∎

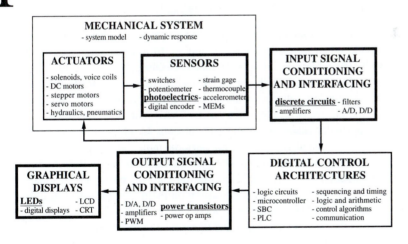

CHAPTER OBJECTIVES

After you read, discuss, study, and apply ideas in this chapter, you will:

1. Comprehend the basic physics of semiconductor devices

2. Be aware of the different types of diodes and how they are used

3. Know the similarities and differences between bipolar junction transistors and field-effect transistors

4. Understand how a transistor can be used to switch current to a load

5. Be able to design circuits using diodes, voltage regulators, bipolar transistors, and field-effect transistors

6. Be able to select semiconductor components for your designs

3.1 INTRODUCTION

We will examine some extraordinary materials that scientists and engineers have transformed into inventions that affect all aspects of life in the 21st century and beyond. To understand these inventions, we need to understand the physical characteristics of a class of materials known as semiconductors, which are used extensively in electronic circuits today. We examine the physics of semiconductors, discuss how electronic components are designed using different types of semiconductor materials, learn the circuit schematic symbols for different semiconductor diodes and transistors, and use the devices in circuit design.

3.2 SEMICONDUCTOR PHYSICS AS THE BASIS FOR UNDERSTANDING ELECTRONIC DEVICES

Metals have a large number of weakly bound electrons in what is called their conduction band. When an electric field is applied to a metal, the electrons migrate freely producing a current through the metal. Because of the ease by which large currents can flow in metals, they are called **conductors.** In contrast, other materials have atoms with valence electrons that are tightly bound, and when an electric field is applied, the electrons do not move easily. These materials are called **insulators** and do not normally sustain large electric currents. In addition, a very useful class of materials, elements in group IV of the periodic table, have properties somewhere between conductors and insulators. They are called **semiconductors.** Semiconductors such as silicon and germanium have current-carrying characteristics that depend on temperature or the amount of light falling on them. As illustrated in Figure 3.1, when a voltage is applied across a semiconductor, some of the valence electrons easily jump to the conductance band and then move in the electric field to produce a current, although smaller than that which would be produced in a conductor.

In a semiconductor crystal, a valence electron can jump to the conduction band, and its absence in the valence band is called a hole. A valence electron from a nearby atom can move to the hole, leaving another hole in its former place. This chain of events can continue, resulting in a current that can be thought of as the movement of holes in one direction or electrons in the other. The net effect is the same, so perhaps Ben Franklin wasn't completely wrong when he thought currents were the movement of positive charges, the common convention used today.

The properties of pure semiconductor crystals can be significantly changed by inserting small quantities of elements from group III or group V of the periodic table into the crystal lattice of the semiconductor. These elements, known as **dopants,** can

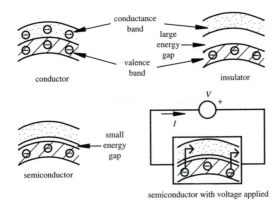

Figure 3.1 Valence and conduction bands of materials.

be diffused or implanted into semiconductors. A thin crystal of silicon, often called a chip, can have a minute pattern of dopants deposited on and diffused into its surface, resulting in devices that are the basis of all modern electronics.

Properties really get interesting when different amounts and different types of dopants are added to semiconductors. Consider what happens if dopants are embedded in the crystal lattice of silicon. Silicon has four valence electrons that form symmetrical electron bonds in the crystal lattice. However, if arsenic or phosphorous from group V is added to the crystal lattice, one of the five valence electrons in each dopant atom remains freer to move around. In this case, the dopant is called a **donor** element because it enhances the electron conductivity of the semiconductor. The resulting semiconductor is called **n-type** silicon due to the electrons available in the crystal lattice as charge carriers. Conversely, if the silicon is doped with boron or gallium from group III, **holes** form due to missing electrons in the lattice where the so-called **acceptor** dopant atoms have replaced silicon atoms. This is because the dopant atom only has three valence electrons. A hole can jump from atom to atom, effectively producing a positive current. What really happens is that electrons move to occupy the holes, and this effectively looks like holes moving. The resulting semiconductor is called **p-type** silicon due to the holes, which are effectively positive charge carriers. In summary, the purpose for doping a semiconductor such as silicon is to elevate and control the number of charge carriers in the semiconductor. In an n-type semiconductor, the charge carriers are electrons, and in a p-type semiconductor, they are holes. As we will see shortly, the interaction between n-type and p-type semiconductor materials is the basis for most semiconductor electronic devices.

3.3 JUNCTION DIODE

Contemporary electronic devices are produced by creating microscopic interfaces between differently doped areas within semiconductor material. If a p-type region of silicon is created adjacent to an n-type region, a **pn junction** is the result. The p-type side of the diode is referred to as the **anode,** and the n-type side is called the **cathode.** As illustrated in the top of Figure 3.2, at the pn junction, electrons from the n-type silicon can diffuse to occupy the holes in the p-type silicon, creating what is called a **depletion region.** A small electric field develops across this thin depletion region due to the diffusion of electrons. This results in a voltage difference across the depletion region called the **contact potential.** For silicon, the contact potential is on the order of 0.6–0.7 V. The positive side of the contact potential is in the n-type region, and the negative side is in the p-type region due to the diffusion of the electrons. Note that we still have not connected the junction to an external circuit.

Now, as shown in the bottom left of Figure 3.2, if a voltage source is connected to the pn junction with the positive side of the voltage source connected to the anode and the negative side connected to the cathode forming a complete circuit, the diode is said to be **forward biased.** The applied voltage overcomes the contact potential and shrinks the depletion region. The anode in effect becomes a source of holes and the cathode becomes a source of electrons so that holes and electrons are continuously replenished at the junction. As the applied voltage approaches the value of the contact potential (0.6–0.7 V for silicon), the current increases exponentially. This effect is quantitatively described by the **diode equation:**

$$I_D = I_0 \left(e^{\frac{qV_D}{kT}} - 1 \right) \qquad (3.1)$$

where I_D is the current through the junction, I_0 is the reverse saturation current, q is the charge of one electron (1.60×10^{-19} C), k is Boltzmann's constant (1.381×10^{-23} J/K), V_D is the forward bias voltage across the junction, and T is the absolute temperature of the junction in Kelvin.

If, as shown in the bottom right of Figure 3.2, the anode is connected to the n-type silicon and the cathode to the p-type silicon, the depletion region is enlarged, inhibiting diffusion of electrons and thus current; and we say the junction is **reverse biased.** A **reverse saturation current** (I_0) does flow, but it is extremely small (on the order of 10^{-9} to 10^{-15} A).

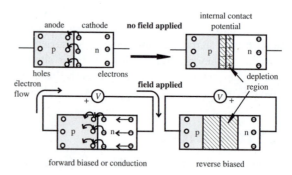

Figure 3.2 pn junction characteristics.

Therefore, a pn junction passes current in only one direction. It is known as a silicon **diode** and is sometimes referred to as a **rectifier.** The schematic symbol for the silicon diode is included in Figure 3.3. Figure 3.4 and Video Demo 3.1 show examples of various common diodes. Included are a small signal diode, a small power diode, and various types of light-emitting diodes. LEDs are described more in Section 3.3.3, and seven-segment displays are presented in Section 6.12.1. The diode is analogous to a fluid check valve, which allows fluid to flow only in one direction as illustrated in Figure 3.5. We will soon see that pn junctions also occur in more advanced devices like transistors and integrated circuits. As we will see, the on-off action of the pn junction provides the basis for all digital devices.

Video Demo

3.1 Diodes

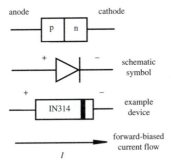

Figure 3.3 Silicon diode.

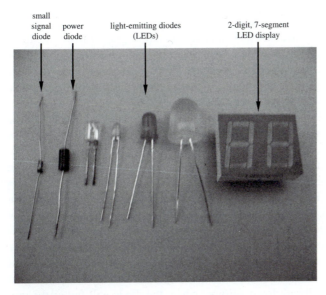

Figure 3.4 Examples of common diodes.

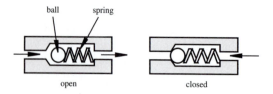

Figure 3.5 Diode check valve analogy.

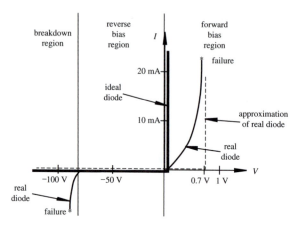

Figure 3.6 Ideal, approximate, and real diode curves.

As described by Equation 3.1, the current-voltage characteristic curve for a semiconductor diode is exponential and is shown graphically in the first quadrant of Figure 3.6 (the curve labeled "real diode"). There is a dramatic nonlinear increase in current as the forward bias voltage approaches 0.7 V. Note the different scales used on the positive and negative sides of the voltage axis. In a first analysis, we approximate the behavior of the semiconductor diode using what we call an **ideal diode** model. The current-voltage characteristic curve for the ideal diode is shown by the dark solid lines in Figure 3.6. This model implies that the diode is fully on for any voltage greater than or equal to 0. Also, when reverse biased, the reverse saturation current is assumed to be 0. Later in actual circuit design, a good first approximation for the **real diode** is given by the dashed lines as this replicates the real voltage drop of 0.6 to 0.7 V measured across the silicon diode when it is forward biased. In summary, an ideal diode has zero resistance when forward biased and infinite resistance when reverse biased. For analytic purposes, it can be replaced by a short circuit if it is forward biased and an open circuit if it is reverse biased. A real diode requires about 0.7 V of forward bias to enable significant current flow. When a real diode is reverse biased, it can withstand a reverse voltage up to a limit known as the **breakdown voltage,** where the diode will fail as the reverse current increases precipitously. We will see in the next section that there is also a class of diodes, called zener diodes, designed for use in the reverse bias region for special applications.

A diode is useful as a rectifier, where it passes only the positive half or the negative half of an AC signal. Example 3.1 illustrates how to analyze a simple ideal diode circuit called a half-wave rectifier. Rectifier circuits are used in the design of power supplies, where AC power must be transformed into DC power for use in electronic devices and digital circuits.

The important specifications that differentiate diodes are the maximum forward current and the maximum reverse bias voltage where breakdown occurs. The instantaneous surge current and average current are usually both specified, and the values calculated for a circuit must not exceed these limits. You must also confirm that reverse bias voltages in your circuit do not exceed the specified breakdown value. Rectifier and power diodes are capable of carrying very large currents. They are designed to be attached to heat sinks in order to efficiently dissipate heat produced in the junction. Diodes require nanoseconds to switch between their on and off states. This switching time is fast enough for most applications, but when designing high-speed circuits it may pose a constraint.

Half-Wave Rectifier Circuit Assuming an Ideal Diode   EXAMPLE 3.1

Given the following circuit containing a diode, we will illustrate how to determine the output voltage V_{out} given a sinusoidal input V_{in}.

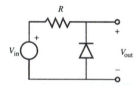 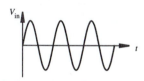

A good approach to solving this problem is to analyze separately the response when $V_{in} > 0$ and then when $V_{in} < 0$. When V_{in} is positive, the diode is reverse biased and therefore equivalent to an open circuit. No current flows through the resistor, and the output V_{out} equals V_{in}. When V_{in} is negative, the diode is forward biased, and it is equivalent to a short circuit. Therefore, there is no voltage drop across the diode and V_{out} is 0 V. Combining these two cases, the resulting output waveform retains the positive peaks in the sine wave and loses the negative peaks (see the following figure). Because only the positive half of the wave remains, this circuit is known as a **half-wave rectifier.** Question 3.6 at the end of the chapter deals with a full-wave rectifier.

■ CLASS DISCUSSION ITEM 3.1
Real Silicon Diode in a Half-Wave Rectifier

In Example 3.1, we assumed that the diode was ideal. The first approximation to a real diode assumes that 0.7 V is required to forward bias the diode. Using the current-voltage relation shown by the dashed curve in Figure 3.6, show how the output of the half-wave rectifier would be different.

■ CLASS DISCUSSION ITEM 3.2
Inductive "Kick"

The following inductive circuit illustrates a common application of a diode to reduce current arcs (sparks) between the switch contacts when the switch is opened. Diodes used in this way are called **flyback, freewheeling,** or **snubber** diodes. Arcs can damage the switch and can create electromagnetic interference (EMI) that can affect surrounding circuits. Why does a switch arc when it is used to open an inductive circuit? What is the purpose of the diode? Consider the current flow in the inductor and how it changes as a function of time. Start with the switch closed and then describe what happens when it is opened.

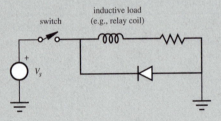

■ CLASS DISCUSSION ITEM 3.3
Peak Detector

The following circuit is known as a **peak detector.** When a time-varying signal V_{in} is applied at the input, the output V_{out} retains the maximum positive value of the input signal. Under what condition does the capacitor charge? Sketch an arbitrary input signal and the resulting ideal output. What behavior would you expect from an actual circuit where the capacitor is "leaky"; that is, the capacitor's charge gradually dissipates? Draw the resulting V_{out} for the "real" (nonideal) capacitor.

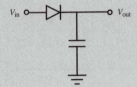

Lab Exercise 4 introduces diodes and how they are used in basic circuits. The Lab also shows the differences between signal diodes and LEDs.

3.3.1 Zener Diode

Reflect back on the current-voltage relationship for a diode shown in Figure 3.6. Note that when a diode is reverse biased with a large enough voltage, the diode allows a large reverse current to flow. This is called diode **breakdown.** For most diodes the breakdown value is at least 50 V and may extend to kilovolts. A special class of diodes is designed to exploit this characteristic. They are known as **zener,** avalanche, or voltage-regulator **diodes.** This family of diodes exhibits steep breakdown curves with well-defined breakdown voltages; thus, they can maintain a nearly constant voltage over a wide range of currents (see Figure 3.7). This characteristic makes them good candidates for building simple voltage regulators, because they can maintain a stable DC voltage in the presence of a variable supply voltage and variable load resistance.

To properly use the zener diode in a circuit, the zener should be reverse biased with a voltage kept in excess of its breakdown or **zener voltage** V_z. Using a zener diode in series with a resistor as shown in Figure 3.8 results in a simple circuit known as a **voltage regulator.** The output voltage of the circuit V_{out} is maintained or regulated by the zener diode at the zener voltage V_z. Even when the current through the zener diode changes (ΔI_z in the figure), the output voltage remains relatively constant (i.e., ΔV_z is small). The narrowness of the voltage range for a given current change is a measure of the voltage regulation of the circuit. If the input voltage and load do not change much, this circuit is effective in obtaining steady and lower DC voltage values from a source, even if the source is not well regulated.

Because the load applied to the voltage regulator will change with time in most applications and the voltage source will exhibit fluctuations, careful consideration

Lab Exercise

Lab 4 Bandwidth, filters, and diodes

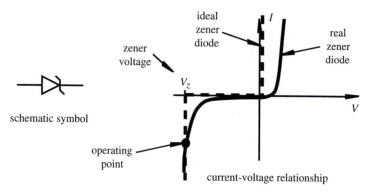

Figure 3.7 Zener diode symbol and current-voltage relationship.

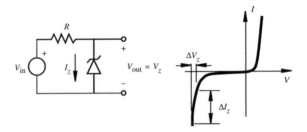

Figure 3.8 Zener diode voltage regulator.

must be paid to the effect on the regulated voltage V_z. For the circuit shown in Figure 3.8, the zener current is related to the circuit voltages according to

$$I_z = \frac{(V_{in} - V_z)}{R} \tag{3.2}$$

To determine how changes in current are related to changes in voltage, we take the finite differential of Equation 3.2, which yields

$$\Delta I_z = \frac{1}{R}(\Delta V_{in} - \Delta V_z) \tag{3.3}$$

The zener diode is a nonlinear circuit element, and therefore ΔV_z is not directly proportional to ΔI_z. However, it is useful to define a dynamic resistance R_d that is the slope of the zener characteristic curve at a particular operating point. This allows us to express the zener current change in terms of the zener voltage change:

$$\Delta I_z = \frac{\Delta V_z}{R_d} \tag{3.4}$$

Normally a manufacturer specifies the nominal zener current I_{zt} and the maximum dynamic impedance (R_d) at the nominal zener current. In a circuit design using a zener diode, the zener current must exceed I_{zt}; otherwise, the zener may operate near the "knee" of the characteristic curve where regulation is poor (i.e., where there is a large change in voltage with a small change in zener current).

By substituting Equation 3.4 into Equation 3.3 and solving for ΔV_z, we can express changes in the regulator output voltage ΔV_{out} in terms of fluctuations in the source voltage ΔV_{in}:

$$\Delta V_{out} = \Delta V_z = \frac{R_d}{R_d + R} \Delta V_{in} \tag{3.5}$$

Therefore, the circuit acts like a voltage divider (for a change in voltage) with the zener diode represented by its dynamic resistance at the operating current of the circuit.

Zener Regulation Performance

EXAMPLE 3.2

We wish to determine the regulation performance of the zener diode circuit shown in Figure 3.8 for a voltage source V_{in} whose value ranges between 20 and 30 V. For the zener diode, we select a 1N4744A manufactured by National Semiconductor from the family 1N4728A to 1N4752A (having different zener voltage values). It is a 15 V, 1 W zener diode. We select a value of R based on the specifications of this diode.

To limit the maximum power dissipation to less than 1 W, the current through the diode must be limited to

$$I_{z_{max}} = 1 \text{ W}/15 \text{ V} = 66.7 \text{ mA}$$

Therefore, using Equation 3.2, the value for resistance R should be chosen to be at least

$$R_{min} = (V_{in_{max}} - V_z)/I_{z_{max}} = (30 \text{ V} - 15 \text{ V})/66.7 \text{ mA} = 225 \text{ }\Omega$$

The closest acceptable standard resistance value is 240 Ω. From the manufacturer's specifications for this zener diode, its dynamic resistance R_d is 14 Ω at 17 mA. The current I_z in this example is larger than this value, so the operating point of the zener diode is on the well-regulated portion of the characteristic curve. Using the given value for R_d in Equation 3.5, we can approximate the resulting output voltage range:

$$\Delta V_{out} = \Delta V_z = \frac{R_d}{R_d + R} \Delta V_{in} = \frac{14}{14 + 240} (30 - 20) \text{ V} = 0.55 \text{ V}$$

which is a measure of regulation of this circuit. This can be expressed as a percentage of the output voltage for a relative measure:

$$\frac{\Delta V_{out}}{V_{out}} 100\% = \frac{0.55 \text{ V}}{15 \text{ V}} 100\% = 3.7\%$$

■ CLASS DISCUSSION ITEM 3.4
Effects of Load on Voltage Regulator Design

Example 3.2 ignored the current that would be drawn by a load. What effect would a load have on the results of the analysis?

Figure 3.9 illustrates a simple voltage regulator circuit where R_L is a load resistance and V_{in} is an unregulated source whose value exceeds the zener voltage V_z. The purpose of this circuit is to provide a constant DC voltage V_z across the load with a corresponding constant current through the load. Providing a stable regulated voltage to a system containing digital integrated circuits is a common application.

If we assume the zener diode is ideal (i.e., its breakdown current-voltage curve is vertical), we can draw some conclusions about the regulator circuit. First, the load

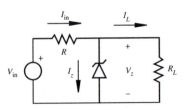

Figure 3.9 Zener diode voltage regulator circuit.

voltage will be V_z as long as the zener diode is subject to reverse breakdown. Therefore, the load current I_L is

$$I_L = \frac{V_z}{R_L} \qquad (3.6)$$

Second, the load current will be the difference between the unregulated input current I_{in} and the zener diode current I_z:

$$I_L = I_{in} - I_z \qquad (3.7)$$

As long as V_z is constant and the load does not change, I_L remains constant. This means that the diode current changes to absorb changes from the unregulated source.

Third, the unregulated source current I_{in} is given by

$$I_{in} = \frac{(V_{in} - V_z)}{R} \qquad (3.8)$$

R is known as a current-limiting resistor because it limits the power dissipated by the zener diode. If I_z gets too large, the zener diode fails.

DESIGN EXAMPLE 3.1

Zener Diode Voltage Regulator Design

Suppose we need to design a regulated 15 V DC source to power a mechatronic system, and we would like to use the voltage regulator circuit shown in Figure 3.9. Furthermore, suppose we have access to only a poorly regulated DC source V_{in} whose nominal value is 24 V.

As the load R_L changes, the zener current I_z increases for larger R_L and decreases for smaller R_L. If we know the maximum possible load resistance (assuming that the output never is an open circuit), we can size the zener diode with regard to its power dissipation characteristics and select a current-limiting resistor. Combining Equations 3.6 and 3.7 and using the maximum value of the load $R_{L_{max}}$ gives

$$I_{z_{max}} = \left(I_{in} - \frac{V_z}{R_{L_{max}}} \right)$$

This is the largest current the zener experiences. The power dissipated by the zener diode is

$$P_{z_{max}} = I_{z_{max}} V_z = \left(I_{in} - \frac{V_z}{R_{L_{max}}} \right) V_z$$

I_{in} is controlled by the current-limiting resistor R. Substituting Equation 3.8 yields

$$P_{z_{max}} = \left(\frac{V_{in} - V_z}{R}\right)V_z - \frac{V_z^2}{R_{L_{max}}}$$

Furthermore, for this design problem, we assume that $R_{L_{max}}$ is 240 Ω, and we wish to select a 1 W zener. Therefore,

$$1\ W = \frac{24\ V - 15\ V}{R_{min}}(15\ V) - \frac{225\ V^2}{240\ Ω}$$

We can now solve for the minimum required current-limiting resistance R:

$$R_{min} = 69.7\ Ω$$

The closest acceptable standard resistance value is 75 Ω.

In summary, zener diodes are useful in circuits where it is necessary to derive smaller regulated voltages from a single higher-voltage source. When designing zener diode circuits, one must select appropriate current-limiting resistors given the power limitations of the diodes. In simple mechatronic designs that may be powered by a 9 V battery and require good 5 V DC supplies for digital devices, a well-designed zener regulator is a cheap and effective solution if the current requirements are modest.

3.3.2 Voltage Regulators

Although the zener diode voltage regulator is cheap and simple to use, it has some drawbacks: The output voltage cannot be set to a precise value, and regulation against source ripple and changes in load is limited. Special semiconductor devices are designed to serve as voltage regulators, some for fixed positive or negative values and others easy to adjust to a desired, nonstandard value. One group of regulators that is easy to use is the three-terminal regulator designated as the 78XX, where the last two digits (XX) specify a voltage with standard values: 5 (05), 12, or 15 V. Using a regulator such as the LM7815C, a well-regulated 15 V source is easy to create, as shown in Figure 3.10.

We could use this design instead of the zener regulator shown in Design Example 3.1 (see Class Discussion Item 3.5). The 78XX can deliver up to 1 amp of current and is internally protected from overload. Using this device, the designer

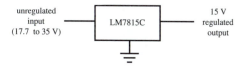

Figure 3.10 15 V regulated DC supply.

need not perform the design calculations shown with the zener diode regulator. The 78XX series of regulators have complementary 79XX series values for the design of $+/-$ voltage supplies.

■ **CLASS DISCUSSION ITEM 3.5**
78XX Series Voltage Regulator

In Design Example 3.1, we used a zener diode to provide a desired DC voltage. Now show how a 78XX voltage regulator can do the same job. Specify the regulator and describe its characteristics.

In some cases, you may need a regulated voltage source with a value not provided in a manufacturer's standard sequence. Then you may use a three-terminal regulator designed to be adjustable by the addition of external resistors. The LM317L can provide an adjustable output with the addition of two external resistors as shown in Figure 3.11. The output voltage is given by

$$V_{\text{out}} = 1.25\left(1 + \frac{R_2}{R_1}\right) \text{ V} \tag{3.9}$$

These adjustable regulators are available in higher current and voltage ratings.

Three-terminal voltage regulators are accurate, reject ripple on the input, reject voltage spikes, have roughly a 0.1% regulation, and are quite stable, making them useful in mechatronic system design.

■ **CLASS DISCUSSION ITEM 3.6**
Automobile Charging System

The typical automobile has a 12 V DC electrical system where a lead-acid battery is charged by a belt-driven AC alternator whose frequency and voltage vary with engine speed. What type of signal conditioning must be performed between the alternator and the battery, and how can this be done?

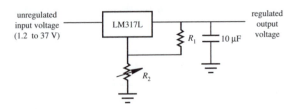

Figure 3.11 1.2 to 37 V adjustable regulator.

3.3.3 Optoelectronic Diodes

Light-emitting diodes are diodes that emit photons when forward biased. The typical LED and its schematic symbol are illustrated in Figure 3.12. The positive lead, or anode, is usually the longer of the two leads. The LED is usually encased in a colored plastic material that enhances the wavelength generated by the diode and sometimes helps focus the light into a beam. The intensity of light is related to the amount of current flowing through the device. LEDs are manufactured to produce a variety of colors, but red, yellow, and green are usually the most common and least expensive. It is important to remember that an LED has a voltage drop of 1.5 to 2.5 V when forward biased, somewhat more than small signal silicon diodes. It takes only a few milliamps of current to dimly light the diode. It is important to include a series current-limiting resistor in the circuit to prevent excess forward current, which can quickly destroy the diode. Usually a 330 Ω resistor is included in series with an LED when used in digital (5 V) circuit designs. Figure 3.13 shows a typical LED circuit. Note that the current is limited to about 9 mA (3 V/330 Ω), which is enough to fully light the LED and is well within the current limit reported by manufacturers for most LEDs. Lab Exercise 4 demonstrates how to build LED circuits properly and shows how much forward bias voltage and current is required to fully light an LED.

Earlier we said that a pn junction is sensitive to light. Special diodes, called **photodiodes,** are designed to detect photons and can be used in circuits to sense light as shown in Figure 3.14. Note that it is the reverse current that flows through the diode when sensing light. It takes a considerable number of photons to provide detectable voltages with these devices. The phototransistor (see Section 3.4.6) can

Lab Exercise

Lab 4 Bandwidth, filters, and diodes

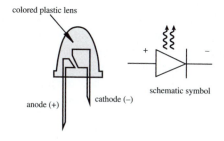

Figure 3.12 Light-emitting diode.

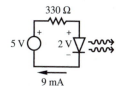

Figure 3.13 Typical LED circuit in digital systems.

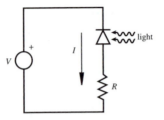

Figure 3.14 Photodiode light detector circuit.

be a more sensitive device, although it is slower to respond. The photodiode is based on quantum effects. If photons excite carriers in a reverse-biased pn junction, a very small current proportional to the light intensity flows. The sensitivity depends on the wavelength of the light.

3.3.4 Analysis of Diode Circuits

Although most of your analyses include single isolated diodes in circuits, there are situations when you design a circuit containing multiple diodes. Because the diode is a nonlinear device, one has to be careful not to naively apply the linear circuit analysis methods discussed so far.

DC circuits that contain many diodes may not be easy to analyze by inspection. The following procedure is a straightforward method to determine voltages and currents in these circuits. First, assume current directions for each circuit element. Next, replace each diode with an equivalent open circuit if the assumed current is in a reverse bias direction or a short circuit if it is in the forward bias direction. Then compute the voltage drops and currents in the circuit loops using KVL and KCL. If the sign on a resulting current is opposite to the assumed direction through an element, you have made the wrong assumption and must change its direction and reanalyze the circuit. Repeat this procedure with different combinations of current directions until there are no inconsistencies between assumed and calculated voltage drops and currents.

EXAMPLE 3.3   Analysis of Circuit with More Than One Diode

This example illustrates the application of the procedure just outlined to a circuit containing two ideal diodes. In the following circuit, we want to determine all currents and voltages. We begin by arbitrarily assuming the current directions as shown.

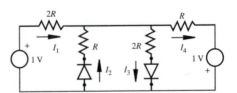

Having assumed the current directions, we replace each diode with a short, because each is assumed to be forward biased. The equivalent circuit follows.

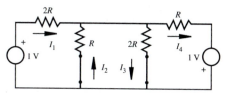

By applying KVL to the loop containing I_2 and I_3, we find that $I_2 = -2I_3$. We conclude that one of the current directions was incorrectly assumed. Therefore, we need to change one of our initial assumptions. Assume I_2 is in the direction opposite to that first chosen. With this assumption, the diode must be replaced with an open circuit because it is reverse biased. The equivalent circuit is shown next.

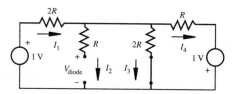

Note that $I_2 = 0$ in this circuit and V_{diode} is the voltage across the reverse-biased diode. By applying KVL to the loop containing I_3 and I_4, the result is that $I_3 < 0$. Therefore, our assumed direction for I_3 is incorrect. We must reverse I_3 and replace the diode with an open circuit. The resulting circuit is shown.

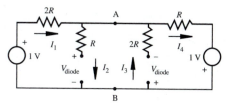

Note that I_2 and I_3 are both 0 in this circuit and each reverse-biased diode has a nonzero voltage across its terminals. The diode branches are in parallel, so they must have the same voltage across them, which is the voltage across nodes A and B. Because the diode currents are assumed to be zero, there would be no drops across the diode-branch resistors. Therefore, the voltage across each diode would need to be the same (magnitude and polarity). This is not the case for this circuit, so the assumed current directions are again incorrect.

The next choice for assumed current directions, which is the only combination we have not investigated, follows.

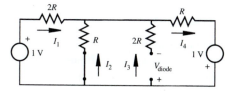

(continued)

(concluded)   If we analyze this circuit (Question 3.9), we find that $I_2 > 0$ and $V_{\text{diode}} > 0$ as assumed. Therefore, there are no inconsistencies, and our results are correct.

In this example, we performed an exhaustive search to check every possible combination of diode biasing. If we had chosen the correct combination earlier in the procedure, by luck or by an educated guess, the exhaustive search would not have been required.

The procedure and example above assumed that when a diode is forward biased, it can be replaced by a short circuit, which implies a 0 forward bias voltage. The procedure has to be modified to model real diodes accurately. To account for the forward bias voltage, instead of replacing the diode with a short, it must be replaced by a small voltage source whose voltage is equal to the forward bias value of the diode.

VOLTAGE LIMIT

##

■ CLASS DISCUSSION ITEM 3.7
Voltage Limiter

The diode portion of the following circuit is called a **voltage limiter.** Explain why. Sketch some input and output waveforms that illustrate the circuit's behavior. Note: $V_H > V_L$.

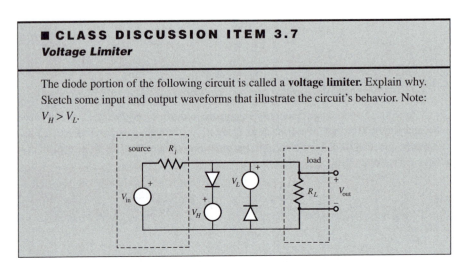

3.4 BIPOLAR JUNCTION TRANSISTOR

The bipolar junction transistor was the salient invention that led to the electronic age, integrated circuits, and ultimately the entire digital world. The transistor has truly revolutionized human existence by impacting practically everything in our everyday lives. We begin this section by providing the physical foundations necessary to understand the function of the transistor. Then we show how it can be used to build some important circuits.

3.4.1 Bipolar Transistor Physics

We saw earlier that a semiconductor diode consists of adjacent regions of p-type and n-type silicon, each connected to a lead. A **bipolar junction transistor** (BJT), in contrast, consists of three adjacent regions of doped silicon, each of which is connected to an external lead. There are two types of BJTs: npn and pnp transistors.

The most common type is the npn BJT, which we discuss in detail and use in our examples. As shown in Figure 3.15, it consists of a thin region or layer of p-type silicon sandwiched between two regions or layers of n-type silicon. Three leads are connected to the three regions, and they are called the **collector, base,** and **emitter.** As denoted by the bold **n** in Figure 3.15, the n-type silicon in the emitter is more heavily doped than the collector, so the collector and emitter are not interchangeable. The corresponding circuit schematic symbol is also shown in the figure with currents and voltages defined and labeled. The construction, schematic, and notation for the pnp BJT is shown in Figure 3.16. The remainder of this section focuses on the npn bipolar junction in Figure 3.15.

V_{CE} is the voltage between the collector and emitter, and V_{BE} is the voltage between the base and emitter. The relationships involving the transistor currents and voltages follow:

$$I_E = I_C + I_B \tag{3.10}$$

$$V_{BE} = V_B - V_E \tag{3.11}$$

$$V_{CE} = V_C - V_E \tag{3.12}$$

For the transistor to be on, the base-to-emitter junction must be forward biased ($V_{BE} = 0.7\text{V}$, so $V_B = V_E + 0.7\text{V}$). When this is the case, a large collector current can flow ($I_C > 0$) with a small base current ($I_B \ll I_C$), and there will be a small voltage drop (V_{CE}) in the collector-to-emitter circuit ($V_E = V_C - V_{CE}$).

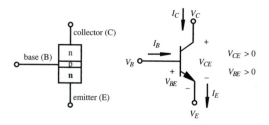

Figure 3.15 npn bipolar junction transistor.

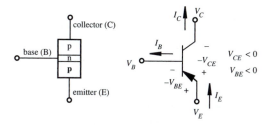

Figure 3.16 pnp bipolar junction transistor.

To understand how the npn BJT functions, we begin by considering the base-to-emitter junction. Because this junction is forward biased ($V_B > V_E$), electrons diffuse from the emitter n-type region to the base p-type region. Because the base-to-collector junction is reverse biased ($V_C > V_B$), there is a depletion region that would ordinarily prevent the flow of electrons from the base region into the collector region. However, because the base region is manufactured to be very thin and the emitter n-type region is more heavily doped than the base, most of the electrons from the emitter accelerate through the base region with enough momentum to cross the depletion region into the collector region without recombining with holes in the base region. Remembering that conventional current is in the opposite direction of electron motion, the result is that a small base current I_B flows from the base to the emitter and a larger current I_C flows from the collector to the emitter. The small base current controls a larger collector current, and therefore the BJT functions as a current amplifier. This characteristic can be approximated with the following equation:

$$I_C = \beta I_B \tag{3.13}$$

which states that the collector current is proportional to the base current with an amplification factor known as the **beta** (β) for the transistor. Manufacturers often use the symbol h_{FE} instead of β. For typical BJTs, beta is on the order of 100, but it can vary significantly among transistors. Beta is also temperature and voltage dependent; therefore, a precise relationship should not be assumed when designing specific transistor circuits.

Because of the BJT's base-collector current characteristics, it can be used to amplify current or to simply switch current on and off. This on-off switching is the basis for most digital computers because it allows easy implementation of a two-state binary representation. We focus on switch design and not amplifier design in our mechatronic applications. Amplifier design requires a more in-depth study of BJTs and is covered thoroughly in electrical engineering microelectronics textbooks.

3.4.2 Common Emitter Transistor Circuit

If a BJT's emitter is grounded and an input voltage is applied to the base, the result is the **common emitter** circuit shown in Figure 3.17. As the base current is gradually increased, the base-to-emitter diode of the transistor begins to conduct when V_{BE} is about 0.6 V. At this point I_C begins to flow and is roughly proportional to I_B ($I_C = \beta I_B$). As I_B is further increased, V_{BE} slowly increases only to about 0.7 V while I_C rises exponentially. As I_C rises, the voltage drop across R_C increases and V_{CE} drops toward ground. The collector cannot drop completely to ground; otherwise, the base-to-collector pn junction would also be forward biased. When V_{CE} reaches its minimum, the transistor is said to go into saturation. In this mode, the collector current is determined by R_C, and the linear relation between I_C and I_B no longer holds.

The characteristics of the common emitter transistor circuit can be summarized by plotting the collector current I_C versus the collector-emitter voltage V_{CE} for different values of base current I_B. The resulting family of curves (see Figure 3.18) describes the **common emitter characteristics** for the transistor.

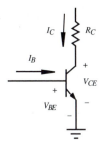

Figure 3.17 Common emitter circuit.

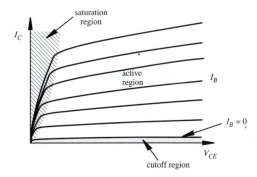

Figure 3.18 Common emitter characteristics for a transistor.

The transistor has a **cutoff region** (where no collector current flows), an **active region** (where collector current is proportional to base current), and a **saturation region** (where collector current is strictly controlled by the collector circuit, assuming sufficient base current). When designing a transistor switch, we need to guarantee that the transistor is fully saturated when it is on. In full saturation, V_{CE} is at its minimum, which is about 0.2 V for a BJT. So in saturation, the base-to-emitter junction is forward biased ($V_{BE} = 0.7$ V), there is a small drop from the collector to the emitter ($V_{CE} = 0.2$ V), and the voltages at the leads of the transistor are related as:

$$V_B = V_E + 0.7 \text{ V} \tag{3.14}$$

$$V_C = V_E + 0.2 \text{ V} \tag{3.15}$$

Example 3.4 shows how to determine how much base current and input voltage are required to saturate a transistor. The power dissipated by the transistor ($I_C V_{CE}$) is smallest, for a given collector current, when it is fully saturated. If the transistor is not fully saturated, it gets hot faster and can fail.

Probably the best way to understand the function of the BJT is to perform voltage and current measurements on actual circuits and plot the results. We've done

EXAMPLE 3.4	Guaranteeing That a Transistor Is in Saturation

The 2N3904 is a small-signal transistor manufactured by many companies as a general purpose amplifier and switch. If you examine the specifications online or in a discrete transistor handbook, you can find a complete list of ratings and electrical characteristics. Here is some of the information provided:

■ Maximum collector current (continuous) = 200 mA

■ V_{CE} (sat) = 0.2 V

■ $h_{FE} = \beta = 100$ (depending on collector current and many other things)

In the following circuit, what minimum input voltage V_{in} is necessary to saturate the transistor?

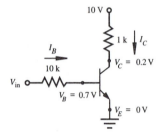

Because V_{CE}(sat) for the 2N3904 is 0.2 V, when the transistor is fully saturated the collector current is

$$I_C = (10 \text{ V} - 0.2 \text{ V})/1 \text{ k}\Omega = 9.8 \text{ mA}$$

Because the DC current gain h_{FE} is about 100, I_B must be at least $I_C/100$ or 0.098 mA. Because $V_{BE} = 0.7$ V, the base current can be related to the input voltage with

$$I_B = 0.098 \text{ mA} = (V_{in} - 0.7 \text{ V})/10 \text{ k}\Omega$$

Therefore, the minimum required input voltage for saturation is

$$V_{in_{min}} = 0.98 \text{ V} + 0.7 \text{ V} = 1.68 \text{ V}$$

Normally you would use a voltage larger than this (e.g., 2 to 5 times larger) to ensure that the transistor is fully saturated, even with variances in parameters.

Lab Exercise

Lab 5 Transistors and photoelectric circuits

just that for the two circuits shown in Figure 3.19, using a 2N3904 small-signal transistor. Lab Exercise 5 also includes some experiences to help develop an understanding of how transistor circuits function. The first circuit (Figure 3.19a) is the common emitter configuration that has been the topic of this section. The second circuit (Figure 3.19b) has an additional resistor in the emitter circuit that results in what is called emitter degeneration.

Figure 3.20 shows the results for the common emitter circuit (Figure 3.19a). In Figure 3.20a, notice how the base-to-emitter forward bias voltage V_{BE} and the

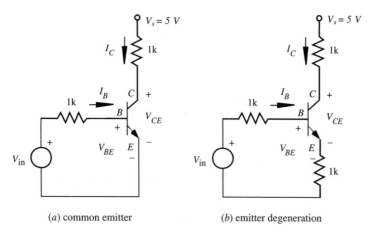

(a) common emitter   (b) emitter degeneration

Figure 3.19 Transistor experiments.

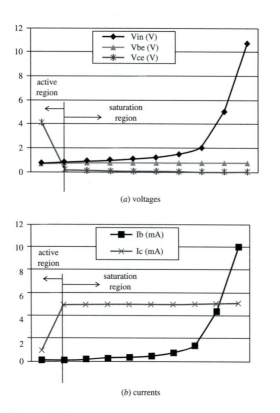

(a) voltages

(b) currents

Figure 3.20 Common emitter experimental results.

collector-to-emitter voltage drop V_{CE} do not change much after the transistor is saturated, even when input voltage V_{in} is increased well above the minimum required for saturation. In Figure 3.20b, notice how the collector current I_C does not increase in the saturation region as the base current I_B is increased above what is required for saturation. As demonstrated in Example 3.4, it is important to fully saturate a transistor by ensuring enough base current flows (by choosing an appropriate base resistor for the available input voltage). However, it is also clear that increasing the input voltage and base current significantly above the minimum required for saturation does not provide any meaningful increase in collector current and will only result in extra heat and energy loss in the base-to-emitter circuit.

Figure 3.21 shows the results for the emitter degeneration circuit (Figure 3.19b). In Figure 3.21a, notice how, just as with the common-emitter circuit, the base-to-emitter forward-bias voltage V_{BE} and the collector-to-emitter voltage drop V_{CE} do not change much after the transistor is saturated, even when input voltage V_{in} is increased well above the minimum required for saturation. In Figure 3.21b, notice how the collector current I_C response is very different from the response in the common-emitter circuit. As before, saturation and maximum collector current occur at a fairly low base current. However, as the base current is increased, the collector

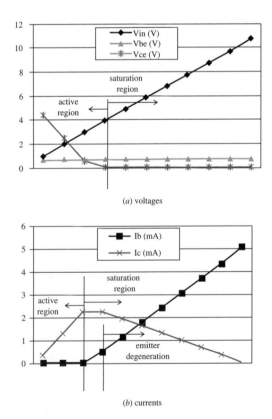

(a) voltages

(b) currents

Figure 3.21 Emitter degeneration experimental results.

current drops. This effect is called **emitter degeneration.** The reason for this is that the base current adds to the collector current, causing an increase in the emitter current ($I_E = I_B + I_C$), creating a larger voltage drop across the emitter resistor ($I_E R_E$). This reduces the voltage difference across the collector resistor, resulting in less collector current. The collector current actually decreases to zero as the voltage at the emitter ($V_E = I_E R_E$) approaches the collector supply voltage V_s. With no voltage difference across the collector resistor, no collector current will flow. For typical mechatronics switching applications, the common-emitter configuration is more appropriate because it is easy to saturate the transistor and ensure maximum collector current over a wide range of circuit parameters.

3.4.3 Bipolar Transistor Switch

Figure 3.22 illustrates a simple transistor switch circuit. When V_{in} is less than 0.7 V, the BE junction of the transistor is not forward biased ($V_{BE} < 0.7$ V), and the transistor does not conduct ($I_C = I_E = 0$). You can therefore assume that the collector-to-emitter circuit can be replaced by a very high impedance or, for all practical purposes, an open circuit. This state, illustrated in Figure 3.23a, is referred to as the **cutoff** or OFF state of the transistor. In cutoff, the output voltage V_{out} is V_C because there is no current through or voltage drop across R_C.

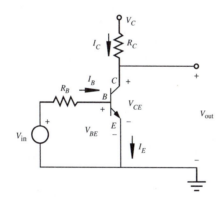

Figure 3.22 Transistor switch circuit.

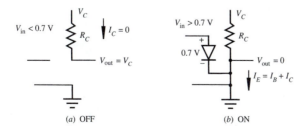

Figure 3.23 Models for transistor switch states.

When the *BE* junction is forward biased ($V_{BE} = 0.7$ V), the transistor conducts. Current passes through the *CE* circuit, and V_{out} is close to ground potential (0.2 V for a saturated BJT). This state, modeled by the forward-biased diode illustrated in Figure 3.23b, is referred to as the **saturated** or ON state of the transistor. We assume that there is enough base current to saturate the transistor. The resistor R_B (see Figure 3.22) is required in this circuit to limit the base current because the *BE* junction essentially behaves like a diode. The relationship between the base current and R_B is given by

$$I_B = (V_{in} - V_{BE})/R_B \qquad\qquad (3.16)$$

When $V_{in} < 0.7$, $I_B = 0$ and $V_{BE} = V_{in}$.

The circuit in Figure 3.22 can serve as a semiconductor switch to turn on or off an LED, electric motor, solenoid, electric light, or some other load (represented by R_C in the figure). These loads require large currents, ranging from milliamps to many amps, to function properly. When the input voltage and current are increased enough to saturate the transistor, a large collector current flows through the load R_C. The magnitude of the collector current is determined by the load resistance R_C and the collector voltage V_C. When the base-to-emitter voltage is below 0.7 V, the transistor is off, and no current flows through the load. The transistors used in power applications, called **power transistors,** are designed to conduct large currents and dissipate more heat. Power transistors are the basis for interfacing low-output current devices such as integrated circuits and computer ports to other devices requiring large currents.

Relays, which mechanically make and break connections, are an alternative to transistors. They cannot switch as fast as transistors and don't last as long, but they are very easy to use and can switch DC as well as AC power. For more information, see Section 10.3. AC current can also be switched with a **TRIAC** (triode for alternating current), which is a semiconductor device. For more information, see Internet Link 3.1.

Internet Link

3.1 TRIAC - triode for alternating current

DESIGN EXAMPLE 3.2

LED Switch

Our objective is to turn a dashboard LED on or off with a digital device having an output voltage of either 0 V or 5 V and a maximum output current of 5 mA. The LED requires 20–40 mA to provide a bright display and has a 2 V voltage drop when forward biased.

We use a transistor switch circuit employing a small-signal transistor (e.g., 2N3904 npn) to provide sufficient current to the LED. The required circuit follows.

5 V

100 Ω

LED

digital device output

10 kΩ

When the digital output is 0 V, the transistor is in cutoff, and the LED is OFF. When the digital output is 5 V, the transistor is in saturation, and the base current is

$$(5 \text{ V} - 0.7 \text{ V})/10 \text{ k}\Omega = 0.43 \text{ mA}$$

which is within the specifications. The 100 Ω collector resistance limits the LED current to a value within the desired range for the LED to be bright (20–40 mA):

$$(5 \text{ V} - 2 \text{ V} - 0.2 \text{ V})/100 \text{ }\Omega = 28 \text{ mA}$$

Lab Exercise 5 demonstrates how to wire and use various types of diode and transistor circuits. Included are a basic transistor switch circuit, a motor driver circuit with flyback protection (see Video Demo 3.2), and a photo-interrupter.

Let us summarize the guidelines for designing a transistor switch. The collector must be more positive than the base or emitter ($V_C > V_B > V_E$). To be ON, the base-to-emitter voltage (V_{BE}) must be 0.7 V. The collector current I_C is independent of base current I_B when the transistor is saturated, as long as there is enough base current to ensure saturation. The minimum base current required can be estimated by first determining the collector current I_C and then applying $I_{B_{\min}} \approx I_C/\beta$. For a given input voltage, the input resistance must be chosen so that the base current exceeds this value by a conservative margin (e.g., 5–10 times larger). The reasons for this are that beta may vary among components, with temperature, and with voltage; and the load resistance may change as current flows through it. It is also important to calculate the maximum values of I_C and I_B to ensure that they fall within the manufacturer's specifications, and add or change series resistors if the currents are too large.

3.4.4 Bipolar Transistor Packages

Transistor manufacturers offer their devices in a number of packages as illustrated in Figure 3.24. The small-signal transistor packages are often the TO-92, and the power transistor packages are the TO-220. Surface mount technology is becoming increasingly popular for use on production printed circuit boards, but such devices are less useful for prototyping because of their small size. Figure 3.25 and Video Demo 3.3 illustrate various common transistor packages. Included are BJTs, metal-oxide-semiconductor field-effect transistors (MOSFETs, which are covered in Section 3.5), and a photo-interrupter (covered in Section 3.4.6).

Lab Exercise

Lab 5 Transistors and photoelectric circuits

Video Demo

3.2 Turning a motor on and off with a transistor

Video Demo

3.3 Transistors

Figure 3.24 Bipolar transistor packages.

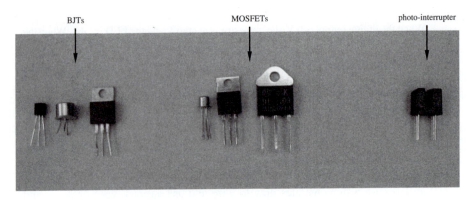

Figure 3.25 Various common transistor packages.

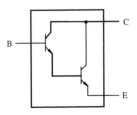

Figure 3.26 Darlington pair.

3.4.5 Darlington Transistor

The schematic in Figure 3.26 represents a transistor known as a **Darlington pair,** which usually comes in a single package. The advantage of this combination is that the current gain is the product of the two individual transistor gains and can exceed 10,000. They may often be found in power circuits for mechatronic systems.

3.4.6 Phototransistor and Optoisolator

A special class of transistor is the **phototransistor,** whose junction between the base and emitter acts as a photodiode (see Section 3.3.3). LEDs and phototransistors are often found in pairs, where the LED is used to create the light, and this light in turn biases the phototransistor. The pair can be used to detect the presence of an object that may partially or completely interrupt the light beam between the LED and transistor (see Lab Exercise 5).

Lab Exercise

Lab 5 Transistors and photoelectric circuits

An **optoisolator** is composed of an LED and a phototransistor separated by a small gap as illustrated in Figure 3.27. The light emitted by the LED causes current to flow in the phototransistor circuit. This output circuit can have a different ground reference, and the supply voltage V_s can be chosen to establish a desired output voltage range. With no common ground, the optoisolator creates a state of electrical isolation between the input and output circuits by transmitting the signal optically rather than through an electrical connection. A benefit of this isolation is that the

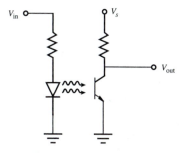

Figure 3.27 Optoisolator.

output is protected from any excessive input voltages that could damage components in the output circuit. Also, because the supplies and grounds are separate, any fluctuations or disturbances that might occur in the output circuit have no effect on the control signals on the input side.

Angular Position of a Robotic Scanner

DESIGN EXAMPLE 3.3

This design example illustrates an application of semiconductor optoelectronic components. Suppose, in the design of an autonomous robot, you wish to include a laser scanning device to sweep the environment to detect obstacles. The head of the scanner is rotated through 360° by a DC motor. Your problem here is to track the angular position of the scan head. How could you do this if you want an on-board computer to use the sensed values?

The solution requires a sensor that provides a digital output, that is, one that can be handled by a digital computer. We will learn more about digital interfacing in Chapter 6. To keep the solution simple at this point, we choose a device that produces a 5 V digital output. An LED-phototransistor pair, also known as a **photo-interrupter,** is at the heart of the design, which is illustrated in the following figure. The pair, which is readily available in a single package, produces a beam of light that can be broken or interrupted. A slotted disk must be designed to attach to the shaft of the motor driving the scan head and to pass through the gap in the photo-interrupter pair. Each slot in the disk provides a digital pulse as it interrupts the light beam during rotation.

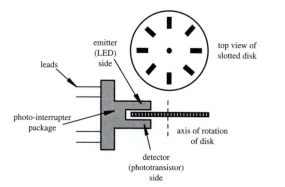

(continued)

(concluded)   In order for the sensor to function properly, we must add the external components shown in the following figure to provide a digital pulse each time a slot is encountered. The emitter LED and its current limiting resistor R_1 are powered by a 5 V DC source. The phototransistor detector and external resistor R_2 provide output signal V_{out}. R_2 is called a **pull-up resistor** because it pulls the output voltage (V_{out}) up from ground (0 V) to 5 V when the transistor is in cutoff. When the transistor is saturated, the output voltage is near 0 V.

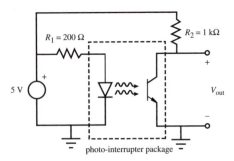

photo-interrupter package

As the slotted disk rotates, light passes through each slot producing a 0 V output and then returns to 5 V when the segments between the slots interrupt the light. The result is a train of pulses. The number of pulses produced provides the measure of angular rotation as a digital approximation. For example, if the disk has 360 slots, each pulse would correspond to 1° of rotation.

3.5 FIELD-EFFECT TRANSISTORS

Using what you have learned so far, you can design circuits for mechatronic systems using BJTs and other discrete components. We now examine the **field-effect transistor** (FET) that operates on a different principle than the BJT but serves a similar role in mechatronic system design. As we will see in Chapter 6, it is also an important component in the design of digital integrated circuits.

Both the BJT and FET are three-terminal devices allowing us to draw analogies between their function and how they are used in circuits. Before we look at the details of how FETs work, we describe their general characteristics. Both BJTs and FETs operate by controlling current between two terminals using a voltage applied to a third terminal. In Section 3.4, we saw that the forward bias of the base-to-emitter junction of the BJT allows charge carriers to enter a thin base region from the emitter, where they are attracted to the collector, resulting in a large collector current controlled by the much smaller base current. We concluded that the BJT is a current amplifier. In contrast, with a FET, the electric field produced by a voltage on one electrode controls the availability of charge carriers in a narrow region, called a channel, through which a current can be made to flow. Therefore, a FET can be described as a transconductance amplifier, which means the output current is controlled by an input voltage.

The nomenclature describing the FET is as follows. The control electrode in the FET, called the **gate,** is analogous to the base of the BJT. In contrast to the BJT base, the FET gate draws no direct current (DC) because it is insulated from the substrate to which it is attached. A conducting channel, whose conductivity is controlled by the gate, lies between the **drain,** which is analogous to the BJT collector, and the **source,** which is analogous to the BJT emitter. There are three families of FETs: enhancement-mode **metal-oxide-semiconductor FETs** (MOSFETs), depletion mode MOSFETs, and **junction field-effect transistors** (JFETs). Each of these families is available in p-channel and n-channel varieties. Understanding the different families and varieties of FETs is somewhat complicated when encountering them for the first time, so we focus primarily on the widely used n-channel enhancement-mode MOSFET. We will see that it is a close analogy to the npn BJT.

The cross section and schematic symbol for an n-channel enhancement-mode MOSFET is illustrated in Figure 3.28. This MOSFET has a p-type substrate and an n-type source and drain that form pn junctions with the substrate. There is a thin silicon dioxide layer insulating the gate from the substrate. As illustrated in Figure 3.29, when a positive DC voltage is applied to the gate, an electric field formed in the substrate below the gate repels holes in the p-type substrate leaving a narrow layer or **channel** in the substrate in which electrons predominate. This is referred to as an **n-channel** in the p-type substrate. The substrate is usually connected to the source internally so that the substrate-source pn junction is not forward biased. In the schematic circuit symbol (see the right side of Figure 3.28), the arrowhead indicates the direction between the p-type substrate and the n-channel.

3.5.1 Behavior of Field-Effect Transistors

Using an n-channel enhancement-mode MOSFET as our example, we explain the details of its operation and discuss the characteristic curves analogous to the BJT. If the gate is grounded ($V_g = 0$), no drain-to-source current I_d flows for a positive

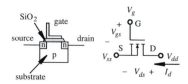

Figure 3.28 n-channel enhancement-mode MOSFET.

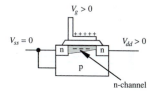

Figure 3.29 Enhancement-mode MOSFET n-channel formation.

drain voltage V_{dd} because the drain pn junction is reverse biased and no conducting channel has formed. In this state, the MOSFET mimics a very large resistor ($\sim 10^8 - 10^{12}\ \Omega$), and no current flows between the drain and source. The MOSFET is said to be in **cutoff.**

As V_{gs} is gradually increased beyond a gate-to-source **threshold voltage** V_t, the n-channel begins to form. V_t depends on the particular MOSFET considered but a typical value is about 2 V. Then as V_{ds} is increased from 0, conduction occurs in the n-channel due to a flow of electrons from source to drain. The drain current I_d, by convention, is shown in the direction opposite to electron flow. As shown in Figure 3.29, a subtle feature of the n-channel is that it is wider near the source than at the drain because the electric field is larger due to the larger difference between V_g and ground at the source end and the smaller difference between V_g and V_{dd} at the drain end.

With a positive V_{gs} larger than V_t, as V_{ds} is increased from 0, we enter the **active region,** also called the **ohmic region,** of the MOSFET. In this region, as V_{gs} is further increased, the conduction channel grows correspondingly, and the MOSFET appears to function like a variable resistor whose resistance is controlled by V_{gs}. However, when $V_{gs} - V_t$ reaches V_{dd}, there is no longer an electric field at the drain end of the MOSFET. Therefore, the width of the n-channel shrinks to a minimum value close to the drain resulting in what is called **pinch-off.** This pinch-off limits a further increase in drain current, and the MOSFET is said to be in **saturation.** In saturation, the current is almost constant with further increases in V_{ds}. The drain-to-source resistance, called R_{on}, is minimal (usually less than 5 Ω) as it enters the saturation region.

Figure 3.30 shows the characteristic family of curves for the n-channel enhancement-mode MOSFET, which graphically illustrates the features just described. The analogous npn BJT family of curves was shown in Figure 3.18. By comparing the characteristic curves, the saturation region of the MOSFET corresponds to the active region of the BJT, so one must be careful when using these terms.

As we did with the npn BJT transistor, let's look at some voltage and current measurements from an actual MOSFET circuit, using an IRF620 power MOSFET. The circuit is shown in Figure 3.31a. For the experiment, the voltage on the gate V_g (which is also the gate-to-source voltage $V_{gs,}$ because the source is grounded) was gradually increased from 0 to 10 V, more gradually in the ranges of interest. Figure 3.31b shows the measurement results for the drain-to-source current I_{ds} and the drain-to-source voltage V_{ds}. Notice that for this MOSFET, the threshold voltage,

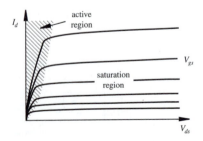

Figure 3.30 n-channel enhancement-mode MOSFET characteristic curves.

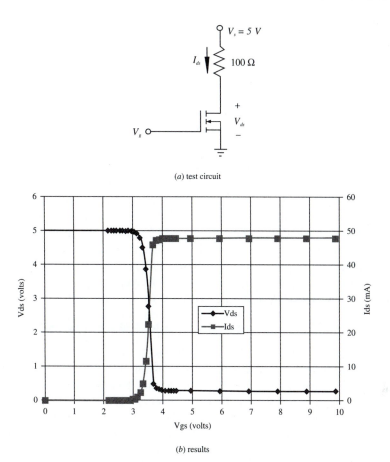

(a) test circuit

(b) results

Figure 3.31 MOSFET experiment.

where conduction begins ($I_{ds} > 0$), is about 3.5 V. Also note that the drain-to-source voltage V_{ds} doesn't drop to zero when the MOSFET is fully on. This is due to the drain-to-source resistance R_{on} of the device (see Question 3.22), which creates a small voltage drop ($V_{ds} = I_{ds}R_{on}$).

The cross section and schematic symbol for a p-channel enhancement-mode MOSFET are illustrated in Figure 3.32. As with the n-channel MOSFET, the arrowhead indicates the direction of the substrate-channel pn junction. If the gate is negative with respect to the source ($V_{sg} > 0$), electrons in the n-type substrate are repelled, forming a p-channel conducting layer beneath the gate. This allows a current to flow from the source to the drain if V_{sd} is positive. The p-channel enhancement-mode MOSFET functions analogously to the pnp BJT.

MOSFETs are very useful in a variety of mechatronic applications. MOSFETs can be used to make excellent high-current voltage-controlled switches. Also, some MOSFETs are designed specifically as analog switches, where signals can be gated

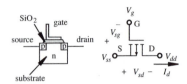

Figure 3.32 p-channel enhancement-mode MOSFET.

(blocked or passed) in circuits. These examples are presented in Section 3.5.3. MOS-FETs are also used in special circuits for driving DC motors. Because of their characteristics, MOSFETs can be used as current sources as a result of the flat characteristics of the saturation region. MOSFETs are also useful in the internal design of integrated circuits (ICs) like microprocessors. The MOSFETs in ICs are often fabricated in complementary (n-channel and p-channel) pairs, and the resulting ICs are known as **complementary metal-oxide-semiconductor** (CMOS) devices. The symmetry of the n-channel and p-channel transistors allows for compact fabrication on a single IC and is useful in the internal design of logic devices (to be presented in Chapter 6).

3.5.2 Symbols Representing Field-Effect Transistors

Because you will come across FETs in numerous circuit designs, it is important to recognize the subtleties of the schematic symbols for them. Because FETs (JFETs and MOSFETs) have two different kinds of channel doping, and because the substrate can be p-type or n-type, there are eight potential FET configurations. The symbols for the four most important classes of FETs are shown in Figure 3.33. The terminal designations are G for gate, S for source, D for drain, and B for substrate. Some of the principal characteristics of the schematic are

1. The direction of the gate or substrate arrow distinguishes between p-channel (arrow out) and n-channel (arrow in).
2. A separation is shown between the gate and the source in the MOSFET but not in the JFET. The separation represents the insulating layer of the metal oxide in the MOSFET.

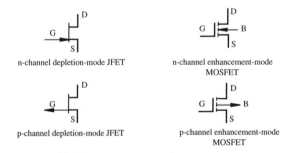

Figure 3.33 Field-effect transistor schematic symbols.

3. A broken line between the source and drain indicates an enhancement-mode device in contrast to a solid line for a depletion-mode device. Enhancement-mode FETs require a gate voltage for conduction, and depletion mode FETs require a gate voltage to reduce the conduction. JFETs are available only in the depletion mode, but MOSFETs are available in both varieties.

4. The gate line is offset toward the source, so the source side can be easily identified. Sometimes the gate line is shown centered; in this case there is no way to distinguish the drain from the source unless they are labeled.

The substrate of a MOSFET may be connected to a separate terminal or internally connected to the source. If there is a separate substrate lead, it must not be biased more positive than the source or drain for an n-channel device and must not be biased more negative than the source or drain for a p-channel device. It should always be connected to something (i.e., it should not be left "floating").

3.5.3 Applications of MOSFETs

The first MOSFET application we consider is switching power to a load. This circuit is analogous to the BJT switch presented in Section 3.4.3. An n-channel enhancement-mode power MOSFET is used with the load on the drain side as shown in Figure 3.34. Note that this MOSFET switch is very easy to design because the gate draws practically no steady state current. We must ensure that $V_g \leq 0$ for the MOSFET to be cutoff so that no current is delivered to the load. When $V_g - V_t \approx V_{dd}$, the MOSFET enters saturation resulting in nearly full voltage V_s across the load (because R_{on} is small). The controlling parameter for the MOSFET is gate voltage V_g. Recall that with the BJT, the controlling parameter is base current I_B. With the BJT, one must ensure adequate base current to saturate the BJT. Using the MOSFET, the current drawn by the gate is essentially 0, so current sourcing is not a concern. However, one needs to calculate the drain current I_d and power dissipation to select a MOSFET capable of switching the desired current for the load. Also, as with a BJT, if the load is inductive, a flyback diode (see Figure 3.34) is necessary to prevent damage to the MOSFET when it is switched off.

The second application we will consider is the use of a MOSFET as an analog switch. Suppose you have a positive analog signal V_{in} and you want to be able to

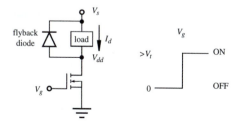

Figure 3.34 MOSFET power switch circuit.

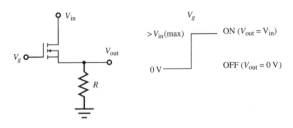

Figure 3.35 MOSFET analog switch circuit.

couple it to another circuit or device, or block it altogether. This is an easy application for a MOSFET using the circuit shown in Figure 3.35. If the control signal V_g is zero, the MOSFET will be cutoff resulting in a huge drain to source impedance (in megaohms) essentially blocking the analog signal ($V_{out} = 0$ V). The pull-down resistor R is required to hold the V_{out} terminal at ground in the off state. When the control signal V_g is larger than the largest value of the analog input signal V_{in} plus threshold voltage V_t, the drain to source channel will conduct with a low resistance, and the output signal will track the input ($V_{out} = V_{in}$).

■ CLASS DISCUSSION ITEM 3.8
Analog Switch Limit

Considering the analog switch circuit shown in Figure 3.35, why does the gate control signal need to be larger than the largest value of the analog signal?

DESIGN EXAMPLE 3.4

Circuit to Switch Power

Among the general problems in a mechatronic system design is delivering electrical power to different portions of the system. MOSFETs are useful devices for this task.

Suppose you have a digital device that produces a binary output, which means its output can be one of two states. For the moment, assume that the output circuit consists of an npn transistor that can be in cutoff or saturation, but with the collector as of yet not connected to anything. As we will see later, this is called an **open-collector-output** device. All you need to know for now is that the output transistor can be turned on and off. Also, it can sink only a very small current, in the milliamp range. How then can we interface the binary output to control the current to a load that may require a current of many amps? A solution to this problem employing an n-channel enhancement-mode MOSFET power transistor is shown in the following figure. The output circuit from the digital device is drawn to the left of the dashed line, and the portion we are designing is to the right.

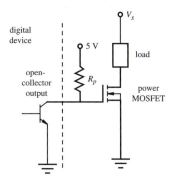

Resistor R_p, connected to the collector lead to complete the digital output circuit, is called a **pull-up resistor** because it "pulls up" the collector voltage to a DC power source (in this case +5 V). It results in 0 V at the MOSFET gate when the output transistor is on, and 5 V at the gate when it is off. To drive a load requiring a current larger than what the digital output can provide, we use a power MOSFET to switch a different power supply V_s.

Now if you are given the specific current and voltage requirements for a load (e.g., a motor), you can refer to manufacturer or supplier data to select the appropriate MOSFET to do the job.

■ **CLASS DISCUSSION ITEM 3.9**
Common Usage of Semiconductor Components

Cite specific examples in your experience where and how each of the following electrical components is used:

- Signal and power diodes
- Light-emitting diodes
- Signal and power transistors

Internet Link

2.4 Electronic component online resources and vendors

3.2 Semiconductor (IC) manufacturers and online resources

3.3 All about circuits – Vol. III – semiconductors

Internet Link 2.4 provides links to various resources and vendors for all types of electronic components, including all of the devices presented in this book. Electronics vendors provide a wealth of online information to make it easy to find data and place orders for their products. Internet Link 3.2 provides links to the largest manufacturers of semiconductor components. The semiconductor manufacturers provide a wealth of useful online information for all sorts of integrated circuits. Internet Link 3.3 is an excellent resource providing a thorough review of semiconductor physics, devices, application circuits, and circuit analysis.

QUESTIONS AND EXERCISES

Section 3.3   Junction Diode

3.1. Sketch the output waveform for V_{out} in the following circuit on axes as shown. Assume the diode is ideal.

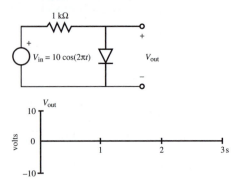

3.2. Sketch the output V_{out} on a set of axes for circuits "a" through "f" with $V_{in} = 1.0 \sin(2\pi t)$ V. Assume the diodes are ideal. Plot the output for one complete cycle of the input ($0 \le t \le 1$s).

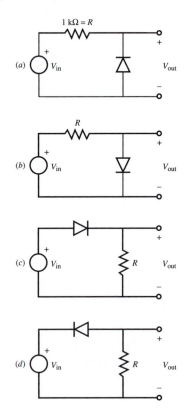

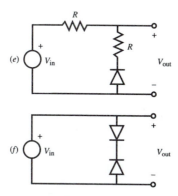

(e) V_{in} V_{out}

(f) V_{in} V_{out}

3.3. Document a complete and thorough answer to Class Discussion Item 3.1.

3.4. Document a complete and thorough answer to Class Discussion Item 3.2.

3.5. Document a complete and thorough answer to Class Discussion Item 3.3.

3.6. Sketch the output waveform for V_{out} in the following circuit on axes as shown. Assume ideal diodes and show your work. Also, explain why this circuit is called a **full-wave rectifier.**

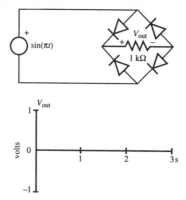

3.7. Document a complete and thorough answer to Class Discussion Item 3.5.

3.8. The following circuits are called **clipping circuits.** Assume ideal diodes and sketch the output voltage V_{out} for two cycles of the input V_{in}.

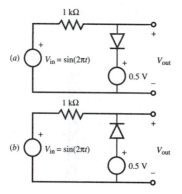

3.9. Compute the currents (I_1, I_2, I_3, I_4) and the diode voltage (V_{diode}) in the final circuit shown in Example 3.3.

3.10. In Example 3.3, for the case where both diodes are assumed to be reverse biased, find I_1, I_4, and V_{AB} (the voltage of node A relative to node B) in the assumed-polarity circuit. Based on voltage V_{AB}, for which diode was the polarity assumption correct?

3.11. Document a complete and thorough answer to Class Discussion Item 3.7.

3.12. For the following circuit, assuming ideal diodes and given $R = 1$ kΩ and $V_{\text{in}} = 10 \sin(\pi t)$ V, plot the output voltage V_{out} on axes with labeled scales for two periods of the input.

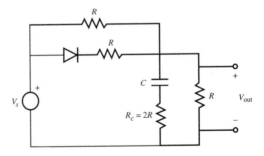

3.13. For the following circuit, find the steady state values for V_{out}, the voltage across and current into the capacitor, and the current through the output resistor for

a. $V_s = 10$ V DC
b. $V_s = -10$ V DC

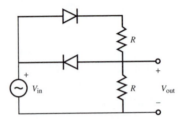

3.14. For Class Discussion Item 3.7 with $V_{\text{in}} = 15 \sin(2\pi t)$ V and $R_i = R_L = 1$ kΩ, draw and label one cycle of V_{in} and V_{out}. Assume ideal diodes.

3.15. Given the following ideal zener diode circuit with breakdown voltage 5.1 V, sketch the output voltage on a set of axes if

a. $V_{\text{in}} = 1.0 \sin(2\pi t)$
b. $V_{\text{in}} = 10.0 + \sin(2\pi t)$

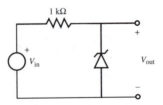

3.16. A digital circuit can output a voltage of either 0 V or 5 V relative to ground. Design a circuit that uses this output to turn an LED on or off assuming the LED

    a. has no forward voltage drop and can carry a maximum of 50 mA

    b. has a forward voltage drop of 2 V and can carry a maximum of 50 mA

Section 3.4 Bipolar Junction Transistor

3.17. In the following circuit, what minimum steady state voltage V_{in} is required to turn the LED on and keep the transistor fully saturated? Assume that the forward bias voltage for the LED is 2 V and there is a 0.2 V collector-to-emitter voltage drop when the transistor is saturated.

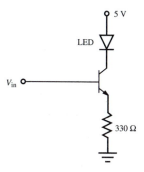

3.18.

    a. Given V_{in} (see the graph that follows) for the following circuit, and assuming the current into the base of the transistor is very small (i.e., assume $I_B = 0$), sketch the LED on-off curve on a graph similar to that shown below the circuit. Assume that the LED has a forward bias voltage of 1 V. Also assume the transistor is in saturation when the LED is on.

    b. Calculate the minimum value of V_{in} required to saturate the transistor assuming a beta of 100. Don't assume $I_B = 0$ in this part.

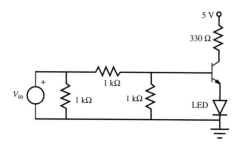

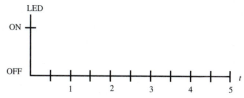

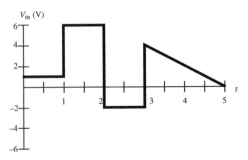

3.19. In the following circuit, find the minimum V_{in} required and the resulting voltage V_{out} to put the transistor in full saturation. Assume that the beta for the transistor is 100 in full saturation.

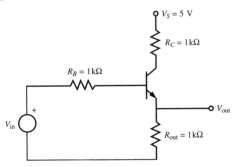

3.20. Consider the design of a solid state switch using an npn power transistor that you plan to control with a digital signal (0 V = off, 5 V = on). Start with the following schematic where components that you must select are labeled with numbers. The series resistor and inductor represents a DC motor that requires 1 A of current at 24 V_{DC}. Replace each of the labeled boxes shown in the figure with the appropriate schematic symbol and then specify the component as completely as you can.

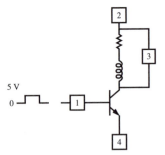

3.21. A photo-interrupter comes in a manufactured package that includes the phototransistor and corresponding LED as shown in the following schematic. What external circuitry must be added to obtain a functioning photo-interrupter? Sketch the resulting schematic.

Section 3.5 Field-Effect Transistors

3.22. For the MOSFET used in the experiment of Figure 3.31, estimate the fully-on drain-to-source resistance R_{on} from the values in the plot.

3.23. Answer Question 3.20 using an n-channel enhancement-mode power MOSFET instead of the npn BJT.

3.24. In Design Example 3.4, suppose $V_s = 15$ V. Replace the MOSFET with a BJT power transistor and specify the type (npn or pnp) and any additional components required. Draw the circuit schematic and describe its characteristics.

3.25. The output of most digital CMOS devices looks like this:

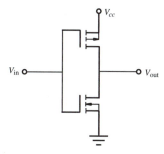

Identify the types of MOSFETs used. What is the value of V_{out} for $V_{in} = 5$ V and for $V_{in} = 0$ V?

3.26. The following table lists various MOSFETs available for a design project. You have a requirement to switch 10 A at 10 V. Select the MOSFET adequate for the requirements and defend your choice.

MOSFET	V_{ds} (V)	R_{ds} (on) (Ω)	I_d cont (A) @ 25°C	P_d (max) (W)
IRF510	100	0.6	16	20
IRF530N	100	0.11	60	63
IRF540	100	0.077	110	150
IRF540N	100	0.052	110	94
IRF610	200	1.5	10	20

3.27. For each state of an n-channel enhancement-mode MOSFET that follows, determine what operating region the MOSFET is in if the threshold voltage V_T is 3 V.

    a. $V_{gs} = 2$ V, $V_{ds} = 5$ V

    b. $V_{gs} = 4$ V, $V_{ds} = 5$ V

    c. $V_{gs} = 6$ V, $V_{ds} = 5$ V

    d. $V_{gs} = -2.5$ V

BIBLIOGRAPHY

Bailar, J. et al., *Chemistry,* Academic Press, New York, 1978.

Gibson, G. and Liu, Y., *Microcomputers for Engineers and Scientists,* Prentice-Hall, Englewood Cliffs, NJ, 1980.

Horowitz, P. and Hill, W., *The Art of Electronics,* 2nd Edition, Cambridge University Press, New York, 1989.

Johnson, D., Hilburn, J., and Johnson, J., *Basic Electric Circuit Analysis,* 2nd Edition, Prentice-Hall, Englewood Cliffs, NJ, 1984.

Lerner, R. and Trigg, G., *Encyclopedia of Physics,* VCH Publishers, New York, 1991.

McWhorter, G. and Evans, A., *Basic Electronics,* Master Publishing, Richardson, TX, 1994.

Millman, J. and Grabel, A., *Microelectronics,* 2nd Edition, McGraw-Hill, New York, 1987.

Mims, F., *Engineer's Mini-Notebook: Basic Semiconductor Circuits,* Radio Shack Archer Catalog No. 276-5013, 1986.

Mims, F., *Engineer's Mini-Notebook: Optoelectronics Circuits,* Radio Shack Archer Catalog No. 276-5012A, 1986.

Mims, F., *Getting Started in Electronics,* Radio Shack Archer Catalog No. 276-5003A, 1991.

Rizzoni, G., *Principles and Applications of Electrical Engineering,* 5th Edition, McGraw-Hill, New York, 2005.

CHAPTER 4

System Response

T his chapter describes how to mathematically model a physical system and characterize its response to dynamic inputs. These topics are important in understanding how actuators, sensors, amplifiers, filters, and other mechatronic system components function. ■

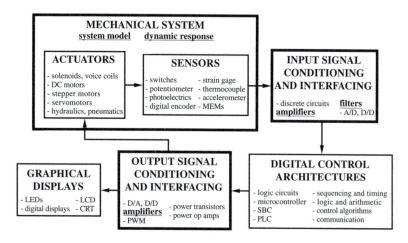

CHAPTER OBJECTIVES

After you read, discuss, study, and apply ideas in this chapter, you will:

1. Understand the three characteristics of a good measurement system: amplitude linearity, phase linearity, and adequate bandwidth

2. Be able to define the Fourier series representation of a signal and use it to show the components of the spectrum of the signal

3. Understand the relationship between an instrument's bandwidth and the spectra of its input and output signals

4. Understand the dynamic response of zero-, first-, and second-order measurement and mechatronic systems

5. Be able to use step and sinusoidal inputs to analyze and characterize the response of measurement and mechatronic systems

6. Understand the analogies among mechanical, electrical, and hydraulic systems.

4.1 SYSTEM RESPONSE

The relationship between the desired output of a mechatronic or measurement system and its actual output is the basis of system response analysis. This chapter deals with analysis techniques that characterize and predict how linear systems respond to specific inputs. We concentrate on measurement systems, which are often integral parts of mechatronic systems.

As we saw in Chapter 1, a measurement system consists of three parts: a transducer, a signal processor, and a recorder. A transducer is a device that usually converts a physical quantity into a time-varying voltage, called an **analog signal.** A signal processor can modify the analog signal, and a recorder provides either a transitory display or storage of the signal. The physical variable we wish to measure is called the input to the measurement system. The transducer transforms the input into a form compatible with the signal processor, which in turn modifies the signal, which then becomes the output of the measurement system. Usually, the recorded output is different from the actual input, as illustrated in Figure 4.1. Generally, we want to have the reproduced output signal match the input as closely as possible unless there is information in the input that we want to eliminate (e.g., electrical noise).

Certain conditions must be satisfied to accomplish adequate reproduction of the input. For a measurement system with time-varying inputs, three criteria must be satisfied in order to ensure that we obtain a quality measurement:

1. Amplitude linearity
2. Adequate bandwidth
3. Phase linearity

We examine each of these criteria in detail in the following sections.

4.2 AMPLITUDE LINEARITY

A good measurement system satisfies the criterion of amplitude linearity. Mathematically, this is expressed as

$$V_{out}(t) - V_{out}(0) = \alpha[V_{in}(t) - V_{in}(0)] \tag{4.1}$$

where α is a constant of proportionality. This means that the output always changes by the same factor times the change in the input. If this does not occur, then the system is not linear with respect to amplitude, and it becomes more difficult to interpret the output. Figure 4.2 displays examples of amplitude linearity and nonlinearity. The first example is linear and $\alpha = 20$. The second two examples are nonlinear because α is not constant. In the third example, the output changes by a factor of 20 on the first pulse and by a factor of 15 on the second pulse.

Normally, a measurement system will satisfy amplitude linearity over only a limited range of input amplitudes. Also, the system usually responds linearly only when the rate of change of the input is within certain limits. This second issue is related to the bandwidth of the system, addressed in Section 4.4. An ideal measurement system will exhibit amplitude linearity for any amplitude or frequency of the input.

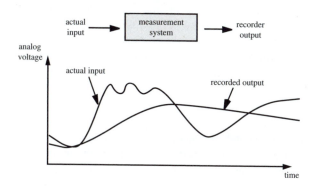

Figure 4.1 Measurement system input-output.

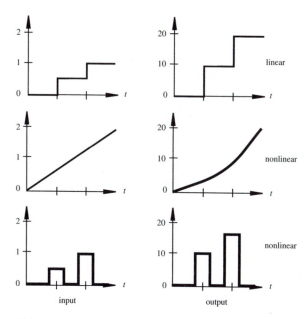

Figure 4.2 Amplitude linearity and nonlinearity.

4.3 FOURIER SERIES REPRESENTATION OF SIGNALS

Before we look at the concepts of bandwidth and phase linearity, which apply to frequency components of an input signal, we first need to review the concept of **Fourier series** representation of a signal. The fundamental premise for the Fourier series representation of a signal is that any periodic waveform can be represented as an infinite series of sine and cosine waveforms of different amplitudes and frequencies. When this infinite series is summed up, it will reproduce the original periodic waveform exactly. What this means is that we can take any complicated but periodic waveform and decompose it into a series of sine and cosine waveforms. In practice, we do not need the entire infinite series because a finite number of the sine and cosine waveforms can adequately represent the original signal.

We define the **fundamental** or first **harmonic** ω_0 as the lowest frequency component of a periodic waveform. It is inversely proportional to the period T:

$$\omega_0 = \frac{2\pi}{T} = 2\pi f_0 \tag{4.2}$$

where f_0 is the fundamental frequency expressed in hertz (Hz). The other sine and cosine waveforms have frequencies that are integer multiples of the fundamental frequency. The second harmonic would be $2\omega_0$, the third harmonic would be $3\omega_0$, and so on. The Fourier series representation of an arbitrary periodic waveform $f(t)$ can be expressed mathematically as

$$F(t) = C_0 + \sum_{n=1}^{\infty} A_n \cos(n\omega_0 t) + \sum_{n=1}^{\infty} B_n \sin(n\omega_0 t) \tag{4.3}$$

where the constant C_0 is the DC component of the signal, and the two summations are infinite series of sine and cosine waveforms. The coefficients of the sine and cosine terms are defined by

$$A_n = \frac{2}{T} \int_0^T f(t) \cos(n\omega_0 t)\, dt \tag{4.4}$$

$$B_n = \frac{2}{T} \int_0^T f(t) \sin(n\omega_0 t)\, dt \tag{4.5}$$

where $f(t)$ is the waveform being represented and T is the period of the waveform. The DC term C_0 represents the average value of the waveform over its period; therefore, it can be expressed as

$$C_0 = \frac{1}{T} \int_0^T f(t)\, dt = \frac{A_0}{2} \tag{4.6}$$

where A_0 is given by substituting $n = 0$ into Equation 4.4.

In the Fourier series representation given by Equation 4.3, there are two different amplitudes (A_n and B_n). However, the cosine and sine terms can be combined with a trigonometric identity (Question 4.4) to create an alternative representation described by a single amplitude and phase. This alternative representation is

$$F(t) = C_0 + \sum_{n=1}^{\infty} C_n \cos(n\omega_0 t + \phi_n) \tag{4.7}$$

where the total amplitude for each harmonic is given by

$$C_n = \sqrt{A_n^2 + B_n^2} \tag{4.8}$$

and the phase for each harmonic is given by

$$\phi_n = -\tan^{-1}\left(\frac{B_n}{A_n}\right) \tag{4.9}$$

To illustrate the application and meaning of a Fourier series, consider an ideal square wave as an example of a periodic waveform. The square wave illustrated in Figure 4.3 is defined mathematically as

$$f(t) = \begin{cases} 1 & 0 \le t < T/2 \\ -1 & T/2 \le t < T \end{cases} \tag{4.10}$$

where T is the period. There are discontinuities at $t = 0$ and $T/2$.

For the square wave defined by Equation 4.10, the coefficients A_n, including A_0, are 0 (see Question 4.5). The coefficients B_n can be found from Equation 4.5 :

$$B_n = \frac{2}{T}\left(\int_0^{T/2} \sin(n\omega_0 t)\, dt - \int_{T/2}^{T} \sin(n\omega_0 t)\, dt\right) \tag{4.11}$$

Integration results in

$$B_n = \frac{2}{T}\left(-\frac{1}{n\omega_0}[\cos(n\omega_0 t)]_0^{T/2} + \frac{1}{n\omega_0}[\cos(n\omega_0 t)]_{T/2}^{T}\right) \tag{4.12}$$

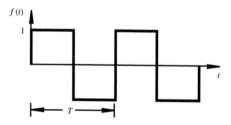

Figure 4.3 Square wave.

Using Equation 4.2 in evaluating the bracketed terms at each of the limits in Equation 4.12 gives

$$B_n = \frac{1}{n\pi}\{-\cos(n\pi) + 1 + 1 - \cos(n\pi)\} \tag{4.13}$$

This can be written as

$$B_n = \frac{2}{n\pi}[1 - \cos(n\pi)] = \begin{cases} \dfrac{4}{n\pi} & n:\ \text{odd} \\[2mm] 0 & n:\ \text{even} \end{cases} \tag{4.14}$$

Therefore, the Fourier series representation of a square wave of amplitude 1 is

$$F(t) = \frac{4}{\pi}\sin(\omega_0 t) + \frac{4}{3\pi}\sin(3\omega_0 t) + \frac{4}{5\pi}\sin(5\omega_0 t) + \cdots \tag{4.15}$$

or, using an infinite sum representation,

$$F(t) = \sum_{n=1}^{\infty} \frac{4}{(2n-1)\pi}\sin[(2n-1)\omega_0 t] \tag{4.16}$$

MathCAD Example

4.1 Fourier series representation of a square wave

Figure 4.4 shows the effects of combining the individual harmonics of the signal incrementally. MathCAD Example 4.1 includes the analysis used to create the figure. Shown on the left side of the figure are the individual harmonics along with their frequencies and amplitudes. Note that, as the harmonic frequency increases, the amplitudes of the harmonics decrease. Illustrated on the right are the superpositions of the successive harmonics, illustrating how the addition of higher harmonics improves the representation of the square wave. If we take the first, third, and fifth harmonics and add them together, we obtain a waveform that begins to look similar to a square wave. As we add additional harmonics to this signal, the quality of the reproduction of the waveform improves and sharp changes are better approximated. If an infinite number of harmonics were used, the result would be a square wave.

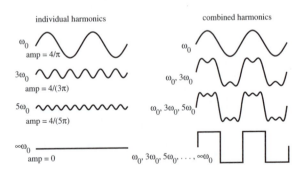

Figure 4.4 Harmonic decomposition of a square wave.

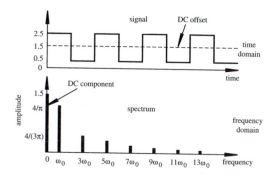

Figure 4.5 Spectrum of a square wave.

Waveforms such as a square wave, which have very sharp or rapid changes associated with them (e.g., at discontinuities), require a large number of higher harmonics for good reproduction.

The top half of Figure 4.5 illustrates a plot of a unity amplitude square wave with a DC offset of 1.5. This plot is known as the **time-domain** representation of the signal. The bottom half of the figure is a plot of the Fourier series amplitudes vs. frequency, which is called the signal's **spectrum.** Since the harmonics are represented as bars or lines on the graph, it is also called a **line spectrum.** Note that the DC offset is the zero frequency component. The spectrum is the signal's **frequency-domain** representation.

When plotting the spectrum for a signal represented by a Fourier series, one should use the single-amplitude version given by Equation 4.7. For the square wave Fourier series given in Equation 4.15, the A_n coefficients are 0, and the C_n coefficients are the same as the B_n coefficients (see Equation 4.8). Each bar in Figure 4.5 represents a different amplitude value. The first bar represents a DC offset of $C_0 = B_0 = 1.5$. The second bar represents the first term in Equation 4.15: $C_1 = B_1 = 4/\pi$. The remaining bars represent the higher nonzero harmonics (B_3, B_5, . . .).

Video Demo 4.1 shows spectra for various output signals from a function generator. A digital oscilloscope makes it very easy to display the spectrum of any electrical signal. Observe how signals with fast transitions (e.g., sharp corners) contain larger amplitude harmonics at higher frequencies.

Internet Link 4.1 provides various resources and demonstrations relating to how signal spectrum concepts can be applied to sound and music theory. Sound consists of pressure waves. A microphone can acquire a pressure signal and convert it into a voltage signal that can be processed and displayed (e.g., on an oscilloscope). The spectral content of different sounds is what differentiates the characters (timbres) of sound (see Class Discussion Item 4.1). Furthermore, a musical note (i.e., its pitch) is defined by its fundamental frequency and its signal amplitude determines its volume (i.e., loudness). Video Demos 4.2 through 4.7 show different examples of these concepts.

4.4   BANDWIDTH AND FREQUENCY RESPONSE

It is important to estimate the spectrum of a signal when choosing a system to measure the signal. Ideally, a measurement system should replicate all frequency components of a signal. Real systems, however, have limitations in their ability to reproduce all frequencies. A scale commonly used to measure the degree of fidelity of a measurement system's reproduction at different frequencies is the **decibel** scale. It is a logarithmic scale that allows comparison of the change in amplitude of a component of a signal when it passes through a measurement system. The decibel (dB) is defined as

$$dB = 20 \log_{10}\left(\frac{A_{out}}{A_{in}}\right) \tag{4.17}$$

where A_{in} is the input amplitude and A_{out} is the output amplitude of a particular harmonic.

The graph shown in Figure 4.6 is an example of a **frequency response curve** for a system. This graph is also called a **Bode plot.** It is a plot of the **amplitude ratio,** A_{out}/A_{in}, vs. the input frequency. It characterizes how com ponents of an input signal are amplified or attenuated by the system. The term **bandwidth** is used to quantify

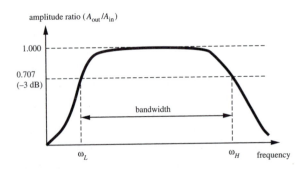

Figure 4.6 Frequency response and bandwidth.

the range of frequencies a system can adequately reproduce. The bandwidth of a system is defined as the range of frequencies where the input of the system is not **attenuated** by more than −3 dB. As illustrated in the figure, a system usually has two frequencies at which the attenuation of the system is −3 dB. They are defined as the low and high **corner** or **cutoff frequencies** ω_L and ω_H. These two frequencies define the bandwidth of the system:

$$\text{bandwidth} = \omega_L \text{ to } \omega_H \tag{4.18}$$

Measurement systems often exhibit no attenuation at low frequencies (i.e., $\omega_L = 0$), and the amplitude ratio degrades only at high frequencies. For these systems, the bandwidth extends from 0 (DC) to ω_H.

The −3 dB cutoff is the decibel value when the power of the output signal (P_{out}) is attenuated to half of its input value (P_{in}):

$$\frac{P_{out}}{P_{in}} = \frac{1}{2} \tag{4.19}$$

For this reason, the cutoff frequencies are referred to as the half-power points. The power of a sinusoidal signal is proportional to the square of the signal's amplitude, thus at the cutoff value,

$$\frac{A_{out}}{A_{in}} = \sqrt{\frac{P_{out}}{P_{in}}} = \sqrt{\frac{1}{2}} \approx 0.707 \tag{4.20}$$

Therefore, at the cutoff frequencies, the amplitude of the signal is attenuated by 29.3% (to 70.7% of its original value), which is approximately −3 dB:

$$\text{dB} = 20 \log_{10}\sqrt{\frac{1}{2}} \approx -3 \text{ dB} \tag{4.21}$$

At first it may seem illogical to define the bandwidth to exclude signal components that exist outside the range of the bandwidth. The half-power points are admittedly somewhat arbitrary, but if applied consistently, they allow us to compare a variety of instruments and system responses. All signal amplitudes of components outside the bandwidth are attenuated by more than 3 dB. Components that lie within the bandwidth, especially those close to the cutoff frequencies, may also be attenuated but by less than 3 dB.

The frequency response of an ideal measurement system has an amplitude ratio of 1, extending from 0 to infinite frequency. An ideal system reproduces all harmonics in a signal without amplification or attenuation. A real measurement system, however, has a limited bandwidth. The bandwidth of a system is influenced by such factors as capacitance, inductance, and resistance in electrical systems and mass, stiffness, and damping in mechanical systems. Through careful design, these parameters can be selected to achieve a desired bandwidth. A properly designed measurement system reproduces all frequency components in a typical input signal. When it does, the system is said to exhibit high **fidelity.**

The proper design or selection of a measurement system requires an understanding of measurement system bandwidth and signal spectrum. Figure 4.7 illustrates an

> ### ■ CLASS DISCUSSION ITEM 4.2
> ### *Measuring a Square Wave with a Limited Bandwidth System*
>
> Assume you have a measurement system whose bandwidth is 0 to 5.1 ω_0, with no attenuation below the cutoff frequency and complete attenuation above. If the input to the system is a square wave with a fundamental frequency of ω_0, describe the difference between the input and the output.

input signal spectrum, a measurement system's frequency response, and the resulting spectrum of the output signal, all using the same frequency scale. The measurement system has a limited bandwidth, so not all input signal frequency components are reproduced satisfactorily. Therefore, the output signal differs from the input signal. Given the input signal spectrum in the figure, the signal can be written as

$$V_{in}(t) = A_1 \sin(\omega_0 t) + A_2 \sin(2\omega_0 t) + A_3 \sin(3\omega_0 t) + \cdots \tag{4.22}$$

Given the measurement system frequency response curve, the output amplitude A_i' can be determined for each input frequency amplitude A_i using the following equation:

$$A_i' = (A_{out}/A_{in})_i \times A_i \tag{4.23}$$

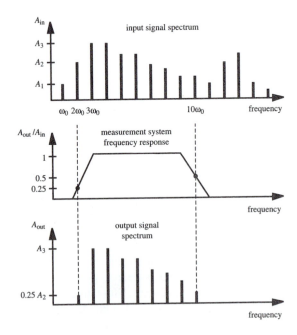

Figure 4.7 Effect of measurement system bandwidth on signal spectrum.

where i is the frequency component number. For example, for $i = 2$ corresponding to frequency $2\omega_0$,

$$A'_2 = (A_{out}/A_{in})_2 \times A_2 = 0.25A_2 \qquad (4.24)$$

Applying Equation 4.23 to each frequency component results in the spectrum shown at the bottom of the figure, and the output signal can be written as

$$V_{out}(t) = A'_2 \sin(2\omega_0 t) + A'_3 \sin(3\omega_0 t) + \cdots + A'_9 \sin(9\omega_0 t) + A'_{10} \sin(10\omega_0 t) \quad (4.25)$$

or

$$V_{out}(t) = 0.25A_2 \sin(2\omega_0 t) + A_3 \sin(3\omega_0 t) + \cdots + A_9 \sin(9\omega_0 t) + 0.5A_{10} \sin(10\omega_0 t) \quad (4.26)$$

Because $A_{out}/A_{in} = 1$ for all frequencies between and including $3\omega_0$ and $9\omega_0$, $A'_i = A_i$ for $i = 3$ through 9. Because $A_{out}/A_{in} = 0$ for the frequency component ω_0 and for all components above $10\omega_0$, these components are completely attenuated and no longer exist in the output signal (i.e., $A'_1 = A'_{11} = A'_{12} = \ldots = 0$). The frequency components at $2\omega_0$ and $10\omega_0$ still exist in the output signal, but their amplitudes are partially attenuated ($0 < A_{out}/A_{in} < 1$).

When designing or choosing a measurement system for an application, it is important that the bandwidth of the system be large enough to adequately reproduce the important frequency components present in the input signal. A measurement system that does not reproduce high frequencies cannot accurately reproduce signals that have rapid changes associated with them.

To experimentally determine the bandwidth of a system, it is necessary to systematically apply pure sinusoidal inputs and to determine the output-to-input amplitude ratio for the desired range of frequencies. The sweep feature on a function generator, which generates a linearly increasing frequency over a selected time, provides a convenient method to perform this for systems with electrical inputs. Lab Exercise 4 applies this technique to filter circuits (see the demonstration in Video Demo 4.8). A method for theoretically determining the frequency response of a system model is presented in Section 4.10.2. Example 4.1 illustrates the basics of the technique, applied to a simple RC filter.

Lab Exercise

Lab 4 Bandwidth, filters, and diodes

Video Demo

4.8 Filter (RC circuit) frequency response

Bandwidth of an Electrical Network | EXAMPLE 4.1

The bandwidth of an electrical system can be readily determined analytically with the help of the steady state AC circuit analysis technique presented in Section 2.6. Consider the following *RC* circuit as an example.

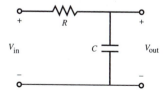

(continued)

(continued)

Using the voltage divider rule for the complex impedances of the resistor ($Z_R = R$) and capacitor ($Z_C = 1/j\omega C$), the output voltage for the network can be written as

$$V_{out} = \frac{\dfrac{1}{j\omega C}}{\dfrac{1}{j\omega C} + R} V_{in}$$

Therefore, the output-to-input ratio as a function of frequency is

$$\frac{V_{out}}{V_{in}} = \frac{1}{j\omega RC + 1}$$

The magnitude of this complex number gives us the amplitude ratio as a function of frequency:

$$\left| \frac{V_{out}}{V_{in}} \right| = \frac{1}{\sqrt{1 + (\omega RC)^2}}$$

The cutoff frequency ω_c for the circuit is

$$\omega_c = \frac{1}{RC}$$

because

$$\left| \frac{V_{out}}{V_{in}} \right| = \frac{1}{\sqrt{2}} = 0.707$$

when $\omega = \omega_c$.

Using ω_c, the amplitude ratio can also be expressed as

$$\left| \frac{V_{out}}{V_{in}} \right| = \frac{1}{\sqrt{1 + (\omega/\omega_c)^2}}$$

MathCAD Example

4.2 Low-pass filter

The frequency response curve representing this relationship follows, where $\omega_r = \omega/\omega_c$ and $A_r = |V_{out}/V_{in}|$. Note that, as ω_r approaches 0, A_r approaches 1 and as ω_r approaches ∞, A_r approaches 0. MathCAD Example 4.2 includes the analysis used to generate the frequency response plot.

This circuit is called a **low-pass filter** because lower frequencies are "passed" to the output with little attenuation and higher frequencies are significantly attenuated (i.e., not "passed").

$$\omega_r := 0, 0.01 \,..\, 2.5 \qquad\qquad A_r(\omega_r) := \frac{1}{\sqrt{1 + \omega_r^2}}$$

If the resistor and capacitor were reversed, the resulting circuit would be called a **high-pass filter** because it would attenuate low frequencies. Two other useful filters are the notch filter and band-pass filter. The **notch filter,** sometimes called the band-reject filter, passes all frequencies except for a narrow band of frequencies that are highly attenuated. A common use for this filter is to eliminate 60 Hz interference often found on signal lines. The **band-pass filter,** on the other hand, passes a narrow band of frequencies and significantly attenuates all others.

4.5 PHASE LINEARITY

The third criterion for a good measurement system is phase linearity, which expresses how well a system preserves the phase relationship between frequency components of the input.

Consider the relationship between phase angle and time displacement between two signals as illustrated in Figure 4.8. Signal 2 lags signal 1 because it occurs later on the time axis. The time displacement between the signals t_d is $T/4$, where T is the period of the signals. Since a cycle of a signal corresponds to 2π radians or 360°, the phase angle between signal 1 and signal 2 is

$$\phi = 360\, t_d\, /T \text{ degrees} = 2\pi\, t_d\, /T \text{ radians} \qquad (4.27)$$

So for $t_d = T/4$, the phase angle is 90° or $\pi/2$ radians.

Measurement systems can cause a delay or time displacement between the input and output signals. For a given frequency f, where $f = 1/T$, Equation 4.27 can be expressed as

$$\phi = 360f \cdot t_d \text{ degrees} = 2\pi f \cdot t_d \text{ radians} \qquad (4.28)$$

Therefore, for a given time displacement, the phase shift for a signal depends on its frequency. Since an input signal may be composed of many frequency components, it is important that all the individual components are displaced by the same amount of time; otherwise, the output of the measurement system would be distorted. For equal time displacement t_d for all frequency components, the following must hold:

$$\phi = k \cdot f \qquad (4.29)$$

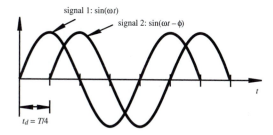

Figure 4.8 Relationship between phase and time displacement.

where k is a constant equal to $360t_d$ degrees or $2\pi t_d$ radians. Therefore, the phase angle must be linear with frequency for equal time displacement of frequency components. When a system functions in this way, it is said to exhibit **phase linearity.** k is usually negative for a measurement system, implying that the recorded output lags behind the actual input signal.

4.6 DISTORTION OF SIGNALS

MathCAD Example

4.3 Amplitude distortion

4.4 Phase distortion

When a system does not exhibit amplitude linearity, the amplitudes of the output frequency components are attenuated. As a result, the output suffers from **amplitude distortion** as illustrated in Figure 4.9 for a square wave. In the figure, the coefficients of the harmonics contain exponential functions whose magnitudes decrease (a) or increase (b) with frequency. Note the change (distortion) in the resulting outputs. MathCAD Example 4.3 includes the analysis used to create the plots in Figure 4.9.

When a system does not exhibit phase linearity, the output frequency components may be of the proper amplitude but are displaced in time with respect to one another. As a result, the output exhibits **phase distortion** as illustrated in Figure 4.10 for a square wave. MathCAD Example 4.4 includes the analysis used to create the plots in Figure 4.10.

A high-fidelity measurement system must exhibit amplitude linearity to prevent amplitude distortion, have adequate bandwidth to pass all frequency components contained in an input signal, and exhibit phase linearity to prevent phase distortion.

When designing or analyzing a measurement system, we want to predict the system's performance. Therefore, we need to be able to model the system and express its behavior in mathematical terms. The remainder of the chapter presents systems analysis tools that let us do this.

$$t := 0, 0.01 .. 2 \qquad n := 1 .. 50$$

$$B_n := \frac{4}{\pi \cdot (2 \cdot n - 1)} \exp[-0.1(2n-1)] \qquad C_n := \frac{4}{\pi(2n-1)}\{1 - \exp[-(2n-1)]\}$$

$$F_{\text{high}}(t) := \sum_n B_n \sin[(2n-1)2\pi t] \qquad F_{\text{low}}(t) := \sum_n C_n \sin[(2n-1)2\pi t]$$

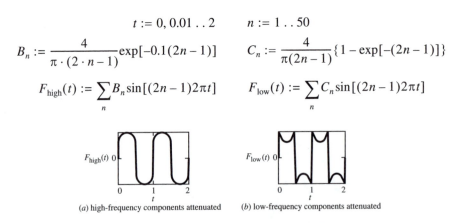

(a) high-frequency components attenuated   (b) low-frequency components attenuated

Figure 4.9 Amplitude distortion of a square wave *(generated with MathCAD).*

$$t := 0, 0.01 \ldots 2 \quad n := 1 \ldots 50$$

$$B_n := \frac{4}{\pi(2n-1)} \qquad dt_n := 0.05\{1 - \exp[-(2n-1)]\}$$

$$F_{\text{lag}}(t) := \sum_n B_n \sin[(2n-1)2\pi(t-dt_n)] \quad F_{\text{lead}}(t) := \sum_n B_n \sin[(2n-1)2\pi(t+dt_n)]$$

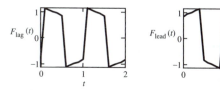

(a) high-frequency components lagging (b) high-frequency components leading

Figure 4.10 Phase distortion of a square wave *(generated with MathCAD)*.

■ CLASS DISCUSSION ITEM 4.3
Analytical Attenuation

In Figure 4.9, the following exponential terms were used to attenuate amplitude components at either low or high frequencies:

$$e^{-0.1(2n-1)} \quad \text{and} \quad 1 - e^{-(2n-1)}$$

Explain how these exponential terms cause the resulting distortion of the square wave. (Hint: Plot the exponential functions and describe how they change the amplitudes of the components of the square wave spectrum.)

4.7 DYNAMIC CHARACTERISTICS OF SYSTEMS

As we will see, many measurement systems can be modeled as linear ordinary differential equations with constant coefficients. This is also true of many mechatronic systems or their component subsystems. These linear models have the following general form:

$$\sum_{n=0}^{N} A_n \frac{d^n X_{\text{out}}}{dt^n} = \sum_{m=0}^{M} B_m \frac{d^m X_{\text{in}}}{dt^m} \tag{4.30}$$

where X_{out} is the output variable, X_{in} is the input variable, and A_n and B_n are constant coefficients. N defines the **order** of the system independent of M. The approach used to determine the differential equation model depends on the type of system being analyzed. For example, Newton's laws and free-body diagrams can be applied to mechanical systems, and KVL and KCL equations can be applied to electric circuits.

As we will see, the constant coefficients A_n represent physical properties of the system of interest.

Many electromechanical systems exhibit nonlinear behavior and cannot be accurately modeled as linear systems. However, a nonlinear system may often exhibit linear behavior over a limited range of inputs and a linear model can be derived that provides an adequate approximation over this range. This process of modeling a nonlinear system with a linear model is called **linearization.**

The next three sections present methods to characterize the simplest and most common forms of Equation 4.30, where $M = 0$ and $N = 0$, 1, or 2. These cases are referred to as zero-order, first-order, and second-order systems, respectively.

4.8 ZERO-ORDER SYSTEM

When $N = M = 0$ in Equation 4.30, the model represents a **zero-order** system whose behavior is described by

$$A_0 X_{\text{out}} = B_0 X_{\text{in}} \tag{4.31}$$

or

$$X_{\text{out}} = \frac{B_0}{A_0} X_{\text{in}} = K X_{\text{in}} \tag{4.32}$$

where $K = B_0/A_0$ is a constant referred to as the **gain** or **sensitivity** of the system, because it represents a scaling between the input and the output. When the sensitivity is high, a small change in the input can result in a significant change in the output. Note that the output of a zero-order system follows the input exactly with no time delay or distortion.

An example of a zero-order measurement system is a potentiometer used to measure displacement. A potentiometer is a variable resistance device whose output resistance changes as an internal wiper moves across a resistive surface. As illustrated in Figure 4.11, it produces an output voltage V_{out} that is directly proportional to the wiper displacement X_{in}. This is a result of the voltage division rule, which gives

$$V_{\text{out}} = \frac{R_x}{R_p} V_s = \left(\frac{V_s}{L}\right) X_{\text{in}} \tag{4.33}$$

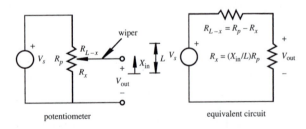

potentiometer   equivalent circuit

Figure 4.11 Displacement potentiometer.

where R_x is the resistance between the potentiometer leads, R_p is the maximum resistance of the potentiometer, X_{in} is the displacement of the potentiometer wiper, and L is the maximum amount of wiper travel.

■ **CLASS DISCUSSION ITEM 4.4**
Assumptions for a Zero-Order Potentiometer

What approximating assumptions must be made regarding the potentiometer to ensure that it is a zero-order measurement system?

THREADED DESIGN EXAMPLE

DC motor power-op-amp speed controller — potentiometer interface **A.2**

The figure below shows the functional diagram for Threaded Design Example A (see Section 1.3 and Video Demo 1.6). The portion described here is highlighted.

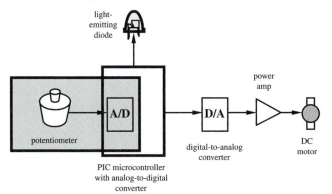

Video Demo

1.6 DC motor power-op-amp speed controller

To interface a potentiometer to a microcontroller, the analog voltage output (see Equation 4.30) must be converted to digital form. Fortunately, a PIC microcontroller can be selected that has a built-in analog-to-digital (A/D) converter. This allows the PIC to sense the potentiometer position, as a voltage, using the simple circuit shown below. The potentiometer has three leads, wired as shown, with the center "wiper" lead connected to the PIC analog input (pin 18, designated as AN1). The voltage on the wiper line varies from 0 V (ground) to 5 V as the knob is turned.

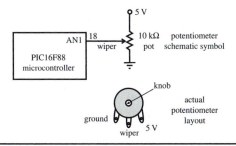

4.9 FIRST-ORDER SYSTEM

When $N = 1$ and $M = 0$ in Equation 4.30, the equation models a **first-order** system whose behavior is described by

$$A_1 \frac{dX_{out}}{dt} + A_0 X_{out} = B_0 X_{in} \tag{4.34}$$

or

$$\frac{A_1}{A_0} \frac{dX_{out}}{dt} + X_{out} = \frac{B_0}{A_0} X_{in} \tag{4.35}$$

As with the zero-order system, the coefficient ratio on the right-hand side is called the sensitivity or **static sensitivity,** defined as

$$K = \frac{B_0}{A_0} \tag{4.36}$$

The coefficient ratio on the left side of Equation 4.35 has a special name and meaning. It is called the **time constant** and is defined as

$$\tau = \frac{A_1}{A_0} \tag{4.37}$$

Why we chose this name will become clear in the subsequent analysis. With these definitions, the first-order system equation can be written as

$$\tau \frac{dX_{out}}{dt} + X_{out} = KX_{in} \tag{4.38}$$

Note that in this standard form, the coefficent of the X_{out} term must be 1.

To characterize how a system responds to various types of inputs, we apply standard inputs to the model, including step, impulse, and sinusoidal functions. A **step input** changes instantaneously from 0 to a constant value A_{in} and is stated mathematically as

$$X_{in} = \begin{cases} 0 & t < 0 \\ A_{in} & t \geq 0 \end{cases} \tag{4.39}$$

The output of the system in response to this input is called the **step response** of the system. For a first-order system, we can find the step response by solving Equation 4.38 using Equation 4.39 with the initial condition

$$X_{out}(0) = 0 \tag{4.40}$$

Assuming a solution of the form Ce^{st} for the homogeneous form of differential Equation 4.38, the **characteristic equation** is

$$\tau s + 1 = 0 \tag{4.41}$$

Since the root of this equation is $s = -1/\tau$, the **homogeneous** or **transient solution** is

$$X_{out_h} = Ce^{-t/\tau} \tag{4.42}$$

where C is a constant determined later by applying initial conditions. A **particular** or **steady state solution** resulting from the step input is

$$X_{out_p} = KA_{in} \tag{4.43}$$

The **general solution** is the sum of the homogeneous and particular solutions (Equation 4.42 + Equation 4.43):

$$X_{out}(t) = X_{out_h} + X_{out_p} = Ce^{-t/\tau} + KA_{in} \tag{4.44}$$

Applying the initial condition (Equation 4.40) to this equation gives

$$0 = C + KA_{in} \tag{4.45}$$

thus,

$$C = -KA_{in} \tag{4.46}$$

so the resulting step response is

$$X_{out}(t) = KA_{in}(1 - e^{-t/\tau}) \tag{4.47}$$

As illustrated in Figure 4.12, this represents an exponential rise in the output toward an asymptotic value of KA_{in}. The rate of the rise depends only on the time constant τ. The response is faster for a smaller time constant.

Note that after one time constant, the output reaches 63.2% of its final value, and from Equation 4.46,

$$X_{out}(\tau) = KA_{in}(1 - e^{-1}) = 0.632KA_{in} \tag{4.48}$$

After four time constants, the step response is

$$X_{out}(4\tau) = KA_{in}(1 - e^{-4}) = 0.982KA_{in} \tag{4.49}$$

Since this value is more than 98% of the steady state value KA_{in}, we usually assume that a first-order system has reached its steady state value within four time constants.

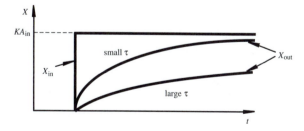

Figure 4.12 First-order response.

Video Demo

4.9 RC circuit charging and discharging

When designing a first-order measurement system, look at quantities that affect τ and try to reduce them if possible. The larger τ is, the longer the measurement system takes to respond to an input.

An important example of a first-order system is an *RC* circuit (see Example 4.1), which has a time constant of $\tau = RC$ (see Question 4.19). *RC* circuits are very common in timing, filter, and other applications (e.g., see Example 4.1, Section 6.12.3, and Example 7.6). Video Demo 4.9 illustrates the step response of an *RC* circuit, showing how the voltage across the capacitor builds over time. The speed of charging and discharging the capacitor is directly related to the time constant of the circuit.

4.9.1 Experimental Testing of a First-Order System

To characterize and evaluate an existing first-order system, we need methods to experimentally determine the time constant τ and the static sensitivity *K*. *K* may be obtained by static calibration, where a known static input is applied, and the output is observed. A common method to determine the time constant τ is to apply a step input to the system and determine the time for the output to reach 63.2% of its final value (see Equation 4.48). An alternative method to determine a value for τ follows.

We can rearrange Equation 4.47, expressing it as

$$\frac{X_{\text{out}} - KA_{\text{in}}}{KA_{\text{in}}} = -e^{-t/\tau} \tag{4.50}$$

Simplifying, we get

$$1 - \frac{X_{\text{out}}}{KA_{\text{in}}} = e^{-t/\tau} \tag{4.51}$$

If we take the natural logarithm of both sides, we get

$$\ln\left(1 - \frac{X_{\text{out}}}{KA_{\text{in}}}\right) = -\frac{t}{\tau} \tag{4.52}$$

If we define the left-hand side as *Z*, then

$$Z = -t/\tau \tag{4.53}$$

and a plot of *Z* vs. *t* is a straight line with the slope

$$\frac{dZ}{dt} = -\frac{1}{\tau} \tag{4.54}$$

Therefore, if we collect experimental data from a step response and plot *Z* vs. *t* as illustrated in Figure 4.13, we can determine τ from the slope of the line:

$$\tau = -\frac{\Delta t}{\Delta Z} \tag{4.55}$$

Note that if experimental data for *Z* vs. *t* deviates from a straight line, then the system is not first order. If this is the case, the system is either of higher order or nonlinear.

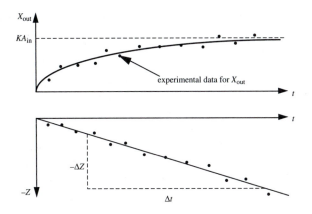

Figure 4.13 Experimental determination of τ.

4.10 SECOND-ORDER SYSTEM

An example of a **second-order** system, where $N = 2$ in Equation 4.30, is the mechanical spring-mass-damper system illustrated in Figure 4.14. Applying Newton's law of motion (in this case, $\Sigma F_x = ma_x$) to the free-body diagram gives the second-order differential equation, which is the mathematical model for the system:

$$m\frac{\mathrm{d}^2 x}{\mathrm{d}t^2} + b\frac{\mathrm{d}x}{\mathrm{d}t} + kx = F_{\text{ext}}(t) \tag{4.56}$$

where m is the mass, b is the damping coefficient, k is the spring constant, and x is the displacement of the mass from the equilibrium (rest) position of the mass measured positively in the downward direction as shown. $F_{\text{ext}}(t)$ represents the resultant of all applied external forces (inputs) in the direction of x. The weight of the mass is not included as a force in the free-body diagram because the displacement x is measured from the equilibrium position. In the equilibrium position, the spring is stretched an amount δ from its unstretched length so that the effect of gravity

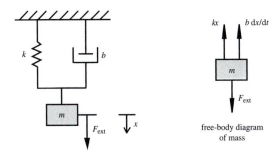

Figure 4.14 Second-order mechanical system and free-body diagram.

is balanced: $k\delta = mg$. Note that the spring force (kx) is in the opposite direction from the mass displacement, and the damper force ($b\,dx/dt$) is in the opposite direction from the velocity of the mass.

As we will see in Section 4.11, the governing equations for many second-order systems other than mechanical spring-mass-damper systems (e.g., mechanical rotational systems and hydraulic systems) have the same form as Equation 4.56.

A good example of a second-order measurement system is the strip chart recorder illustrated in Figure 4.15. It consists of a continuously moving roll of paper on which a pen is moved back and forth. A spring is attached to the pen carriage to keep it centered in a neutral position when an applied force is removed. The strip chart recorder can be thought of as a mechanical oscilloscope. It was used much more before the digital age whenever a large amount of time-varying data needed to be stored (e.g., for later analysis). They are still used in some applications (e.g., in older remote seismograph stations and polygraph machines). The applied force F_{ext} is provided by an electromagnetic coil whose core is attached to the pen carriage. The spring keeps the pen carriage centered in the zero position when the input is 0. The pen carriage bearings and the pen-paper interface result in the damping force. The carriage and the pen constitute the mass to which the forces in the system are applied.

To characterize the unforced response of a second-order system, we need to solve the differential Equation 4.56 with $F_{ext} = 0$. The characteristic equation is

$$ms^2 + bs + k = 0 \tag{4.57}$$

This quadratic equation has two roots for s:

$$s_1 = -\frac{b}{2m} + \sqrt{\left(\frac{b}{2m}\right)^2 - \frac{k}{m}} \tag{4.58}$$

$$s_2 = -\frac{b}{2m} - \sqrt{\left(\frac{b}{2m}\right)^2 - \frac{k}{m}} \tag{4.59}$$

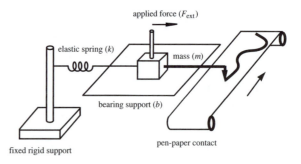

Figure 4.15 Strip chart recorder as an example of a second-order system.

If there were no damping in the system (i.e., $b = 0$), the roots would be

$$s_1 = j\sqrt{\frac{k}{m}} \tag{4.60}$$

$$s_2 = -j\sqrt{\frac{k}{m}} \tag{4.61}$$

and the corresponding homogeneous solution would be

$$x_h(t) = A\cos\left(\sqrt{\frac{k}{m}}\,t\right) + B\sin\left(\sqrt{\frac{k}{m}}\,t\right) \tag{4.62}$$

where A and B are constants determined from the initial conditions $x(0)$ and $dx/dt(0)$. This motion represents pure **undamped** oscillatory motion with radian frequency:

$$\omega_n = \sqrt{\frac{k}{m}} \tag{4.63}$$

This is called the **natural frequency** of the system, because it is the frequency at which the undamped system would naturally oscillate if the spring were stretched and the mass released and allowed to move without any external force ($F_{ext} = 0$).

If there is damping in the system (i.e., $b \neq 0$) and the radicand in Equations 4.58 and 4.59 is zero, the roots are double real roots, and the resulting transient homogeneous solution is

$$x_h(t) = (A + Bt)e^{-\omega_n t} \tag{4.64}$$

This represents an exponentially decaying motion. A system with this behavior is said to be **critically damped,** because it is just on the verge of damped oscillatory motion. The damping constant that results in critical damping is called the **critical damping constant** b_c. It is the value of b that makes the radicand 0, so

$$b_c = 2\sqrt{km} = 2m\omega_n \tag{4.65}$$

The **damping ratio** ζ *(zeta)* for a noncritically damped system is defined as

$$\zeta = \frac{b}{b_c} = \frac{b}{2\sqrt{km}} \tag{4.66}$$

It is a measure of the proximity to critical damping. A critically damped system has a damping ratio of 1.

With the definitions of natural frequency and damping ratio, the roots of the characteristic equations (Equations 4.58 and 4.59) can be written as

$$s_1 = -\zeta\omega_n + \omega_n\sqrt{\zeta^2 - 1} \tag{4.67}$$

$$s_2 = -\zeta\omega_n - \omega_n\sqrt{\zeta^2 - 1} \tag{4.68}$$

If there is damping in the system (i.e., $b \neq 0$) and the radicand in Equations 4.67 and 4.68 is negative (i.e., $\zeta < 1$), the roots are complex conjugates, and the resulting transient homogeneous solution is

$$x_h(t) = e^{-\zeta\omega_n t}[A\cos(\omega_n\sqrt{1-\zeta^2}\ t) + B\sin(\omega_n\sqrt{1-\zeta^2}\ t)] \qquad (4.69)$$

This motion represents damped oscillation consisting of sinusoidal motion with exponentially decaying amplitude. A system with these characteristics is said to be **underdamped,** because it has less than critical damping ($\zeta < 1$). The frequency of oscillation is

$$\omega_d = \omega_n\sqrt{1-\zeta^2} \qquad (4.70)$$

which is called the **damped natural frequency** of the system.

If there is damping in the system (i.e., $b \neq 0$) and the radicand in Equations 4.67 and 4.68 is positive (i.e., $\zeta > 1$), the roots are distinct real roots and the resulting transient homogeneous solution is

$$x_h(t) = A\,e^{(-\zeta+\sqrt{\zeta^2-1})\omega_n t} + B\,e^{(-\zeta-\sqrt{\zeta^2-1})\omega_n t} \qquad (4.71)$$

This represents an exponentially decaying output. A system with these characteristics is said to be **overdamped,** because its damping exceeds critical damping ($\zeta > 1$).

Examples of transient responses for all three cases of damping (underdamped, critically damped, and overdamped) are illustrated in Figure 4.16. The curves represent unforced motion of a second-order system with different amounts of damping, when the system is released from rest ($dx/dt(0) = 0$) at $x(0) = 1$.

Based on the definitions above, a more standard form for the differential equation representing a second-order system can be written as

$$\ddot{x} + 2\zeta\omega_n\dot{x} + \omega_n^2 x = \frac{\omega_n^2}{k}F_{\text{ext}} \qquad (4.72)$$

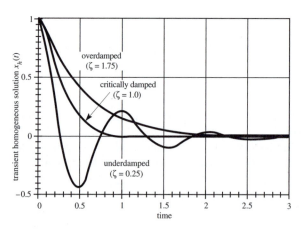

Figure 4.16 Transient response of second-order systems.

Equations 4.63 and 4.66 define how the terms in this equation relate to the coefficient parameters in Equation 4.56.

■ **CLASS DISCUSSION ITEM 4.5**
Spring-Mass-Damper System in Space

Would a spring-mass-damper system behave any differently inside a space station orbiting the earth than it does on the surface of the earth? Why or why not? How could you use a spring to measure the mass of an astronaut in the orbiting space station?

4.10.1 Step Response of a Second-Order System

As we found when analyzing a first-order system, an important input used to study the dynamic characteristics of a system is a step function. The step response is a good measure of how fast and smoothly a system responds to abrupt changes in input. The step response consists of two parts: a transient homogeneous solution $x_h(t)$, which is of the form presented in Section 4.10 for the unforced response, plus a steady state particular solution $x_p(t)$, which is a result of the forcing function. For a step input given by

$$F_{ext}(t) = \begin{cases} 0 & t < 0 \\ F_i & t \geq 0 \end{cases} \tag{4.73}$$

it is clear from Equation 4.56 that a particular solution is

$$x_p(t) = \frac{F_i}{k} \tag{4.74}$$

The general solution for the step response is then

$$x(t) = x_h(t) + x_p(t) \tag{4.75}$$

where the constants in $x_h(t)$ are determined by applying the initial conditions $x(0)$ and $dx/dt(0)$ to the general solution $x(t)$. As with the unforced case, there are three distinctly different types of response based on the amount of damping in the system, as illustrated in Figure 4.17.

Figure 4.18 illustrates the step response of an underdamped system and defines several terms used when describing the step response. The **steady state value** is the value the system reaches after all transients dissipate. The **rise time** is the time required for the system to go from 10% to 90% of the steady state value. The **overshoot** is a measure of the amount the output exceeds the steady state value before settling, usually specified as a percentage of the steady state value. The **settling time** is the time required for the system to settle to within an amplitude band whose width is a specified ±percentage of the steady state value. A settling time tolerance band of ±10% is shown in Figure 4.18, although, ±2% is more commonly used when calculating and reporting values. These terms can be used to characterize the step

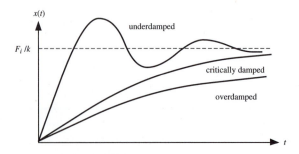

Figure 4.17 Second-order step responses.

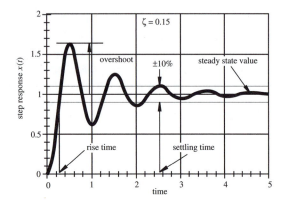

Figure 4.18 Features of an underdamped step response.

response of a second-order system with different amounts of damping: An under-damped system exhibits a fast rise time, but it overshoots and may take a while to settle, although, with the right amount of damping, it can settle faster than critically damped or overdamped systems. For cases where no overshoot is allowed or desired, a critically damped system has the fastest rise and settle times. An overdamped system also has no overshoot, but has slower rise and settle times.

The step response may be useful in identifying the order of a system, because the shape of the response curve can help indicate the order. Also, if the system is second order, we can learn whether it is underdamped, overdamped, or critically damped.

■ **CLASS DISCUSSION ITEM 4.6**
Good Measurement System Response

Describe and sketch what you think is a "good" step response of a measurement system. Is overshoot bad? What advantage does it provide? Does critical damping provide optimal response?

4.10.2 Frequency Response of a System

To determine the frequency response of a linear system, we apply a sinusoidal input and determine the output response for different input frequencies. For the second-order system described in Section 4.10, the sinusoidal forcing function input can be represented as

$$F_{ext}(t) = F_i \sin(\omega t) \tag{4.76}$$

where F_i is the amplitude of the external force and ω is the input frequency.

When the sinusoidal input is first applied, the system exhibits a combined transient and steady state response. Once the transients dissipate, the steady state output of the system is sinusoidal with the same frequency as the input but possibly out of phase with the input. We can represent this steady state output in the following general form:

$$x(t) = X_o \sin(\omega t + \phi) \tag{4.77}$$

where X_o is the output magnitude and ϕ is the phase difference between the input and the output.

The procedure to determine analytically the frequency response of a linear system follows. The linear system could model a measurement system or a more general mechatronic system of any order. Each step in the procedure is shown in its generic form and then applied to a second-order system.

Analytical Procedure to Determine the Frequency Response of a System

1. *Find the Laplace transform of the system differential equation assuming initial conditions are zero:* $x(0) = dx/dt\ (0) = 0$. *The Laplace transform converts the differential equation into an algebraic equation that is related to the frequency response of the system.*

   The governing equation for a second-order system (from Equation 4.72) can be expressed as

   $$\frac{d^2x(t)}{dt^2} + 2\zeta\omega_n\frac{dx(t)}{dt} + \omega_n^2x(t) = \frac{\omega_n^2}{k}F_{ext}(t) \tag{4.78}$$

   Applying the Laplace transform to both sides of the equation, assuming zero initial conditions, gives

   $$(s^2 + 2\zeta\omega_ns + \omega_n^2)X(s) = \frac{\omega_n^2}{k}F_{ext}(s) \tag{4.79}$$

   where $F_{ext}(s)$ and $X(s)$ are the Laplace transforms of the input forcing function and the output displacement, respectively.

2. *Find the **transfer function** of the system, which is the ratio of the output and input Laplace transforms.*

   For this system, the transfer function is

   $$G(s) = \frac{X(s)}{F_{ext}(s)} = \frac{\frac{\omega_n^2}{k}}{(s^2 + 2\zeta\omega_ns + \omega_n^2)} \tag{4.80}$$

3.   *To simulate a harmonic input, replace s with jω in the transfer function. This yields the frequency response behavior of the system.*

For this system, replacing s with jω in Equation 4.80 and dividing the numerator and denominator by ω_n^2 gives

$$G(j\omega) = \frac{1/k}{\left[1 - \left(\dfrac{\omega}{\omega_n}\right)^2\right] + j\left(2\zeta\dfrac{\omega}{\omega_n}\right)} \qquad (4.81)$$

4.   *Find the desired **amplitude ratio** between the output and input by determining the magnitude of the complex transfer function:*

$$\text{mag}[G(j\omega)] = |G(j\omega)| \qquad (4.82)$$

For this system, taking the magnitude of the denominator, which is the square root of the sum of the squares of the real and imaginary components, and moving the k term to the left side of the equation, results in the following dimensionless ratio:

$$\frac{X_o}{F_i/k} = \frac{1}{\left\{\left[1 - \left(\dfrac{\omega}{\omega_n}\right)^2\right]^2 + 4\zeta^2\left(\dfrac{\omega}{\omega_n}\right)^2\right\}^{1/2}} \qquad (4.83)$$

5.   *Find the **phase angle** ϕ between the output and input by determining the argument of the complex transfer function:*

$$\phi = \arg[G(j\omega)] = \angle G(j\omega) \qquad (4.84)$$

From Equation 4.81, recalling that the argument of a complex number is the arctangent of the ratio of its imaginary and real parts and that the argument of a ratio is the difference of the numerator and denominator arguments, the phase angle for this system is found as

$$\phi = 0 - \tan^{-1}\left\{\frac{2\zeta\dfrac{\omega}{\omega_n}}{\left[1 - \left(\dfrac{\omega}{\omega_n}\right)^2\right]}\right\} = -\tan^{-1}\left\{\frac{2\zeta}{\dfrac{\omega_n}{\omega} - \dfrac{\omega}{\omega_n}}\right\} \qquad (4.85)$$

MathCAD Example

4.5 Second-order system frequency response

A plot of the frequency response of the second-order system as a function of the damping ratio ζ is shown in Figure 4.19. MathCAD Example 4.5 includes the analysis used to create the plot. When the input frequency ω equals the natural frequency ω_n (i.e., $\omega/\omega_n = 1$), **resonance** occurs. Often ω_n is called the resonant frequency. For small damping ratios, the output amplitude becomes quite large at the resonant frequency. For input frequencies less than the natural frequency (i.e., $\omega/\omega_n < 1$), the amplitude ratio is close to 1. The damping ratio ζ affects the range of frequencies

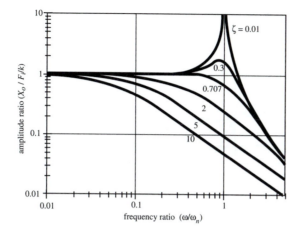

Figure 4.19 Second-order system amplitude response.

for which this is true. Recall that, for a good measurement system, we require the amplitude ratio to be nearly constant (within ±3 dB) over the range of frequencies we wish to measure. A damping ratio of 0.707 provides the best amplitude linearity over the largest bandwidth.

A plot of the phase angle as a function of frequency ratio and damping ratio is shown in Figure 4.20. The phase angle is negative, implying that the output lags behind the input, and the magnitude ranges from 0°, which represents no phase difference, to 180°, which represents a half-cycle phase difference. As ω approaches the natural frequency ω_n (i.e., $\omega/\omega_n = 1$), the output lags the input by 90° (a quarter of a cycle). When the damping in a system is very small (e.g., $\zeta = 0.01$) and the value of the excitation frequency passes through the natural frequency (ω_n), the phase shift changes abruptly by 180°. A damping ratio of 0.707 provides the best approximation of phase linearity for a second-order system.

A very large number of mechanical systems can be approximated by second-order linear ordinary differential equations. Also, complex systems or their component subsystems can often be reduced to second-order systems for a first approximation analysis. Therefore, all engineers should completely understand all implications of Figures 4.19 and 4.20. Video Demos 4.10 through 4.12 show examples of how the second-order system response concepts apply to some basic systems. Video Demo 4.13 shows a classic failure that resulted from engineers not having a firm understanding of some of these concepts. In this case, there were poorly understood wind-induced oscillating loads and inadequate damping, leading to catastrophic failure of this famous suspension bridge.

The output of second-order measurement systems always differs from the input in both amplitude and phase. In designing a measurement system, the goal is to minimize these effects when possible, so the choice of 0.707 for the damping ratio is optimal because it results in the best combination of amplitude linearity

Video Demo

4.10 Spring-mass second-order system frequency response

4.11 Pendulum second-order system frequency response

4.12 Torsional system frequency response and control

4.13 Tacoma Narrows bridge failure

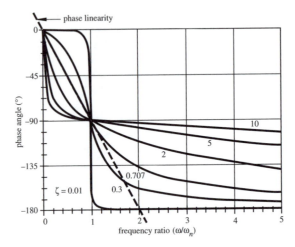

Figure 4.20 Second-order system phase response.

and phase linearity over the widest range of frequencies. When designing other types of systems (e.g., see Design Example 4.1, which addresses the design of a car suspension), damping ratios other than 0.707 might be more appropriate for the application.

In general, whenever a periodic input is applied to a second-order system, resonance, attenuation, and phase shift may occur. In order to predict how a system will respond, it is important to be able to determine the system's frequency response characteristics.

■ **CLASS DISCUSSION ITEM 4.7**
Slinky Frequency Response

Using a Slinky with a mass suspended from one end, demonstrate the frequency response characteristics illustrated in Figures 4.19 and 4.20.

DESIGN
EXAMPLE 4.1

Automobile Suspension Selection

A design team has been asked to select a spring and shock absorber for the front suspension of a new automobile. The team's task is to choose a suspension that will provide the best response to varying road conditions. Since many US manufacturers still specify their products in English units, the design team has reluctantly agreed to perform all analyses using the English system.

After preliminary analyses and meetings with other design teams, the suspension team has narrowed its choices down to three alternative suspension designs. The three alternatives are characterized by the following spring constants k and damping constants b:

Alternative	k (lbf/in)	b (lbf sec /in)
1	500	10
2	200	20
3	120	10

To evaluate the alternatives under different road conditions, the design team decides to simulate bumps or pot holes as a ground force step input of magnitude F_0 and regular bumps or waviness in the road as a sinusoidal ground force input of magnitude B_0 and frequency ω. To subject each alternative to similar input displacement levels, the following fixed ratios are used in the comparison:

$$F_0/k = 6 \text{ in} \quad B_0/k = 3 \text{ in}$$

Each alternative is subjected to the following three frequencies to simulate different speeds or bump spacing:

$$\omega_1 = 10 \text{ rad/sec} \quad \omega_2 = 20 \text{ rad/sec} \quad \omega_3 = 30 \text{ rad/sec}$$

The team decides on a single degree of freedom system that models only a single wheel and neglects the response of the other wheels. The resulting model is a second-order system described by Equation 4.56, where m is the mass supported ("sprung") by the suspension and F_{ext} is the ground force. The weight of the entire vehicle is 2000 lbf; therefore, the sprung mass supported by a single wheel is assumed to be 1/4 of the total mass:

$$m = 500 \text{ lbm} = 15.53 \text{ slugs}$$

Note that the wheel mass and tire dynamics are not included in this simplified model (see Class Discussion Item 4.8).

Using Equations 4.63 and 4.66, the natural frequency ω_n and the damping constant ζ are calculated for each alternative:

Alternative	ω_n	ζ
1	19.6	0.20
2	12.4	0.62
3	9.6	0.40

Since $\zeta < 1.0$ for each alternative, the suspensions will exhibit an underdamped step response described by Equation 4.69.

The step response and sinusoidal response analyses are performed in MathCAD (see MathCAD Example 4.6). The input variables (for Alternative 1) are defined as

$$k := 500 \frac{\text{lbf}}{\text{in}} \qquad b := 10 \text{ lbf} \frac{\text{sec}}{\text{in}} \qquad m := 500 \text{ lb}$$

$$F_0 := (6 \text{ in})k \qquad B_0 := (3 \text{ in})k$$

$$\omega_1 := 10 \frac{\text{rad}}{\text{sec}} \qquad \omega_2 := 20 \frac{\text{rad}}{\text{sec}} \qquad \omega_3 := 30 \frac{\text{rad}}{\text{sec}}$$

$$t := 0 \text{ sec}, 0.01 \text{ sec} \ . \ . \ 2 \text{ sec}$$

MathCAD Example

4.6 Suspension design

(continued)

(continued)

The step response is defined by (see Question 4.28)

$$\omega_n := \sqrt{\frac{k}{m}} \qquad \zeta := \frac{b}{2\sqrt{km}} \qquad \omega_d := \omega_n\sqrt{1-\zeta^2}$$

$$x(t) := \frac{F_0}{k}\left[1 - e^{-\zeta\omega_n t}\cos(\omega_d t)\right]$$

and the sinusoidal response is defined by

$$X_0(\omega) := \frac{\dfrac{B_0}{k}}{\sqrt{\left[1 - \left(\dfrac{\omega}{\omega_n}\right)^2\right]^2 + 4\zeta^2\left(\dfrac{\omega}{\omega_n}\right)^2}} \qquad \phi(\omega) := -\text{angle}\left[\left(\frac{\omega_n}{\omega} - \frac{\omega}{\omega_n}\right), 2\zeta\right]$$

$$x(t, \omega) := X_0(\omega)[\sin(\omega t + \phi(\omega))]$$

The results for the three alternatives follow.

■ Alternative 1

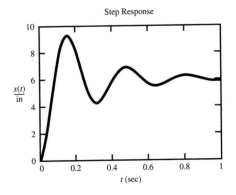

Step Response

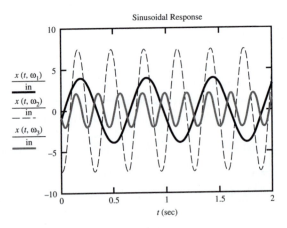

Sinusoidal Response

■ **Alternative 2**

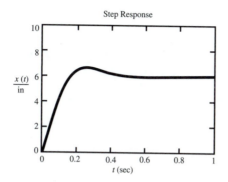

Step Response

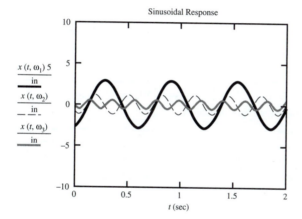

Sinusoidal Response

■ **Alternative 3**

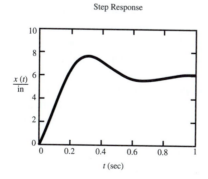

Step Response

(continued)

(concluded)

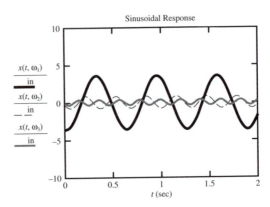

A good suspension has a fast response (short settling time), resulting in good handling, and prevents excessive bounce (oscillation) or large displacement, which improves rider comfort. From the preceding results, it appears that Alternative 2 is the best of the three designs. The step response settles fast with little bounce and the sinusoidal responses have low amplitude.

■ **CLASS DISCUSSION ITEM 4.8**
Suspension Design Results

Were the step response and sinusoidal response graphs required to draw conclusions about the three alternatives? Also, why was the amplitude of the sinusoidal response excessive for Alternative 1 and least for Alternative 2? What would it take to include the wheel mass and tire dynamics in the system model?

4.11 SYSTEM MODELING AND ANALOGIES

Any system—whether it be mechanical, electrical, hydraulic, thermal, or some combination—can be modeled by ordinary linear differential equations that relate the output responses of the system to the inputs. These differential equations are all very similar mathematically and differ only in the constants that appear in front of the derivative terms. These constants represent the physical parameters of the system, and there are analogies for these parameters among the different system genres. For instance, a resistor in an electrical system is analogous to a dashpot or damper in a mechanical system or to a valve or flow restriction in a hydraulic system. Similarly, mass or inertia in a mechanical system is analogous to inductance in an electrical system and fluid inertance in a hydraulic system. The generic terms used to describe the analogous system parameters and variables are **effort, flow, displacement, momentum, resistance, capacitance, and inertia.** Table 4.1 summarizes the generic quantities along with specific analogies for mechanical, electrical, and hydraulic systems. Also included are equations relating variables for energy storage and energy dissipation elements in each system.

Table 4.1 Second-order system modeling analogies

Generic quantity	Mechanical translation	Mechanical rotation	Electrical	Hydraulic
Effort (E)	Force (F)	Torque (T)	Voltage (V)	Pressure (P)
Flow (F)	Speed (v)	Angular speed (ω)	Current (i)	Volumetric flow rate (Q)
Displacement (q)	Displacement (x)	Angular displacement (θ)	Charge (q)	Volume ($\forall$)
Momentum (p)	Linear momentum ($p = mv$)	Angular momentum ($h = J\omega$)	Flux linkage ($I = N\Phi = Li$)	Momentum/area ($\Gamma = IQ$)
Resistor (R)	Damper (b)	Rotary damper (B)	Resistor (R)	Resistor (R)
Capacitor (C)	Spring ($1/k$)	Torsion spring ($1/k$)	Capacitor (C)	Tank (C)
Inertia (I)	Mass (m)	Moment of inertia (J)	Inductor (L)	Inertance (I)
Inertia energy storage (special case)	$F = \dot{p}$ ($F = ma$)	$T = \dot{h}$ ($T = J\alpha$)	$V = \dot{\lambda}$ ($V = L\, di/dt$)	$P = \dot{\Gamma}$ ($P = I\, dQ/dt$)
Capacitor energy storage	$F = kx$	($T = k\theta$)	$V = (1/C)q$	$P = (1/C)\forall$
Dissipative	$F = bv$	$T = B\omega$	$V = Ri$	$P = RQ$

The only problematic analogy in the table is hydraulic resistance, because this quantity is usually not constant. It is typically a nonlinear function of flow rate and geometry. Analogies can also be extended to other phenomena such as heat transfer, where temperature difference is the "effort," heat flow is the "flow," the amount of heat transferred is the "displacement," heat capacity due to mass and specific heat is "capacitance," and thermal resistance due to convection and conduction is "resistance." However, heat transfer and many other phenomena cannot be adequately modeled by second-order linear differential equations, so there are no direct analogies for the relations given in the table.

Figure 4.21 illustrates an example of electrical, mechanical, and hydraulic system analogies along with their governing equations. Each of these systems is described by differential equations of exactly the same form. The only differences are the constant parameters that describe the analogous elements. Any quantity in one system is directly analogous to a quantity in each of the other two systems. For example, the capacitor in the electrical system is analogous to the spring in the mechanical system and to the tank in the hydraulic system, and the current through the capacitor ($I_2 = I_1 - I_3$) is analogous to the rate of compression of the spring ($v_2 = v_1 - v_3$) in the mechanical system and to the flow rate into the tank ($Q_2 = Q_1 - Q_3$) in the hydraulic system.

Knowledge of these analogies helps one gain a deeper understanding and intuitive feel for how different systems respond. It also provides a framework to model and analyze a great variety of systems in a generic form. An understanding of the characteristics and analysis of one type of system can directly apply to any other system.

In the past, before the advent of digital computers, modeling analogies were essential in the application of analog computers to simulating and analyzing the

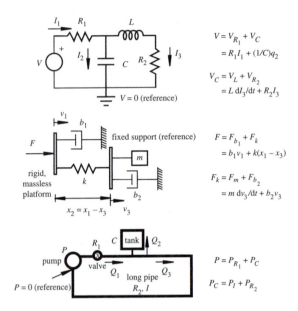

$$V = V_{R_1} + V_C$$
$$= R_1 I_1 + (1/C)q_2$$

$$V_C = V_L + V_{R_2}$$
$$= L \, dI_3/dt + R_2 I_3$$

$$F = F_{b_1} + F_k$$
$$= b_1 v_1 + k(x_1 - x_3)$$

$$F_k = F_m + F_{b_2}$$
$$= m \, dv_3/dt + b_2 v_3$$

$$P = P_{R_1} + P_C$$

$$P_C = P_I + P_{R_2}$$

Figure 4.21 Example of system analogies.

behavior of various types of systems. The procedure was to develop an electrical system analogous to the physical system of interest; build the analogous electrical system with the aid of an analog computer, which provided scaling, integration, differentiation, mixing, and other functions; and perform simulations by providing inputs in the form of voltage signals resulting in voltage or current outputs analogous to quantities of interest in the physical system being simulated. As an example, the behavior of the mechanical system shown in Figure 4.21 could be studied by building the electrical circuit in the figure with $C = s/k$, $R_1 = s \cdot b_1$, $R_2 = s \cdot b_2$, and $L = s \cdot m$, where s is some constant scaling factor. Predicting the force on and speed of the mass m given some applied force $F(t)$ would be a simple matter of applying $V(t) = s \cdot F(t)$ volts to the circuit and monitoring the voltage across and the current through the inductor. The force and speed that would be experienced by the mass in the actual mechanical system would be $F_m = (V_L/s)$ and $v_m = (I_L/s)$. Before the availability of numerical integration simulation software on digital computers, this was the only method for simulating the behavior of complex physical systems.

■ **CLASS DISCUSSION ITEM 4.9**
Initial Condition Analogy

What is the electrical analogy for an initial displacement of a spring in a mechanical system? In other words, what is the corresponding initial condition in the electrical system?

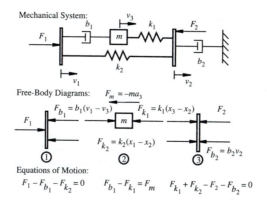

Figure 4.22 Mechanical system analogy example.

The following procedure facilitates converting from a system in one form to an analogous system in some other form. As an illustrative example, the mechanical system shown in Figure 4.22 is converted to its analogous electrical system.

Steps in Converting from One System to an Analogous System

1. *Label the flows with the appropriate sign for each element in the system and its analogy.*

   The algebraic sign given to a flow depends on the direction of the effort on the element. For the example in Figure 4.22, the applied load F_1 is in the direction of speed v_1, so its flow is v_1, but applied load F_2 is in the opposite direction to v_2, so its flow is $-v_2$. For the spring and damper flows, because the forces are drawn assuming compression, the flow for b_1 is $(v_1 - v_3)$ and the flow for k_1 is $(v_3 - v_2)$. The quantities for the analogous system are obtained by direct substitution with quantities based on Table 4.1. The flows and analogies for the entire example are shown in the table that follows.

Mechanical element	Mechanical flow	Electrical flow	Electrical element
F_1	v_1	I_1	V_1
b_1	$v_1 - v_3$	$I_1 - I_3$	R_1
m	v_3	I_3	L
k_1	$v_2 - v_2$	$I_3 - I_2$	C_1
k_2	$v_2 - v_2$	$I_1 - I_2$	C_2
b_2	v_2	I_2	R_2
F_2	$-v_2$	$-I_2$	V_2

2. *Formulate the system equations at each node using the equations of motion ($\Sigma F = ma$) for mechanical systems or KVL loop equations for electrical systems.*

   For the example, see the free-body diagrams and system equations in Figure 4.22. Note that for the massless, rigid platforms, the equation is of the form $\Sigma F = 0$, because they are assumed to have negligible mass. For the mass m, the form is $\Sigma F = ma$. Or, alternatively, the inertial term (ma) can be considered as an inertial force ($Fm = -ma$) as shown in Figure 4.22. In this case, the system equation is again of the form $\Sigma F = 0$.

3. *Pick an element to start with and begin constructing the schematic for the analogy using the flows for guidance. It is best to start with an element that is embedded in the system (i.e., its flow affects many other elements), although any element would suffice.*

    Figure 4.23 shows how the schematic is started, beginning with inductor L (mass m) and recursively branching to other elements that involve its flow (I_3). Note that the free-body diagram (FBD) or KVL equations and the element flow relationships need to be enforced as the schematic is constructed. The completed circuit is shown in Figure 4.24.

4. *Verify the graph with the analogous system equations.*

    For the example, the KVL loop equations indeed are the same form as the free-body diagram equations of motion (see Figures 4.22 and 4.24).

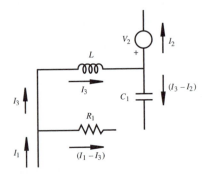

Figure 4.23 Beginning the analog schematic.

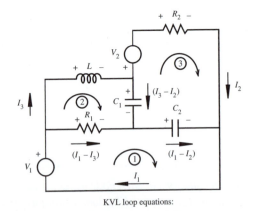

KVL loop equations:

① $V_1 = V_{R_1} + V_{C_2}$ ② $V_L = V_{R_1} - V_{C_1}$ ③ $V_2 = V_{C_1} + V_{C_2} - V_{R_2}$

Figure 4.24 Electrical system analogy example.

■ CLASS DISCUSSION ITEM 4.10
Measurement System Physical Characteristics

Inertia, capacitance, and damping are almost always present in a measurement system. Sometimes these characteristics are desirable; other times they are undesirable. Think of examples of measurement systems containing mechanical, electrical, or hydraulic components and discuss the advantages and disadvantages of changing the physical characteristics of the system. Also describe how these changes might be made.

QUESTIONS AND EXERCISES

Section 4.2   Amplitude Linearity

4.1. If the following input (V_{in}) and output (V_{out}) relationships exist for different measurement systems, indicate whether each is linear or nonlinear?

a. $V_{out}(t) = 5\,V_{in}(t)$

b. $V_{out}(t)/V_{in}(t) = 5t$

c. $V_{out}(t) = V_{in}(t) + 5$

d. $V_{out}(t) = V_{in}(t) + V_{in}(t)$

e. $V_{out}(t) = V_{in}(t) \times V_{in}(t)$

f. $V_{out}(t) = V_{in}(t) + 10t$

g. $V_{out}(t) = V_{in}(t) + \sin(5)$

Section 4.3   Fourier Series Representation of Signals

4.2. What is the Fourier series and fundamental frequency (in hertz) of the waveform $f(t) = 5\sin(2\pi t)$?

4.3. What is the Fourier series and fundamental frequency (in rad/sec) of a standard U.S. household AC voltage?

4.4. Verify that Equations 4.7 through 4.9 are equivalent to Equation 4.3.

4.5. Show that the Fourier series coefficients A_n for the square wave defined by Equation 4.10 are 0.

4.6. The discrete Fourier series representation of a half-sine-wave pulse train waveform is mathematically represented as

$$V(t) = \frac{1}{\pi} + \frac{\sin(2\pi t)}{2} - \frac{2}{\pi}\left[\frac{\cos(4\pi t)}{1\cdot 3} + \frac{\cos(8\pi t)}{3\cdot 5} + \frac{\cos(12\pi t)}{5\cdot 7} + \cdots\right]$$

Using a computer plotting application, plot three cycles of $V(t)$ displaying

a. DC component + fundamental (first harmonic)

b. DC component + first 10 harmonics (Note that some harmonics have zero amplitude)

c. DC component + as many harmonics as you think are necessary to provide a good reproduction of the waveform

Section 4.4   Bandwidth and Frequency Response

4.7. Given the following measurement system frequency response curve,

a. What is the bandwidth of the measurement system?

b. If the input signal (V_{in}) is a 2 V, peak-to-peak, square wave of period 1 sec, what will the steady state output signal ($V_{out}(t)$) from the measurement system be? Use the square-wave Fourier series representation given in Equation 4.15 or 4.16 for $V_{in}(t)$.

c. Plot the resulting output on a computer for $t = 0$ to 2 sec.

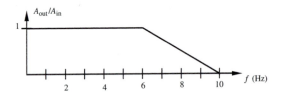

4.8. For an RC low-pass filter with $R = 1$ kΩ and $C = 1$ μF,

a. What is the bandwidth of the filter?

b. If the input is a 2 V, peak-to-peak, square wave of period 1 sec, what will the filter output be? Use the square wave Fourier series representation given in Equation 4.15 or 4.16 for $V_{in}(t)$.

c. Plot the resulting output on a computer for $t = 0$ to 2 sec.

4.9. Solve Question 4.8 with R = 100 kΩ instead.

4.10. What is the bandwidth (in Hz) of a system with the frequency response that follows?

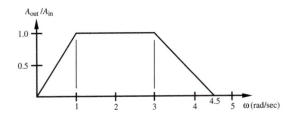

4.11. If a signal is given by

$$F(t) = \sum_{n=1}^{5} n \sin(n\omega t)$$

where $\omega = 1$ rad/sec,

a. What is the range of frequencies of the signal?

b. Plot the spectrum of $F(t)$.

c. Assuming the signal is being measured by a measurement system with the frequency response curve shown in Question 4.10, plot the spectrum of the resulting output of the system.

4.12. Determine and plot the frequency response curve for a high-pass filter (see Example 4.1). Also derive an expression for the cutoff frequency.

4.13. Derive an expression for the phase angle between the output and input voltages of the low-pass filter in Example 4.1 . Plot the result for frequencies between 0 and 2.5 times the cutoff frequency.

4.14. Document a complete and thorough answer to Class Discussion Item 4.2.

Section 4.6   Distortion of Signals

4.15. Using the Fourier series representation of a square wave in Equation 4.15 or 4.16, plot a square wave using 20 harmonics. Then plot it with the first three harmonics attenuated by 1/4. Finally, plot it with the first three harmonics unattenuated and the next 17 attenuated by 1/4. What do you conclude about the influence of the amplitudes of low and high harmonics?

4.16. Plot the frequency response curves for the exponential attenuation terms in Class Discussion Item 4.3.

Section 4.8   Zero-Order System

4.17. A simple static spring scale is an example of a zero-order system where the input is the mass to be measured and the output is the calibrated deflection of the spring. Given this, what is the gain or sensitivity of the scale?

4.18. We generally assume that an oscilloscope is a zero-order measurement system. Why is this so? When will this assumption be in error?

Section 4.9   First-Order System

4.19. Relate the time constant, cutoff frequency, and bandwidth for the low-pass filter circuit shown in Example 4.1 by writing a general first-order differential equation for the circuit. What is the dependent variable in the differential equation? Write an expression for the output voltage as a function of time $V_{out}(t)$ for a step input voltage of amplitude A_i. Assume the capacitor is discharged to begin with.

4.20. A simple glass bulb thermometer is an example of a first-order system where the input is the surrounding temperature (T_{in}) and the output is the temperature of the liquid inside the bulb (T_{out}), which expands to provide a reading on a scale. Using basic heat transfer principles, where the rate of convective heat transfer in equals the rate of change of internal energy of the fluid, derive the system equation and put it in standard form. Identify the time constant and relate the heat transfer parameters and fluid properties to the electrical parameters in an RC circuit (see Question 4.19). The parameters in your equation should include the fluid mass (m) and specific heat (c), and the bulb external area (A) and heat transfer coefficient (h).

4.21. The following set of data were collected from a system. Would it be appropriate to assume that this system is first order? If so, find the time constant of the system. Also, what is an approximate value for the static sensitivity?

t	$X_{out}(t)$
0.0	0.0
0.1	1.4
0.2	2.3
0.3	3.0
0.4	3.6
0.5	4.1
0.6	4.2
0.7	4.6
0.8	4.7
0.9	4.8
1.0	4.9

Section 4.10 Second-Order System

4.22. Is the damped natural frequency greater or smaller than the natural frequency of a second-order system? Explain why.

4.23. Draw a schematic and write the system equation for an example second-order system for each of the following categories:

a. mechanical rotary

b. electrical

c. hydraulic

MathCAD Example

4.7 First-order system frequency response

4.24. Use the procedure presented in Section 4.10.2 to derive the amplitude ratio and phase relationship for a first-order system expressed in standard form. MathCAD Example 4.7 shows plots of the results.

4.25. For a spring-mass-damper system with $F_{ext}(t) = 20 \sin(0.75t)$ N, $m = 10$ kg, $k = 12$ N/m, and $b = 10$ Ns/m, what is the equation for the steady state sinusoidal response of $x(t)$?

4.26. A cheap, mechanically actuated, strip chart recording system consisting of a plotter head of mass m driven by an elastic belt of stiffness k is illustrated in the following figure.

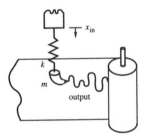

The input displacement x_{in} is coupled mechanically to the pen head through the spring. The spring constant k is 32.4 N/m and the mass of the pen head is 0.10 kg. Derive the differential equation of motion for the pen head, given the displacement input. Assume that the damping constant between the pen and paper is 10 Ns/m. Using the procedure in Section 4.10.2, determine the steady state output displacement magnitude and phase angle for each steady state input that follows. Explain the differences between the output and input displacements and phase angles.

a. $x_{\text{in}} = 0.05 \sin(10t)$

b. $x_{\text{in}} = 0.05 \sin(1000t)$

c. $x_{\text{in}} = 0.05 \sin(10,000t)$

4.27. Document a complete and thorough answer to Class Discussion Item 4.5.

4.28. Derive the expression for the step response $x(t)$ in Design Example 4.1.

4.29. Document a complete and thorough answer to Class Discussion Item 4.8.

Section 4.11   System Modeling and Analogies

4.30. Using the capacitor energy storage equation, derive the expression for the capacitance of a cylindrical tank of diameter D and fluid height h holding bottom-fed fluid of specific weight γ.

4.31. Using the inertia energy storage equation, derive the expression for the inertance of the fluid (of density ρ) in a straight pipe of length L and cross-sectional area A. Start by applying $F = ma$ to a control volume surrounding the fluid.

4.32. Convert the translational mechanical system that follows to an analogous electrical system.

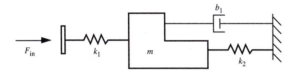

4.33. Convert the following translational mechanical system to an analogous electrical system.

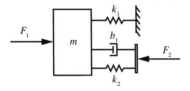

4.34. Convert the following electrical system to an analogous hydraulic system.

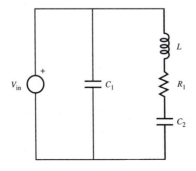

4.35. For the translational mechanical system that follows, draw free-body diagrams for each node, list the flow for each element, and construct (draw) an analogous electrical system with all elements and flows labeled.

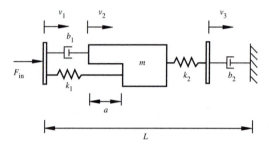

4.36. For the three-degree-of-freedom mechanical system that follows, draw free-body diagrams for each of the components, carefully showing all external forces and moments. Be sure to show the reference positions for the position variables on the free-body diagrams. Directly from the component free-body diagrams, write down the equations of motion.

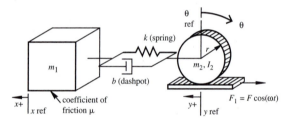

BIBLIOGRAPHY

Beckwith, T., Marangoni, R., and Lienhard, J., *Mechanical Measurement,* 5th Edition, Addison-Wesley, Reading, MA, 1993.

Doebelin, E., *Measurement Systems Application and Design,* McGraw-Hill, New York, 1990.

Figliola, R. and Beasley, D., *Theory and Design for Mechanical Measurements,* John Wiley, New York, 1995.

Holman, J. P., *Experimental Methods for Engineers,* McGraw-Hill, New York, 1994.

Ross, S., *Introduction to Ordinary Differential Equations,* 3rd Edition, John Wiley, New York, 1980.

Shearer, J., Murphy, A., and Richardson, H., *Introduction to System Dynamics,* Addison-Wesley, Reading, MA, 1971.

Thomson, W., *Theory of Vibration with Applications,* 2nd Edition, Prentice-Hall, Englewood Cliffs, NJ, 1981.

CHAPTER 5

Analog Signal Processing Using Operational Amplifiers

T his chapter presents operational amplifier circuits that are important for interfacing analog components in a mechatronic system. ■

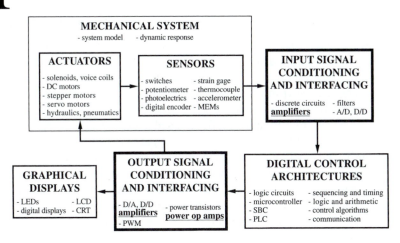

After you read, discuss, study, and apply ideas in this chapter, you will:

1. Understand the input/output characteristics of a linear amplifier

2. Understand how to use the model of an ideal operational amplifier in circuit analysis

3. Know how to design op amp circuits

4. Be able to design an inverting amplifier, noninverting amplifier, summer, difference amplifier, instrumentation amplifier, integrator, differentiator, and sample and hold amplifier

5. Understand the characteristics and limitations of a "real" operational amplifier

5.1   INTRODUCTION

Since electrical circuits occur in virtually all mechatronic and measurement systems, it is essential that engineers develop a basic understanding of the acquisition and processing of electrical signals. Usually these signals come from transducers, which convert physical quantities (e.g., temperature, strain, displacement, flow rate) into currents or voltages, usually the latter. The transducer output is usually described as an **analog signal,** which is continuous and time varying.

Often the signals from transducers are not in the form we would like them to be. They may

■ Be too small, usually in the millivolt range

■ Be too "noisy," usually due to electromagnetic interference

■ Contain the wrong information, sometimes due to poor transducer design or installation

■ Have a DC offset, usually due to the transducer and instrumentation design

Many of these problems can be remedied, and the desired signal information extracted through appropriate analog signal processing. The simplest and most common form of signal processing is amplification, where the magnitude of the voltage signal is increased. Other forms include signal inversion, differentiation, integration, addition, subtraction, and comparison.

Analog signals are very different from digital signals, which are discrete, using only a finite number of states or values. Since computers and microprocessors require digital signals, any application involving computer measurement or control requires analog-to-digital conversion. This chapter covers the basic elements of analog signal processing including the design and analysis of signal processing circuits. The operational amplifier is an integrated circuit used as a building block in many of these circuits. Chapter 6 focuses on digital circuits, and Chapter 8 deals with converting analog signals into a format that can be processed by digital devices such as computers.

5.2   AMPLIFIERS

People have spent their lives studying and writing about amplifiers, so we cannot expect to do justice to the subject in a few pages. However, we will look at the salient features of amplifiers and determine how we may design one using integrated circuits.

Ideally, an **amplifier** increases the amplitude of a signal without affecting the phase relationships of different components of the signal. When choosing or designing an amplifier, we must consider size, cost, power consumption, input impedance, output impedance, gain, and bandwidth. Physical size depends on the components used to construct the amplifier. Prior to the 1960s, vacuum tube amplifiers were common, but they were heavy power consumers with significant heat dissipation. Portable units were large and heavy and required frequent battery replacement. Since its advent, **solid state** technology, where charge carriers move through a solid semiconductor material, has replaced the vacuum tube technology, where bulky tubes enclosed a gas at low pressure through which electrons flowed. Today, solid state transistors and integrated circuits have dramatically changed amplifier design, resulting in small, cool-running amplifiers. They consume little power and are easily made portable using rechargeable batteries.

Generally, we model an amplifier as a two-port device, with an input and output voltage referenced to ground, as illustrated in Figure 5.1. The voltage **gain** A_v of an amplifier defines the factor by which the voltage is changed:

$$V_{out} = A_v V_{in} \tag{5.1}$$

Normally we want an amplifier to exhibit amplitude linearity, where the gain is constant for all frequencies. However, amplifiers may be designed to intentionally amplify only certain frequencies, resulting in a filtering effect. In such cases, the output characteristics are governed by the amplifier's bandwidth and associated cutoff frequencies.

The input impedance of an amplifier, Z_{in}, is defined as the ratio of the input voltage and current:

$$Z_{in} = V_{in}/I_{in} \tag{5.2}$$

Most amplifiers are designed to have a large input impedance so very little current is drawn from the input.

The output impedance is a measure of how much the output voltage drops with output current:

$$Z_{out} = \Delta V_{out}/I_{in} \tag{5.3}$$

where the voltage drop ΔV_{out} is measured relative to the output voltage with no current. Most amplifiers are designed to have a very small output impedance so the output voltage will not change much as the output current changes.

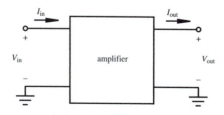

Figure 5.1 Amplifier model.

See Section 2.4 for more information related to input and output impedance. The input impedance of an amplifier is analogous to that of a voltmeter, and the output impedance is analogous to that of a voltage source.

5.3 OPERATIONAL AMPLIFIERS

The operational amplifier, or **op amp,** is a low-cost and versatile integrated circuit consisting of many internal transistors, resistors, and capacitors manufactured into a single chip of silicon. It can be combined with external discrete components to create a wide variety of signal processing circuits. The op amp is the basic building block for

- Amplifiers
- Integrators
- Summers
- Differentiators
- Comparators
- A/D and D/A converters
- Active filters
- Sample and hold amplifiers

We present most of these applications in subsequent sections. The op amp derives its name from its ability to perform so many different operations.

5.4 IDEAL MODEL FOR THE OPERATIONAL AMPLIFIER

Figure 5.2 shows the schematic symbol and terminal nomenclature for an ideal op amp. It is a differential input, single output amplifier that is assumed to have infinite gain. The two inputs are called the **inverting input,** labeled with a minus sign, and the **noninverting input,** labeled with a plus sign. The ∞ symbol is sometimes used in the schematic to denote the infinite gain and the assumption that it is an ideal op amp. The voltages are all referenced to a common ground. The op amp is an

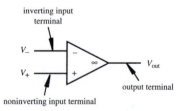

Figure 5.2 Op amp terminology and schematic representation.

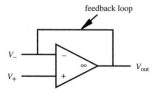

Figure 5.3 Op amp feedback.

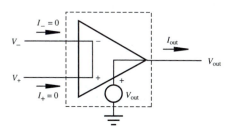

Figure 5.4 Op amp equivalent circuit.

active device requiring connection to an external power supply, usually plus and minus 15 V. The external supply is not normally shown on circuit schematics. Since the op amp is an active device, output voltages and currents can be larger than the signals applied to the inverting and noninverting terminals.

As illustrated in Figure 5.3, an op amp circuit usually includes **feedback** from the output to the negative (inverting) input. This so-called **closed loop** configuration results in stabilization of the amplifier and control of the gain. When feedback is absent in an op amp circuit, the op amp is said to have an **open loop** configuration. This configuration results in considerable instability due to the very high gain, and it is seldom used. The utility of feedback will become evident in the examples presented in the following sections.

Figure 5.4 illustrates an ideal model that can aid in analyzing circuits containing op amps. This model is based on the following assumptions that describe an ideal op amp:

1. *It has infinite impedance at both inputs;* hence, no current is drawn from the input circuits. Therefore,

$$I_+ = I_- = 0 \qquad (5.4)$$

2. *It has infinite gain.* As a consequence, the difference between the input voltages must be 0; otherwise, the output would be infinite. This is denoted in Figure 5.4 by the shorting of the two inputs. Therefore,

$$V_+ = V_- \qquad (5.5)$$

Even though we indicate a short between the two inputs, we assume no current may flow through this short.

3. *It has zero output impedance.* Therefore, the output voltage does not depend on the output current.

Note that V_{out}, V_+, and V_- are all referenced to a common ground. Also, for stable linear behavior, there must be feedback between the output and the inverting input.

These assumptions and the model may appear illogical and confusing, but they provide a close approximation to the behavior of a real op amp when used in a circuit that includes negative feedback. With the aid of this ideal model, we need only Kirchhoff's laws and Ohm's law to completely analyze op amp circuits.

Actual op amps are usually packaged in an eight-pin dual in-line package (DIP) integrated circuit (IC), sometimes called a chip. The designation for a general purpose op amp produced by many IC manufacturers is 741. It is illustrated in Figure 5.5 with its pin configuration (pin-out). As with all ICs, one end of the chip is marked with an indentation or spot, and the pins are numbered counterclockwise (looking at the top of the IC) and consecutively starting with 1 at the left side of the marked end. For a 741 series op amp, pin 2 is the inverting input, pin 3 is the noninverting input, pins 4 and 7 are for the external power supply, and pin 6 is the op amp output. Pins 1, 5, and 8 are not normally used, and no connections are required. Figure 5.6 illustrates the internal design of a 741 IC available from National Semiconductor. Note that the circuits are composed of transistors, resistors, and capacitors that are easily manufactured on a single silicon chip. The most valuable details for the user are the input and output parts of the circuit having characteristics that affect externally connected components.

The manufacture of integrated circuits is very involved, requiring extremely expensive equipment. Video Demo 5.1 shows various common packages in which ICs are made available, and Video Demos 5.2 through 5.4 describe and illustrate all of the steps of the manufacturing process. Fortunately, due to high production volume of components used in consumer and industrial applications, the economy of scale prevails, and ICs can be sold at very affordable prices.

Many different op amp designs are available from IC manufacturers. The input impedances, bandwidth, and power ratings can vary significantly. Also, some require only a single-output (unipolar) power supply. Although the 741 is widely used, another common op amp is the TL071 manufactured by Texas Instruments. Its pin

Video Demo

5.1 Integrated circuits

5.2 Integrated circuit manufacturing process stages

5.3 Chip manufacturing process

5.4 How a CPU is made

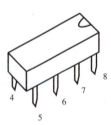
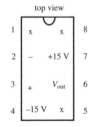

Figure 5.5 741 op amp pin-out.

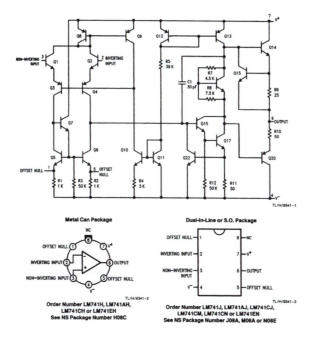

Figure 5.6 741 internal design. *(Courtesy of National Semiconductor, Santa Clara, CA)*

configuration is identical to the 741, but because it has FET inputs, it has a larger input impedance and a wider bandwidth.

IC manufacturers provide complete information on their devices in documents called **data sheets.** Internet Links 5.1 and 5.2 point to the complete data sheets for both the 741 and TL071. If you don't have much experience looking at these data sheets, they can be very overwhelming. However, they contain all the information you would ever need to use the devices. Useful information includes the device pin-out (e.g., see Figure 5.5), input and output current and voltage specifications, voltage source requirements, and impedances. Section 5.14 at the end of this chapter provides some guidance on specific things to look for in an op data sheet.

Internet Link

5.1 μA741 op amp data sheet

5.2 5.2 TL071 FET input op amp data sheet

5.5 INVERTING AMPLIFIER

An **inverting amplifier** is constructed by connecting two external resistors to an op amp as shown in Figure 5.7. As the name implies, this circuit inverts and amplifies the input voltage. Note that the resistor R_F forms the feedback loop. This feedback loop always goes from the output to the inverting input of the op amp, implying negative feedback.

We now use Kirchhoff's laws and Ohm's law to analyze this circuit. First, we replace the op amp with its ideal model shown within the dashed box in Figure 5.8.

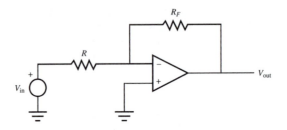

Figure 5.7 Inverting amplifier.

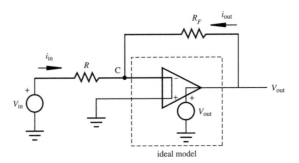

Figure 5.8 Equivalent circuit for an inverting amplifier.

Applying Kirchhoff's current law at node C and utilizing assumption 1, that no current can flow into the inputs of the op amp,

$$i_{in} = -i_{out} \tag{5.6}$$

Also, because the two inputs are assumed to be shorted in the ideal model, C is effectively at ground potential:

$$V_C = 0 \tag{5.7}$$

Because the voltage across resistor R is $V_{in} - V_C = V_{in}$, from Ohm's law,

$$V_{in} = i_{in}R \tag{5.8}$$

and because the voltage across resistor R_F is $V_{out} - V_C = V_{out}$,

$$V_{out} = i_{out}R_F \tag{5.9}$$

Substituting Equation 5.6 into Equation 5.9 gives

$$V_{out} = -i_{in}R_F \tag{5.10}$$

Dividing Equation 5.10 by Equation 5.8 yields the input/output relationship:

$$\frac{V_{out}}{V_{in}} = -\frac{R_F}{R} \tag{5.11}$$

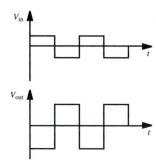

Figure 5.9 Illustration of inversion.

Therefore, the voltage gain of the amplifier is determined simply by the external resistors R_F and R, and it is always negative. The reason this circuit is called an inverting amplifier is that it reverses the polarity of the input signal. This results in a phase shift of $180°$ for periodic signals. For example, if the square wave V_{in} shown in Figure 5.9 is connected to an inverting amplifier with a gain of -2, the output V_{out} is inverted and amplified, resulting in a larger amplitude signal $180°$ out of phase with the input.

■ **CLASS DISCUSSION ITEM 5.1**
Kitchen Sink in an Op Amp Circuit

Consider the following op amp circuit:

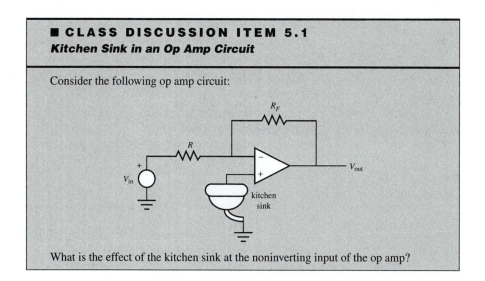

What is the effect of the kitchen sink at the noninverting input of the op amp?

5.6 NONINVERTING AMPLIFIER

The schematic of a **noninverting amplifier** is shown in Figure 5.10. As the name implies, this circuit amplifies the input voltage without inverting the signal. Again, we can apply Kirchhoff's laws and Ohm's law to determine the voltage gain of

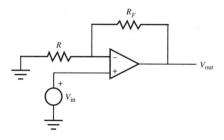

Figure 5.10 Noninverting amplifier.

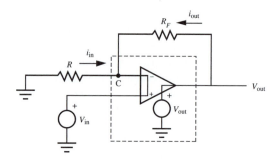

Figure 5.11 Equivalent circuit for a noninverting amplifier.

this amplifier. As before, we replace the op amp with the ideal model shown in the dashed box in Figure 5.11.

The voltage at node C is V_{in} because the inverting and noninverting inputs are at the same voltage. Therefore, applying Ohm's law to resistor R,

$$i_{in} = \frac{-V_{in}}{R} \tag{5.12}$$

and applying it to resistor R_F,

$$i_{out} = \frac{V_{out} - V_{in}}{R_F} \tag{5.13}$$

Solving Equation 5.13 for V_{out} gives

$$V_{out} = i_{out} R_F + V_{in} \tag{5.14}$$

Applying KCL at node C gives

$$i_{in} = -i_{out} \tag{5.15}$$

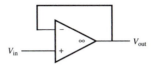

Figure 5.12 Buffer or follower.

so Equation 5.12 can be written as

$$V_{in} = i_{out} R \qquad (5.16)$$

Using Equations 5.14 and 5.16, the voltage gain can be written as

$$\frac{V_{out}}{V_{in}} = \frac{i_{out} R_F + V_{in}}{V_{in}} = \frac{i_{out} R_F + i_{out} R}{i_{out} R} = 1 + \frac{R_F}{R} \qquad (5.17)$$

Therefore, the noninverting amplifier has a positive gain greater than or equal to 1. This is useful in isolating one portion of a circuit from another by transmitting a scaled voltage without drawing appreciable current.

If we let $R_F = 0$ and $R = \infty$ in the noninverting op amp circuit in Figure 5.10, the resulting circuit can be represented as shown in Figure 5.12. This circuit is known as a **buffer** or **follower** because $V_{out} = V_{in}$. It has a high input impedance and low output impedance. This circuit is useful in applications where you need to couple to a voltage signal without loading the source of the voltage. The high input impedance of the op amp effectively **isolates** the source from the rest of the circuit. Lab Exercise 6 explores this characteristic by showing how to tap a voltage in a resistor circuit without affecting the currents in the circuit.

Lab Exercise

Lab 6
Operational
amplifier circuits

■ **CLASS DISCUSSION ITEM 5.2**
Positive Feedback

Ideally, what would happen to the output of the buffer amplifier shown in Figure 5.12 if V_{in} was applied to the inverting input and feedback was from the output to the non-inverting input instead?

■ **CLASS DISCUSSION ITEM 5.3**
Example of Positive Feedback

A good example of positive feedback is the effect Jimi Hendrix used to achieve when he would move his guitar close to the front of his amplifier speaker. Describe the effect of this technique and describe what is going on physically.

THREADED DESIGN EXAMPLE

A.3   *DC motor power-op-amp speed controller—Power amp motor driver*

Video Demo

1.6 DC motor power-op-amp speed controller

The figure below shows the functional diagram for Threaded Design Example A (see Section 1.3 and Video Demo 1.6), with the portion described here highlighted.

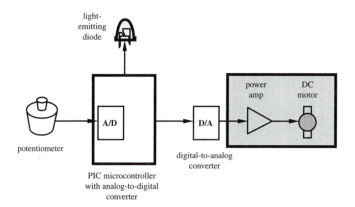

Internet Link

1.4 Threaded design example components

1.5 Digikey electronics supplier

1.6 Jameco electronics supplier

The D/A converter outputs a voltage directly related to the potentiometer position. However, the D/A converter's output current is limited and not enough to drive a motor. A power op amp circuit, configured as a noninverting amplifier, can drive the motor at the higher currents needed. In effect, the power amp will serve as a buffer between the D/A converter and the motor. The circuit below shows the components used along with their interconnections. As shown, the OPA547 can be powered by a bipolar ($\pm$) 9 V supply (instead of a standard bipolar 15 V supply). In Video Demo 1.6, we show how these voltages can be created with either a standard laboratory power supply or with two 9 V batteries wired in series. With an input resistor of 10 kΩ and a feedback resistor of 1 kΩ, from Equation 5.17, the power amp circuit has a gain of 1.1. Therefore, the voltage from the D/A converter is not amplified very much, but the circuit is able to source ample current to the motor. Op amp components that are designed to output significant current are called power op amps. The OP547 is one such op amp. As with all of the design examples, if you want more information about any of the components used, see Internet Links 1.4 through 1.6.

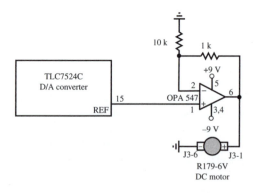

5.7 SUMMER

The **summer** op amp circuit shown in Figure 5.13 is used to add analog signals. By analyzing the circuit with

$$R_1 = R_2 = R_F \qquad (5.18)$$

we can show (Question 5.8) that

$$V_{\text{out}} = -(V_1 + V_2) \qquad (5.19)$$

Therefore, the circuit output is the negative sum of the inputs.

5.8 DIFFERENCE AMPLIFIER

Lab Exercise

Lab 6
Operational
amplifier circuits

The **difference amplifier** circuit shown in Figure 5.14 is used to subtract analog signals (see Lab Exercise 6). In analyzing this circuit, we can use the principle of **superposition,** which states that, whenever multiple inputs are applied to a linear system (e.g., an op amp circuit), we can analyze the circuit and determine the response for each of the individual inputs independently. The sum of the individual responses is equivalent to the overall response to the multiple inputs. Specifically, when the inputs are ideal voltage sources, to analyze the response due to one source, the other sources are shorted. If some inputs are current sources, they are replaced with open circuits.

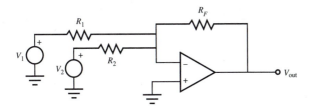

Figure 5.13 Summer circuit.

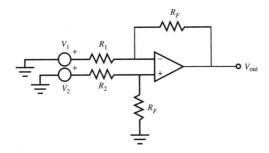

Figure 5.14 Difference amplifier circuit.

The first step in analyzing the circuit in Figure 5.14 is to replace V_2 with a short circuit, effectively grounding R_2. As shown in Figure 5.15, the result is an inverting amplifier (see Figure 5.7 and Class Discussion Item 5.1). Therefore, from Equation 5.11, the output due to input V_1 is

$$V_{out_1} = -\frac{R_F}{R_1} V_1 \tag{5.20}$$

The second step in analyzing the circuit in Figure 5.14 is to replace V_1 with a short circuit, effectively grounding R_1, as shown in Figure 5.16a. This circuit is equivalent to the circuit shown in Figure 5.16b where the input voltage is

$$V_3 = \frac{R_F}{R_2 + R_F} V_2 \tag{5.21}$$

because V_2 is divided between resistors R_2 and R_F.

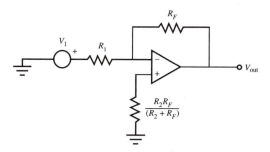

Figure 5.15 Difference amplifier with V_2 shorted.

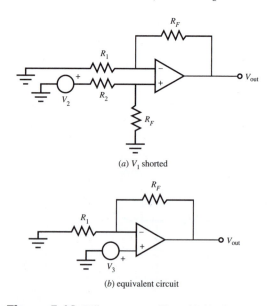

(a) V_1 shorted

(b) equivalent circuit

Figure 5.16 Difference amplifier with V_1 shorted.

The circuit in Figure 5.16b is a noninverting amplifier (see Figure 5.10). Therefore, the output due to input V_2 is given by Equation 5.17:

$$V_{out_2} = \left(1 + \frac{R_F}{R_1}\right)V_3 \tag{5.22}$$

By substituting Equation 5.21, this equation can be written as

$$V_{out_2} = \left(1 + \frac{R_F}{R_1}\right)\left(\frac{R_F}{R_2 + R_F}\right)V_2 \tag{5.23}$$

The principle of superposition states that the total output V_{out} is the sum of the outputs due to the individual inputs:

$$V_{out} = V_{out_1} + V_{out_2} = -\left(\frac{R_F}{R_1}\right)V_1 + \left(1 + \frac{R_F}{R_1}\right)\left(\frac{R_F}{R_2 + R_F}\right)V_2 \tag{5.24}$$

When $R_1 = R_2 = R$, the output voltage is an amplified difference of the input voltages:

$$V_{out} = \frac{R_F}{R}(V_2 - V_1) \tag{5.25}$$

This result can also be obtained using the op amp rules, KCL, and Ohm's law (Question 5.10).

5.9 INSTRUMENTATION AMPLIFIER

The difference amplifier presented in Section 5.8 may be satisfactory for low-impedance sources, but its input impedance is too low for high-output impedance sources. Furthermore, if the input signals are very low level and include noise, the difference amplifier is unable to extract a satisfactory difference signal. The solution to this problem is the **instrumentation amplifier.** It has the following characteristics:

■ Very high input impedance
■ Large **common mode rejection ratio** (CMRR). The CMRR is the ratio of the difference mode gain to the common mode gain. The **difference mode gain** is the amplification factor for the difference between the input signals, and the **common mode gain** is the amplification factor for the average of the input signals. For an ideal difference amplifier, the common mode gain is 0, implying an infinite CMRR. When the common mode gain is nonzero, the output is nonzero when the inputs are equal and nonzero. It is desirable to minimize the common mode gain to suppress signals such as noise that are common to both inputs.
■ Capability to amplify low-level signals in a noisy environment, often a requirement in differential-output sensor signal-conditioning applications
■ Consistent bandwidth over a large range of gains

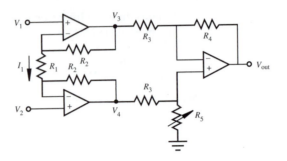

Figure 5.17 Instrumentation amplifier.

Instrumentation amplifiers are commercially available as monolithic ICs (e.g., Analog Devices 524 and 624 and National Semiconductor LM 623). A single external resistor is used to set the gain. This gain can be higher and is more stable than gains achievable with a simple difference amplifier.

An instrumentation amplifier can also be constructed with inexpensive discrete op amps and precision resistors as illustrated in Figure 5.17. We analyze this circuit in two parts. The two op amps on the left provide a high-impedance amplifier stage where each input is amplified separately. This stage involves a moderate CMRR. The outputs V_3 and V_4 are supplied to the op amp circuit on the right, which is a difference amplifier with a potentiometer R_5 used to maximize the overall CMRR.

We first apply KCL and Ohm's law to the left portion of the circuit to express V_3 and V_4 in terms of V_1 and V_2. Using the assumptions and rules for an ideal op amp, it is clear that the current I_1 passes through R_1 and both feedback resistors R_2. Applying Ohm's law to the feedback resistors gives

$$V_3 - V_1 = I_1 R_2 \tag{5.26}$$

and

$$V_2 - V_4 = I_1 R_2 \tag{5.27}$$

Applying Ohm's law to R_1 gives

$$V_1 - V_2 = I_1 R_1 \tag{5.28}$$

To express V_3 and V_4 in terms of V_1 and V_2, we eliminate I_1 by solving Equation 5.28 for I_1 and substituting it into Equations 5.26 and 5.27. The results are

$$V_3 = \left(\frac{R_2}{R_1} + 1\right)V_1 - \frac{R_2}{R_1}V_2 \tag{5.29}$$

and

$$V_4 = -\frac{R_2}{R_1}V_1 + \left(\frac{R_2}{R_1} + 1\right)V_2 \tag{5.30}$$

By analyzing the right portion of the circuit, it can be shown (Question 5.12) that

$$V_{out} = \frac{R_5(R_3 + R_4)}{R_3(R_3 + R_5)} V_4 - \frac{R_4}{R_3} V_3 \tag{5.31}$$

We can substitute the expressions for V_3 and V_4 from Equations 5.29 and 5.30 into Equation 5.31 to express the output voltage V_{out} in terms of the input voltages V_1 and V_2. Assuming $R_5 = R_4$, the result is

$$V_{out} = \left[\frac{R_4}{R_3} \left(1 + 2\frac{R_2}{R_1} \right) \right] (V_2 - V_1) \tag{5.32}$$

A design objective for the instrumentation amplifier is to maximize the CMRR by minimizing the common mode gain. For a common mode input, $V_1 = V_2$, Equation 5.32 yields an output voltage $V_{out} = 0$. Hence, the common mode gain is 0, and the CMRR is infinite if $R_5 = R_4$. In practice, the resistances never match exactly. Also, if the temperature varies within the discrete circuit, resistance mismatches are further exaggerated. By using a potentiometer for R_5, the designer can minimize the mismatch between R_5 and R_4, resulting in a maximum CMRR.

The problems of resistor matching with discrete components are avoided by using a monolithic instrumentation amplifier constructed with laser-trimmed resistors. These amplifiers have a very high CMRR not usually obtainable with discrete components. In addition, the gain is programmable by selecting an appropriate external resistor R_1.

5.10 INTEGRATOR

If the feedback resistor of the inverting op amp circuit is replaced by a capacitor, the result is an **integrator** circuit. It is shown in Figure 5.18. Referring to the analysis for the inverting amplifier, Equation 5.9 is replaced by the relationship between voltage and current for a capacitor:

$$\frac{dV_{out}}{dt} = \frac{i_{out}}{C} \tag{5.33}$$

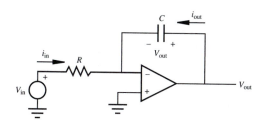

Figure 5.18 Ideal integrator.

Integrating gives

$$V_{out}(t) = \frac{1}{C} \int_0^t i_{out}(\tau)\, d\tau \tag{5.34}$$

where τ is a dummy variable of integration. Since $i_{out} = -i_{in}$ and $i_{in} = V_{in}/R$,

$$V_{out}(t) = -\frac{1}{RC} \int_0^t V_{in}(\tau)\, d\tau \tag{5.35}$$

Therefore, the output signal is an inverted, scaled integral of the input signal.

Newton

■ **CLASS DISCUSSION ITEM 5.4**
Integrator Behavior

If a DC voltage is applied as an input to an ideal integrator, how does the output change over time? What is the output given a sinusoidal input? What would the result be if a small DC offset were added to the sinusoidal input?

A more practical integrator circuit is shown in Figure 5.19. The resistor R_s placed across the feedback capacitor is called a **shunt resistor.** Its purpose is to limit the low-frequency gain of the circuit. This is necessary because even a small DC offset at the input would be integrated over time, eventually saturating the op amp (see Section 5.14 and Class Discussion Item 5.7). The integrator is useful only when the scaled integral always remains below the maximum output voltage for the op amp. As a good rule of thumb, R_s should be greater than $10R_1$.

Because of the impedance and frequency response of the feedback circuit containing R_s and C, the circuit in Figure 5.19 acts as an integrator only for a range of frequencies. At very low frequencies, the circuit behaves as an inverting amplifier because the impedance of the feedback loop is effectively R_s because the impedance

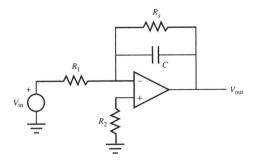

Figure 5.19 Improved integrator.

of C is large at low frequencies. At very high frequencies (i.e., $\omega >> 1/R_1C$), the output is attenuated to zero because the feedback loop is effectively a short. Lab Exercise 6 explores and Video Demo 5.5 demonstrates how the response of the integrator changes over a large frequency range.

Any DC offset due to the input bias current (see item A in Section 5.14.1) is minimized by R_2, which should be chosen to approximate the parallel combination of the input and shunt resistors:

$$R_2 = \frac{R_1 R_s}{R_1 + R_s} \tag{5.36}$$

The reason for this is that the input bias current flowing into the inverting terminal is a result of the currents through R_1 and R_s, and the input bias current flowing into the noninverting terminal flows through R_2. If the voltages generated by the bias current are the same, they have no net effect on the output.

Lab Exercise

Lab 6
Operational
amplifier circuits

Video Demo

5.5 Op amp
integrator circuit
at different
frequencies

5.11 DIFFERENTIATOR

If the input resistor of the inverting op amp circuit is replaced by a capacitor, the result is a **differentiator** circuit. It is shown in Figure 5.20. Referring to the analysis for the inverting amplifier, Equation 5.8 is replaced by the relationship between voltage and current for a capacitor:

$$\frac{dV_{in}}{dt} = \frac{i_{in}}{C} \tag{5.37}$$

Since $i_{in} = -i_{out}$ and $i_{out} = V_{out}/R$,

$$V_{out} = -RC\frac{dV_{in}}{dt} \tag{5.38}$$

Therefore, the output signal is an inverted, scaled derivative of the input signal.

One note of caution concerning using differentiation in signal processing is that any electrical noise in the input signal will be accentuated in the output. In effect, the differentiator amplifies the fast-changing noise. Integration, on the other hand, has a smoothing effect, so noise is not a concern when using an integrator.

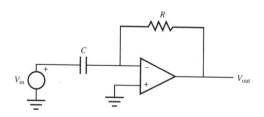

Figure 5.20 Differentiator.

Recommend possible improvements to the differentiator circuit in Figure 5.20. Consider the effects of high-frequency noise in the input signal.

Newton

Think of various applications for the integrator and differentiator circuits. Consider how a differential equation could be solved with an analog computer. Also consider how to convert between sawtooth and square-wave function generator outputs and how to process position and speed sensor signals.

5.12 SAMPLE AND HOLD CIRCUIT

A **sample and hold** circuit is used extensively in analog-to-digital conversion (discussed in Chapter 8), where a signal value must be stabilized while it is converted to a digital representation. The sample and hold circuit illustrated in Figure 5.21 consists of a voltage-holding capacitor and a voltage follower. With switch S closed,

$$V_{out}(t) = V_{in}(t) \tag{5.39}$$

When the switch is opened, the capacitor C holds the input voltage corresponding to the last sampled value, because negligible current is drawn by the follower. Therefore,

$$V_{out}(t - t_{sampled}) = V_{in}(t_{sampled}) \tag{5.40}$$

where $t_{sampled}$ is the time when the switch was last opened. Often, an op amp buffer is also used on the V_{in} side of the switch to minimize current drain from the input voltage source V_{in}.

The type of capacitor used for this application is important. A low-leakage capacitor such as a polystyrene or polypropylene type would be a good choice. An electrolytic capacitor would be a poor choice because of its high leakage. This leakage would cause the output voltage value to drop during the "hold" period.

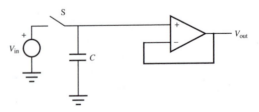

Figure 5.21 Sample and hold circuit.

5.13 COMPARATOR

The comparator circuit illustrated in Figure 5.22 is used to determine whether one signal is greater than another. The **comparator** is an example of an op amp circuit where there is no negative feedback and the circuit exhibits infinite gain. The result is that the op amp **saturates.** Saturation implies that the output remains at its most positive or most negative output value. Video Demo 4.2 shows an interesting example of amplifier saturation where an audio amplifier is clipping the output of a sound wave from a microphone.

Video Demo

4.2 Spectra of whistling and humming, and amplifier saturation

Certain op amps are specifically designed to operate in saturation as comparators. The output of the comparator is defined by

$$V_{out} = \begin{cases} +V_{sat} & V_{in} > V_{ref} \\ -V_{sat} & V_{in} < V_{ref} \end{cases} \tag{5.41}$$

where V_{sat} is the saturation voltage of the comparator and V_{ref} is the reference voltage to which the input voltage V_{in} is being compared. The positive saturation value is slightly less than the positive supply voltage, and the negative saturation value is slightly greater than the negative supply voltage.

Often, comparators (e.g., LM339) have **open-collector outputs,** where the output states are controlled by an output transistor operating at cutoff or saturation. This type of output, illustrated in Figure 5.23, is called an open-collector output because the collector of the output transistor is not connected internally and requires an external powered circuit. The output transistor is ON (at saturation) and the output is effectively grounded when $V_{in} > V_{ref}$, and the output transistor is OFF (at cutoff) and the output is open circuited when $V_{in} < V_{ref}$.

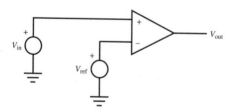

Figure 5.22 Comparator.

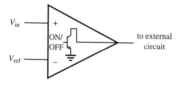

Figure 5.23 Comparator open-collector output.

5.14 THE REAL OP AMP

An actual operational amplifier deviates somewhat in characteristics from an ideal op amp. The best way to familiarize yourself with an IC is to review its specifications in the data book provided by the manufacturer. Complete descriptions of op amps and many other analog ICs are found online or in manufacturers' *Linear* data books. Some of the more important parameters that can be found on op amp data sheets are described in the next section.

As the ideal operational amplifier model implies, real op amps have a very high input impedance, so very little current is drawn at the inputs. At the same time, there is very little voltage difference between the input terminals. However, the input impedance of a real op amp is not infinite, and its magnitude is an important terminal characteristic of the op amp.

Another important terminal characteristic of any real op amp is the maximum output voltage that can be obtained from the amplifier. Consider an op amp circuit with a gain of 100 set by the external resistors in a noninverting amplifier configuration. For a 1 V input, you would expect a 100 V output. In reality, the maximum voltage output will be about 1.4 V less than the supply voltage to the op amp for a large load impedance. So if a ± 15 V supply is being used, the maximum voltage output would be approximately 13.6 V, and the minimum would be -13.6 V. Attempts to drive the output beyond this range result in saturation and clipping at the maximum output-voltage-swing limits.

Two other important characteristics of a real op amp are associated with its response to a square wave input. When you apply a square wave input to an amplifier circuit, you ideally would expect a square wave output. However, as illustrated in Figure 5.24, the output cannot change infinitely fast; instead, it exhibits a ramp from one level to the next. In order to quantify the op amp step response, two parameters are defined:

■ **Slew rate**—The maximum time rate of change possible for the output voltage

$$SR = \frac{\Delta V}{\Delta t} \tag{5.42}$$

■ **Rise time**—The time required for the output voltage to go from 10% to 90% of its final value. This parameter is specified by manufacturers for specific load and input parameters.

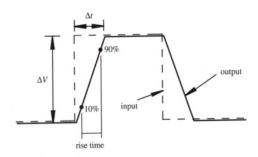

Figure 5.24 Effect of slew rate on a square wave.

Another important characteristic of a real op amp is its frequency response. An ideal op amp exhibits infinite bandwidth. In practice, however, a real op amp has a finite bandwidth, which is a function of the gain established by external components. To quantify this dependence of bandwidth on the gain, another definition is used: the **gain bandwidth product** (GBP). The GBP of an op amp is the product of the open-loop gain and the bandwidth at that gain. The GBP is constant over a wide range of frequencies because, as shown in Figure 5.25, typical op amps exhibit a linear loglog relationship between open-loop gain and frequency. Note how the op amp's gain decreases with input signal frequency. Higher-quality op amps have larger GBPs. The open-loop gain is a characteristic of the op amp without feedback. The closed-loop gain is the overall gain of an op amp circuit with feedback. The closed-loop gain is always limited by the open-loop gain of the op amp. For example, a noninverting amplifier with a closed-loop gain of 100 would have a bandwidth of 0 Hz to approximately 10,000 Hz as illustrated in Figure 5.25. The frequency where the open-loop gain curve first starts to limit the closed-loop gain is called the **fall-off frequency.** As you increase the gain of a circuit, you limit its bandwidth. Likewise, if your application requires only a small bandwidth (e.g., in a low-frequency application), larger gains can be used without signal attenuation or distortion.

5.14.1 Important Parameters from Op Amp Data Sheets

Most of the parameters used to describe the characteristics of actual op amps are listed and described here. These parameters are important when designing and using op amp circuits.

A. Input Parameters

 ■ Input Voltage (V_{icm})—This is the maximum input voltage that can be applied between either input and ground. In general, this voltage is equal to the supply voltage.

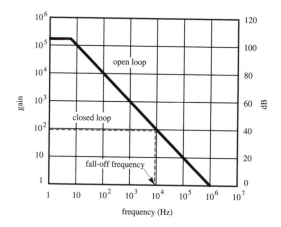

Figure 5.25 Typical op amp open- and closed-loop response.

- Input Offset Voltage (V_{io})—This is the voltage that must be applied to one of the input terminals, with the other input being at 0 V, to give a zero output voltage. Remember, for an ideal op amp, the output voltage offset is 0.

- Input Bias Current (I_{ib})—This is the average of the currents flowing into both inputs when the output voltage is 0. Ideally, the two input currents are 0.

- Input Offset Current (I_{io})—This is the difference between the input currents when the output voltage is 0.

- Input Voltage Range (V_{cm})—This is the range of allowable common mode input voltage, where the same voltage is placed on both inputs.

- Input Resistance (Z_i)—This is the resistance "looking into" either input with the other input grounded.

B. Output Parameters

- Output Resistance (Z_{oi})—This is the internal resistance of the op amp's output circuit (i.e., "looking into" the op amp).

- Output Short Circuit Current (I_{osc})—This is the maximum output current that the op amp can deliver to a load.

- Output Voltage Swing ($\pm V_{omax}$)—This is the maximum peak-to-peak output voltage that the op amp can supply without saturating or clipping.

C. Dynamic Parameters

- Open Loop Voltage Gain (A_{OL})—This is the ratio of the output to the differential input voltage of the op amp without external feedback.

- Large Signal Voltage Gain—This is the ratio of the maximum voltage swing to the change in the input voltage required to drive the output from 0 to a specified voltage.

- Slew Rate (SR)—This is the time rate of change of the output voltage, assuming a step input, with the op amp circuit having a voltage gain of 1.

D. Other Parameters

- Maximum Supply Voltage ($\pm V_s$)—This is the maximum positive and negative voltage permitted to power the op amp.

- Supply Current—This is the current that the op amp draws from the power supply.

- Common Mode Rejection Ratio (CMRR)—This is a measure of the ability of the op amp to reject signals of equal value at the inputs. It is the ratio of the difference mode gain (the output gain corresponding to the difference between the inputs) to the common mode gain (the output gain occurring when the same voltage is applied to both inputs), usually expressed in decibels (dB).

- Channel Separation—Whenever there is more than one op amp in a single package, such as the 747 op amp IC, a certain amount of **cross-talk** is present. That is, a signal applied to the input of one op amp produces a finite output signal in the second op amp, even though there is no direct connection.

Data for each of these parameters are usually provided in IC manufacturers' *Linear* data books, which are available online. Figure 5.26 is a reproduction of the LM741 data sheet from National Semiconductor. It is divided into a maximum ratings section and an electrical characteristics section. This data sheet is typical of

Absolute Maximum Ratings

If Military/Aerospace specified devices are required, please contact the National Semiconductor Sales Office/ Distributors for availability and specifications. (Note 5)

	LM741A	LM741E	LM741	LM741C
Supply Voltage	±22V	±22V	±22V	±18V
Power Dissipation (Note 1)	500 mW	500 mW	500 mW	500 mW
Differential Input Voltage	±30V	±30V	±30V	±30V
Input Voltage (Note 2)	±15V	±15V	±15V	±15V
Output Short Circuit Duration	Continuous	Continuous	Continuous	Continuous
Operating Temperature Range	−55°C to +125°C	0°C to +70°C	−55°C to +125°C	0°C to +70°C
Storage Temperature Range	−65°C to +150°C	−65°C to +150°C	−65°C to +150°C	−65°C to +150°C
Junction Temperature	150°C	100°C	150°C	100°C
Soldering Information				
N-Package (10 seconds)	260°C	260°C	260°C	260°C
J- or H-Package (10 seconds)	300°C	300°C	300°C	300°C
M-Package				
Vapor Phase (60 seconds)	215°C	215°C	215°C	215°C
Infrared (15 seconds)	215°C	215°C	215°C	215°C

See AN-450 "Surface Mounting Methods and Their Effect on Product Reliability" for other methods of soldering surface mount devices.

| ESD Tolerance (Note 6) | 400V | 400V | 400V | 400V |

Electrical Characteristics (Note 3)

Parameter	Conditions	LM741A/LM741E			LM741			LM741C			Units
		Min	Typ	Max	Min	Typ	Max	Min	Typ	Max	
Input Offset Voltage	$T_A = 25°C$ $R_S \leq 10\,k\Omega$ $R_S \leq 50\Omega$		0.8	3.0		1.0	5.0		2.0	6.0	mV mV
	$T_{AMIN} \leq T_A \leq T_{AMAX}$ $R_S \leq 50\Omega$ $R_S \leq 10\,k\Omega$			4.0			6.0			7.5	mV mV
Average Input Offset Voltage Drift				15							μV/°C
Input Offset Voltage Adjustment Range	$T_A = 25°C, V_S = ±20V$	±10				±15			±15		mV
Input Offset Current	$T_A = 25°C$		3.0	30		20	200		20	200	nA
	$T_{AMIN} \leq T_A \leq T_{AMAX}$			70		85	500			300	nA
Average Input Offset Current Drift				0.5							nA/°C
Input Bias Current	$T_A = 25°C$		30	80		80	500		80	500	nA
	$T_{AMIN} \leq T_A \leq T_{AMAX}$			0.210			1.5			0.8	μA
Input Resistance	$T_A = 25°C, V_S = ±20V$	1.0	6.0		0.3	2.0		0.3	2.0		MΩ
	$T_{AMIN} \leq T_A \leq T_{AMAX},$ $V_S = ±20V$	0.5									MΩ
Input Voltage Range	$T_A = 25°C$							±12	±13		V
	$T_{AMIN} \leq T_A \leq T_{AMAX}$				±12	±13					V

Figure 5.26 Example op amp data sheet. *(Courtesy of National Semiconductor, Santa Clara, CA)*

(continued)

Electrical Characteristics (Note 3) (Continued)

Parameter	Conditions	LM741A/LM741E			LM741			LM741C			Units
		Min	Typ	Max	Min	Typ	Max	Min	Typ	Max	
Large Signal Voltage Gain	$T_A = 25°C, R_L \geq 2\,k\Omega$ $V_S = \pm 20V, V_O = \pm 15V$ $V_S = \pm 15V, V_O = \pm 10V$	50			50	200		20	200		V/mV V/mV
	$T_{AMIN} \leq T_A \leq T_{AMAX}$, $R_L \geq 2\,k\Omega$, $V_S = \pm 20V, V_O = \pm 15V$ $V_S = \pm 15V, V_O = \pm 10V$ $V_S = \pm 5V, V_O = \pm 2V$	32 10			25			15			V/mV V/mV V/mV
Output Voltage Swing	$V_S = \pm 20V$ $R_L \geq 10\,k\Omega$ $R_L \geq 2\,k\Omega$	±16 ±15									V V
	$V_S = \pm 15V$ $R_L \geq 10\,k\Omega$ $R_L \geq 2\,k\Omega$				±12 ±10	±14 ±13		±12 ±10	±14 ±13		V V
Output Short Circuit Current	$T_A = 25°C$ $T_{AMIN} \leq T_A \leq T_{AMAX}$	10 10	25	35 40		25			25		mA mA
Common-Mode Rejection Ratio	$T_{AMIN} \leq T_A \leq T_{AMAX}$ $R_S \leq 10\,k\Omega, V_{CM} = \pm 12V$ $R_S \leq 50\Omega, V_{CM} = \pm 12V$	80	95		70	90		70	90		dB dB
Supply Voltage Rejection Ratio	$T_{AMIN} \leq T_A \leq T_{AMAX}$, $V_S = \pm 20V$ to $V_S = \pm 5V$ $R_S \leq 50\Omega$ $R_S \leq 10\,k\Omega$	86	96		77	96		77	96		dB dB
Transient Response Rise Time Overshoot	$T_A = 25°C$, Unity Gain		0.25 6.0	0.8 20		0.3 5			0.3 5		µs %
Bandwidth (Note 4)	$T_A = 25°C$	0.437	1.5								MHz
Slew Rate	$T_A = 25°C$, Unity Gain	0.3	0.7			0.5			0.5		V/µs
Supply Current	$T_A = 25°C$					1.7	2.8		1.7	2.8	mA
Power Consumption	$T_A = 25°C$ $V_S = \pm 20V$ $V_S = \pm 15V$		80	150		50	85		50	85	mW mW
LM741A	$V_S = \pm 20V$ $T_A = T_{AMIN}$ $T_A = T_{AMAX}$			165 135							mW mW
LM741E	$V_S = \pm 20V$ $T_A = T_{AMIN}$ $T_A = T_{AMAX}$			150 150							mW mW
LM741	$V_S = \pm 15V$ $T_A = T_{AMIN}$ $T_A = T_{AMAX}$					60 45	100 75				mW mW

Note 1: For operation at elevated temperatures, these devices must be derated based on thermal resistance, and T_j max. (listed under "Absolute Maximum Ratings"). $T_j = T_A + (\theta_{jA} P_D)$.

Thermal Resistance	Cerdip (J)	DIP (N)	HO8 (H)	SO-8 (M)
θ_{jA} (Junction to Ambient)	100°C/W	100°C/W	170°C/W	195°C/W
θ_{jC} (Junction to Case)	N/A	N/A	25°C/W	N/A

Note 2: For supply voltages less than ±15V, the absolute maximum input voltage is equal to the supply voltage.

Note 3: Unless otherwise specified, these specifications apply for $V_S = \pm 15V$, $-55°C \leq T_A \leq +125°C$ (LM741/LM741A). For the LM741C/LM741E, these specifications are limited to 0°C $\leq T_A \leq +70°C$.

Note 4: Calculated value from: BW (MHz) = 0.35/Rise Time(µs).

Note 5: For military specifications see RETS741X for LM741 and RETS741AX for LM741A.

Note 6: Human body model, 1.5 kΩ in series with 100 pF.

Figure 5.26 (continued)

those from other manufacturers. Figure 5.27 shows the frequency response characteristics of the TL071. These types of graphs are also typically provided on op amp data sheets. Internet Links 5.1 and 5.2 point to the complete data sheets for both the 741 and the TL071. Lab Exercise 6 explores how some of the parameters and information reported in the data sheets are important in understanding how various op amp circuits respond to various inputs.

As you gain practical experience with op amp circuits, you should develop an appreciation of the significance of the many parameters affecting op amp performance.

Internet Link

5.1 741 op amp data sheet

5.2 TL071 FET input op amp data sheet

Lab Exercise

Lab 6
Operational amplifier circuits

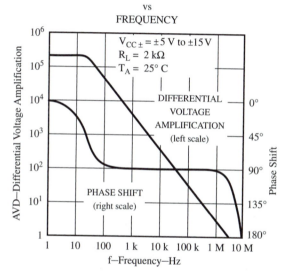

Figure 5.27 TL071 FET input op amp. *(Courtesy of Texas Instruments, Dallas, TX)*

■ **CLASS DISCUSSION ITEM 5.7**
Real Integrator Behavior

Reflecting back on Class Discussion Item 5.4, if the integrator is made with a real op amp, what happens to the output in contrast to that of the ideal integrator?

EXAMPLE 5.1

Sizing Resistors in Op Amp Circuits

The ideal model of the op amp would imply that if you constructed the following two op amp circuits in the laboratory, they would have the same gain. Ideally, each circuit would have a gain of -2. However, the top circuit would be a very poor design and not function as expected. The reason for this can be found by considering the Output Short Circuit Current found on the specification sheet for the op amp. From Figure 5.26, the value for a LM741 is typically 25 mA. This is the largest current that the output can source. But, looking at the circuit, the output current is $V_{out} / 2\ \Omega$, and because $V_{out} = -2V_{in} = -10$ V, the output current would be 5 A! This is far above the current sourcing capability of the op amp. To avoid this problem, larger resistances such as the ones shown in the bottom circuit are used. Here, the output current is 5 mA (10 V / 2 kΩ), which is well within the op amp specification.

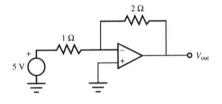

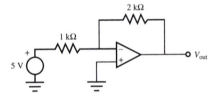

DESIGN EXAMPLE 5.1

Myogenic Control of a Prosthetic Limb

Interfacing prosthetic devices to the human body presents one of the most interesting and challenging problems for engineers. These problems pose a number of medical and engineering challenges in the fields of materials, fluids, electronics, control, and mechanics. Think of the artificial heart, dialysis machine, hip joint replacement, osmotic skin patch, and artificial retina as examples. As we develop improved technological products, we find very important applications in bioengineering. Let us consider one important problem that uses our knowledge of operational amplifiers.

Suppose you would like to design a prosthetic limb (e.g., an artificial arm or leg) that could be controlled by the thoughts of the wearer. Early prosthetic limbs were either purely passive or somewhat mechanically controlled by contracting other muscles. A novel approach however, would be to provide thought control of the limb. Two possibilities present themselves: neural control or myogenic control. For neural control we would have to electrically tap the nervous system, which still is a problem that has not been completely solved technically. Myogenic control is easier to realize. When a muscle is caused to move or twitch, the tiny movement of electrolytes in the muscles below the skin cause an electric field that induces a small voltage on the surface of the skin. This voltage is quite small, otherwise we would shock each other every time we touched. It ranges from microvolts to millivolts and may be mixed with other biopotential signals. The problem then is to sense and isolate this small voltage and convert it into a signal capable of switching something like an electric motor that could be attached to a prosthetic device. So here is our problem: How can we design a mechatronic system that uses the surface skin potential from a muscle as an input to control an actuator such as an electric motor?

Let us begin by outlining what the approach will be. First we have to tap the skin potential with a special surface electrode. Then we have to amplify the signal and filter it to eliminate unwanted noise components and achieve the correct frequency response. Then we need to convert it to a form that allows setting different levels for a control strategy. Finally, we need to drive an electric motor that requires significant current. You currently have the design capability to do all of these things. We start by looking at the transducer to sense the skin potential.

The electric fields that occur in living tissue are caused by charge separations in electrolytes and not by the movement of electrons. In order to sense the voltage at the skin, we need a transducer that converts subdermal (below the skin) electrolyte ion currents to electron currents in our electronic system. Silver–silver chloride electrodes have this property. So if we place a silver chloride electrode on the skin and couple it with a conducting gel, we can sense the body's voltage at that location. The magnitude of the voltage is related to how much a subcutaneous (beneath the skin) muscle contracts. This voltage is the myoelectric signal of interest. The problem that remains is that the electrode produces a very small signal, at best a few millivolts. Also, a considerable amount of 60 Hz background noise and other signals can obscure the signal associated with the muscle. Moreover, the electrode-skin connection has a high impedance.

This is an application where an instrumentation amplifier is necessary to provide the high input impedance, high common mode rejection ratio, and gain necessary to extract the biopotential signal produced by the contracting muscle. The following figure displays the preamplifier stage of our electromyogenic (EMG) detector. For the components shown, it should be easy to create a CMRR in excess of 60 dB and a gain of 125 with an input impedance of 10 MΩ. Note that two active (differential) electrodes 1 and 2 will be mounted close together above the muscle. The third electrode is a ground reference. This circuit is satisfactory in capturing the EMG signal.

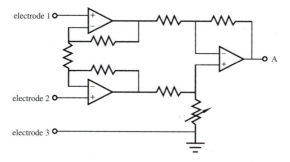

The instrumentation amplifier was chosen because it can extract a very small signal difference between the two signal electrodes (electrodes 1 and 2) while significantly

(continued)

(continued)

attenuating noise common to both electrodes. This eliminates a considerable amount of 60 Hz common mode noise (resulting from electromagnetic interference) and other signals common to both electrodes. However, something called a *motion artifact* can still occur due to relative motion between the electrodes and the tissue. Relative motion can produce voltages sufficient to saturate the second-stage amplifier. The frequencies of the motion artifact are usually at the low end of the bandwidth of the EMG signals. Therefore, a 2 Hz high-pass filter on the input of the second stage of the amplifier that follows can be used to reduce these artifacts. Signal "A" is the output from the instrumentation amplifier circuit and signal "B" is the output of the op-amp-based filter. The filter circuit is a combination of an RC filter and a noninverting amplifier. The capacitor in the op amp feedback loop limits the gain at frequencies higher than those of interest.

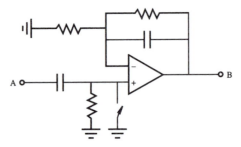

At this point, the EMG signal (signal "B") observed on an oscilloscope would look like the following, where the large amplitude bursts are associated with muscle contractions.

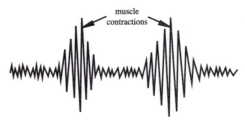

This is a rather high-frequency signal with components between a few Hz and 250 Hz. To make this signal more useful for control purposes, we need to extract the envelope of the signal between 0 V and its maximum positive amplitude. We can accomplish this with a rectifier and a low-pass filter. A normal silicon diode would not be satisfactory to rectify the signal because it requires a 0.7-V turn-on voltage, which is larger than the amplitude of the input signal. Because the signal is very small, we must use a *precision rectifier* circuit (see the figure that follows) that more closely approximates the action of an ideal diode.

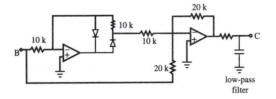

The precision-rectified EMG and the resulting low-pass-filtered signals look like those shown in the next figure. The low-pass-filtered signal is basically the envelope of the rectified signal. Now we have a signal that can be input to a comparator (see the figure that follows next) to provide a binary control signal (signal "D"), which will be on when the muscle is contracted and off when relaxed. The designer or user can adjust the reference voltage with the potentiometer for the desired sensitivity.

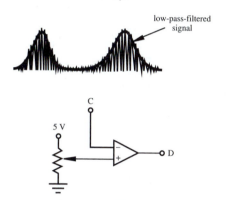

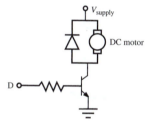

As shown in the final figure, the output of the rectifier can then be input to a power transistor circuit to control the current in a motor.

In summary, we have used a variety of op amp circuits to process an analog signal. It exemplifies the extraction of a very low level signal in the presence of noise, a variety of analog signal processing methods, and the interface to an actuator to control mechanical power. In this case, we have converted the EMG signal into a binary control signal for on-off control of a DC motor that could, for example, flex the elbow in a prosthetic arm. A more complete and detailed solution to a similar problem, where an EMG signal is used to control an industrial robot, is presented in Section 11.4.

■ **CLASS DISCUSSION ITEM 5.8**
Bidirectional EMG Controller

In Design Example 5.1, we discussed turning on and off a motor via an EMG signal. Unfortunately, the controller can actuate the joint in one direction only. Discuss how you might change the design to allow bidirectional movement.

QUESTIONS AND EXERCISES

Section 5.5   Inverting Amplifier

5.1. An inverting op amp circuit is designed with 1/4 W resistors (i.e., they are capable of dissipating up to 1/4 W of energy without failure). If the input voltage is 5 V, what are the minimum required values for the input and feedback resistors if the gain is

   a.   1

   b.   10

5.2. Determine V_{out} as a function of I (provided by a current source) and the resistor values for each of the op amp circuits that follows. Assume ideal op amp behavior.

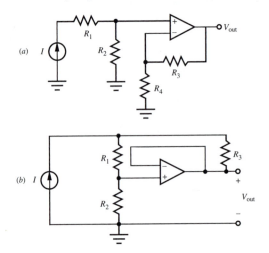

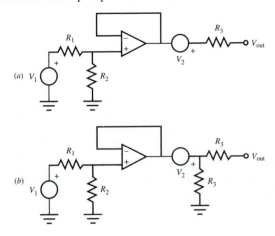

Section 5.6   Noninverting Amplifier

5.3. If resistor R_F shown in Figure 5.10 is replaced with a short (i.e., $R_F = 0$), what is the gain of the circuit?

5.4. Determine V_{out} in the following circuits with $R_1 = R_2 = R_3 = 1$ kΩ, $V_1 = 10$ V, and $V_2 = 5$ V. Assume ideal op amp behavior.

5.5. Determine I_4 in terms of V_{in}, R_1, R_2, R_3, and R_4 in the following circuit. Assume ideal op amp behavior.

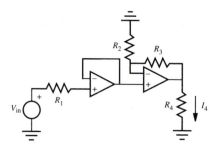

5.6. For the following circuit, express I_{out_1} and V_{out_2} in terms of V_1, V_2, and R.

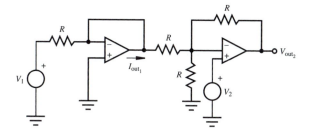

5.7. Explain why $V_{out} \neq V_{in}$ in the following circuit.

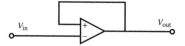

Section 5.7 Summer

5.8. Analyze the summer circuit in Figure 5.13 and determine an equation for the output voltage V_{out} in terms of the input voltages V_1 and V_2 and the resistances R_1, R_2, and R_F. Use this result to verify that Equation 5.19 is correct. Show and explain all work.

5.9. If a voltage source V_3 is inserted between ground and the noninverting input in the summer circuit shown in Figure 5.13, what is the resulting equation for the output voltage V_{out} in terms of the input voltages V_1 and V_2 if $R_1 = R_2 = R_F$?

Section 5.8 Difference Amplifier

5.10. Derive Equation 5.24 for the difference amplifier without using the principle of superposition.

5.11. Use the principle of superposition to derive an expression for the output voltage in the following circuit and explain why the circuit is called a **level shifter.**

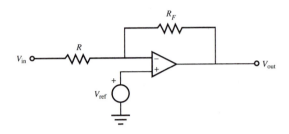

Section 5.9   Instrumentation Amplifier

5.12. Derive Equation 5.31 that expresses V_{out} in terms of the V_3 and V_4 shown in Figure 5.17.

Section 5.10   Integrator

5.13. Solve and generate plots for Class Discussion Item 5.4 for a 100 Hz sine wave with an amplitude of 1 V and a DC offset of 0.1 V. Plot the output for 5 cycles of the sinusoid.

Section 5.11   Differentiator

5.14. Find $V_{out}(t)$ given $V_{in}(t)$ in the op amp circuit that follows.

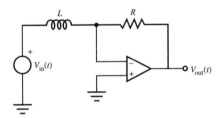

5.15. Derive Equation 5.36 assuming the input and output voltages are both zero and the currents flowing into each op amp input are equal.

5.16. Using the following input waveform, sketch the corresponding output waveform for each op amp circuit (a through d). Assume ideal op amp behavior.

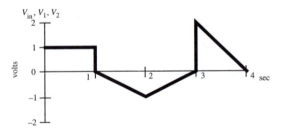

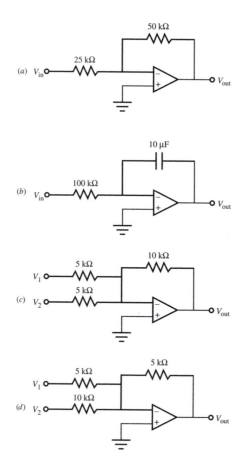

Section 5.13   Comparator

5.17. Using a standard output (not open-collector) comparator, draw a circuit that could be used to turn on an LED when an input voltage exceeds 5 V.

5.18. Using an open-collector output comparator, draw a circuit that could be used to turn on an LED when an input voltage exceeds 5 V.

Section 5.14   The Real Op Amp

5.19. If the short-circuit output current of a real op amp is 10 mA, calculate the minimum resistance required for the feedback resistor in an inverting op amp circuit with a gain of 10 and a maximum output voltage of 10 V.

5.20. Given the following circuit and op amp open loop gain curve, what is the fall-off frequency for the circuit when $R_F = 20$ kΩ and $R = 2$ kΩ?

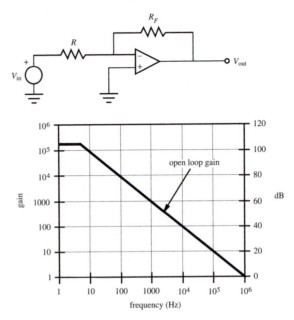

5.21. Document a complete and thorough answer to Class Discussion Item 5.7.

BIBLIOGRAPHY

Coughlin, R. and Driscoll, F., *Operational Amplifiers and Linear Integrated Circuits,* 4th Edition, Prentice-Hall, Englewood Cliffs, NJ, 1991.

Horowitz, P. and Hill, W., *The Art of Electronics,* 2nd Edition, Cambridge University Press, New York, 1989.

Johnson, D., Hilburn, J., and Johnson, J., *Basic Electric Circuit Analysis,* 2nd Edition, Prentice-Hall, Englewood Cliffs, NJ, 1984.

McWhorter, G. and Evans, A., *Basic Electronics,* Master Publishing, Richardson, TX, 1994.

Mims, F., *Engineer's Mini-Notebook: Op Amp IC Circuits,* Radio Shack Archer Catalog No. 276-5011, 1985.

Mims, F., *Getting Started in Electronics,* Radio Shack Archer Catalog No. 276-5003A, 1991.

Texas Instruments, *Linear Circuits Data Book, Volume 1—Operational Amplifiers,* Dallas, TX, 1992.

CHAPTER 6

Digital Circuits

his chapter presents digital electronic devices used for logic, display, sequencing, timing, and other functions in mechatronic systems. The fundamentals presented in this chapter are important in understanding the basic functioning of all digital components and systems used in the control of mechatronic systems. ∎

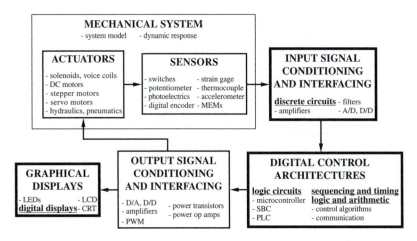

After you read, discuss, study, and apply ideas in this chapter, you will:

1. Be able to define a digital signal

2. Understand how the binary and hexadecimal number systems are used in coding digital data

3. Know the characteristics of different logic gates

4. Know the differences between combinational and sequential logic

5. Be able to draw a timing diagram for a digital circuit

6. Be able to use Boolean mathematics to analyze logic circuits

7. Be able to design logic networks

8. Be able to use a variety of flip-flops for storing data

9. Understand differences between TTL and CMOS logic devices

10. Know how to construct an interface between TTL and CMOS devices

11. Be able to use counters for different counting applications

12. Know how to display numerical data using LED displays

6.1 INTRODUCTION

In contrast to an analog signal that changes in a continuous manner, a digital signal exists only at specific levels or states and changes its level in discrete steps. An analog signal and a digital signal are illustrated in Figure 6.1. Most digital signals have only two states: high and low. A system using two-state signals allows the application of Boolean logic and binary number representations, which form the foundation for the design of all digital devices.

Digital devices are categorized according to their function as **combinational logic** or **sequential logic** devices. Digital devices convert digital inputs into one or more digital outputs. The difference between the two categories is based on signal timing. For sequential logic devices, the timing, or sequencing history, of the input signals plays a role in determining the output. This is not the case with combinational logic devices whose outputs depend only on the instantaneous values of the inputs.

Before we introduce the various digital devices, we will review the binary number system and the application of binary numbers in digital calculations. Then we discuss Boolean algebra, which is the mathematical basis for digital computations. Finally, we discuss a number of specific combinational and sequential logic devices and their applications.

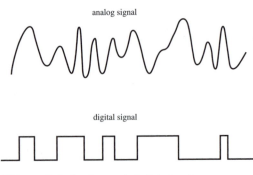

Figure 6.1 Analog and digital signals.

6.2 DIGITAL REPRESENTATIONS

We grow up becoming proficient using the base 10 decimal number system. The **base** of the number system indicates the number of different symbols that can be used to represent a digit. In base 10, the symbols are 0, 1, 2, 3, 4, 5, 6, 7, 8, and 9. Each digit in a decimal number is a placeholder for different powers of 10 according to

$$d_{n-1} \ldots d_3 d_2 d_1 d_0 = d_{n-1} \cdot 10^{n-1} + \cdots + d_2 \cdot 10^2 + d_1 \cdot 10^1 + d_0 \cdot 10^0 \quad (6.1)$$

where n is the number of digits and each digit d_i is one of the ten symbols. Note that the highest power of 10 is $(n - 1)$, 1 less than the number of digits. As an example, the decimal number 123 can be expanded as

$$123 = 1 \times 10^2 + 2 \times 10^1 + 3 \times 10^0 \quad (6.2)$$

Fractions may also be included if digits for negative powers of 10 are included $(d_{-1}, d_{-2}, \ldots)$.

In order to represent and manipulate numbers with digital devices such as computers, we use a base 2 number system called the **binary number system.** The reason for this is that the operation of digital devices is based on transistors that switch between two states: the ON or saturated state and the OFF or cutoff state. These states are designated by the symbols 1 (ON) and 0 (OFF) in the base 2 system. The digits in a binary number, as with the base 10 system, correspond to different powers of the base. A binary number can be expanded as

$$(d_{n-1} \ldots d_3 d_2 d_1 d_0)_2 = d_{n-1} \cdot 2^{n-1} + \cdots + d_2 \cdot 2^2 + d_1 \cdot 2^1 + d_0 \cdot 2^0 \quad (6.3)$$

where each digit d_i is one of the two symbols 0 and 1. The trailing subscript 2 is used to indicate that the number is base 2 and not the normally assumed base 10. As an example of Equation 6.3, the binary number 1101 can be expanded as

$$1101_2 = 1 \cdot 2^3 + 1 \cdot 2^2 + 0 \cdot 2^1 + 1 \cdot 2^0 = 8_{10} + 4_{10} + 1_{10} = 13_{10} \quad (6.4)$$

The digits of a binary number are also called **bits,** and the first and last bits have special names. The first, or leftmost, bit is known as the **most significant bit** (MSB) because it represents the largest power of 2. The last, or rightmost, bit is known as the **least significant bit** (LSB) because it represents the smallest power of 2. A group of 8 bits is called a **byte.**

In general, the value of a number represented in any base can be expanded and computed with

$$(d_{n-1} \ldots d_3 d_2 d_1 d_0)_b = (d_{n-1} \cdot b^{n-1} + \cdots + d_2 \cdot b^2 + d_1 \cdot b^1 + d_0 \cdot b^0) \quad (6.5)$$

where b is the base and n is the number of digits. Often it is necessary to convert from one base system to another. Equation 6.5 provides a mechanism to convert from an arbitrary base to base 10. To convert a number from base 10 to some other base, the procedure is to successively divide the decimal number by the base and record the

Table 6.1 Decimal to binary conversion

Successive Divisions	Remainder	
123/2	1	LSB
61/2	1	
30/2	0	
15/2	1	
7/2	1	
3/2	1	
1/2	1	MSB
Result	1111011	

Video Demo

6.1 Binary counting machine using marbles and wood toggle switches

remainders after each division. The remainders, when written in reverse order from left to right, form the digits of the number represented in the new base. Table 6.1 illustrates this procedure by converting the decimal number 123 to its binary equivalent. You can use Equation 6.3 to verify the binary result by calculating its expansion.

Binary arithmetic is carried out in the same way as the more familiar base 10 arithmetic. Example 6.1 illustrates the similarities. Video Demo 6.1 shows a binary counting machine using marbles as bits and wood toggle switches to store values. It quite nicely illustrates how binary counting and bit-carry work.

EXAMPLE 6.1   **Binary Arithmetic**

This example illustrates the analogy between decimal addition and multiplication and binary addition and multiplication. Note that, when adding two 1 bits (1 + 1), the sum is 0 with a carry of 1 to the next higher order bit. This occurs in the addition example below, where a 1 is carried from bit one to bit two and from bit two to bit three:

$$
\begin{array}{cccc}
 & 11 & & \\
9 & 1001 & 9 & 1001 \\
\underline{+\ 3} & \underline{+\ 0011} & \underline{\times\ 3} & \underline{\times\ 0011} \\
12 & 1100 & 27 & 1001 \\
 & & & +\ 1001 \\
 & & & +\ 0000 \\
 & & & \underline{+\ 0000} \\
 & & & 11011 \\
\end{array}
$$

Because binary numbers can be long and cumbersome to write and display, often the **hexadecimal** (base 16) number system is used as an alternative representation. Table 6.2 lists the symbols for the hexadecimal system along with their binary and decimal equivalents. Note that the letters A through F are used to represent the digits larger than 9.

To convert a binary number to hexadecimal, divide the number into groups of four digits beginning with the least significant bit and replace each group with its hexadecimal equivalent. For example,

$$123_{10} = 0111\ 1011_{2} = 7B_{16} \tag{6.6}$$

Table 6.2 Hexadecimal symbols and equivalents

Binary	Hexadecimal	Decimal
0000	0	0
0001	1	1
0010	2	2
0011	3	3
0100	4	4
0101	5	5
0110	6	6
0111	7	7
1000	8	8
1001	9	9
1010	A	10
1011	B	11
1100	C	12
1101	D	13
1110	E	14
1111	F	15

Another base useful in representing binary numbers in concise form is octal (base 8). To convert a binary number to octal, divide the number into groups of three digits beginning with the least significant bit and replace each group with its octal equivalent. For example,

$$123_{10} = 001\ 111\ 011_2 = 173_8 \tag{6.7}$$

■ CLASS DISCUSSION ITEM 6.1
Nerd Numbers

Why do geeky engineers sometimes confuse Halloween (OCT 31) and Christmas (DEC 25)?

In addition to numbers, alphanumeric characters can also be represented in digital (binary) form with **ASCII** codes. ASCII is short for the *A*merican *S*tandard *C*ode for *I*nformation *I*nterchange. ASCII codes are 7-bit codes used to denote all of the alphanumeric characters. The 7-bit codes are usually stored in an 8-bit byte. There is a unique code for each alphanumeric character. Some example codes are

$$A: 0100\ 0001 = 41_{16} = 65_{10}$$
$$B: 0100\ 0010 = 42_{16} = 66_{10}$$
$$0: 0011\ 0000 = 30_{16} = 48_{10}$$
$$1: 0011\ 0001 = 31_{16} = 49_{10}$$

The complete set of ASCII codes can be found at Internet Link 6.1.

Internet Link

6.1 ASCII codes

Binary coded decimal (BCD) is another type of digital representation sometimes used for input and output of numerical data. With BCD, 4 bits are used to represent a single, base 10 digit. BCD is a convenient mechanism for representing decimal numbers in a binary number format, but it is inefficient for storing or transmitting multiple-digit numbers because only 10 of the 16 (2^4) possible states of the 4-bit number are used. To convert a decimal number to BCD, assemble the 4-bit codes for each decimal digit. For example,

$$123_{10} = 0001\ 0010\ 0011_{bcd} \tag{6.8}$$

Note that this is different from the binary representation:

$$123_{10} = 0111\ 1011_2 \tag{6.9}$$

■ CLASS DISCUSSION ITEM 6.2
Computer Magic

How can a digital computer perform the complex operations it does given that its architecture and operation are based on simply manipulating bits (zeros and ones)?

6.3 COMBINATIONAL LOGIC AND LOGIC CLASSES

Combinational logic devices are digital devices that convert binary inputs into binary outputs based on the rules of mathematical logic. The basic operations, schematic symbols, and algebraic expressions for combinational logic devices are shown in Table 6.3. These devices are also called **gates,** because they control the flow of signals from the inputs to the single output. A small circle at the input or output of a digital device denotes signal inversion; that is, a 0 becomes a 1 or a 1 becomes a 0. For example, the **NAND** and **NOR** gates are AND and OR gates, respectively, with the output inverted, hence the circle is shown on the output. The **truth table** for each device is shown on the right. The truth table is a compact means of displaying all combinations of inputs and their corresponding outputs. Usually the combination of inputs is written as the ascending list of binary numbers whose number of bits corresponds to the number of inputs (e.g., 00, 01, 10, 11 for two inputs). Internet Link 6.2 provides a fun online digital circuit simulator allowing you to drag-and-drop and connect logic gates in different ways to interactively see how combinational logic circuits function.

The standard AND, NAND, OR, NOR, and XOR gates have only two inputs, but other forms are available with more than two inputs. In the case of a multiple input **AND** gate, the output is 1 if and only if all inputs are 1; otherwise, the output is 0. In the case of the **OR** gate, the output is 0 if and only if all inputs are 0; otherwise, it is 1. In the case of the **XOR** gate, the output is 0 if all of the inputs are 0 or if all of

Internet Link

6.2 "Logicly" combinational logic online simulator

Table 6.3 Combinational logic operations

Gate	Operation	Symbol	Expression	Truth Table
Inverter (INV, NOT)	Invert signal (complement)		$C = \overline{A}$	$\begin{array}{cc} A & C \\ 0 & 1 \\ 1 & 0 \end{array}$
AND gate	AND logic		$C = A \cdot B$	$\begin{array}{ccc} A & B & C \\ 0 & 0 & 0 \\ 0 & 1 & 0 \\ 1 & 0 & 0 \\ 1 & 1 & 1 \end{array}$
NAND gate	Inverted AND logic		$C = \overline{A \cdot B}$	$\begin{array}{ccc} A & B & C \\ 0 & 0 & 1 \\ 0 & 1 & 1 \\ 1 & 0 & 1 \\ 1 & 1 & 0 \end{array}$
OR gate	OR logic		$C = A + B$	$\begin{array}{ccc} A & B & C \\ 0 & 0 & 0 \\ 0 & 1 & 1 \\ 1 & 0 & 1 \\ 1 & 1 & 1 \end{array}$
NOR gate	Inverted OR logic		$C = \overline{A + B}$	$\begin{array}{ccc} A & B & C \\ 0 & 0 & 1 \\ 0 & 1 & 0 \\ 1 & 0 & 0 \\ 1 & 1 & 0 \end{array}$
XOR gate	Exclusive OR logic		$C = A \oplus B$ $= A \cdot \overline{B} + \overline{A} \cdot B$	$\begin{array}{ccc} A & B & C \\ 0 & 0 & 0 \\ 0 & 1 & 1 \\ 1 & 0 & 1 \\ 1 & 1 & 0 \end{array}$
Buffer	Increase output signal current		$C = A$	$\begin{array}{cc} A & C \\ 0 & 0 \\ 1 & 1 \end{array}$

the inputs are 1; otherwise, it is 1. The algebraic symbols used to represent the logic functions are: plus (+) for logic OR, dot (·) for the logic AND, and an overbar ($\overline{X}$) for logic **NOT,** denoting inversion.

A **buffer** is used to increase the current supplied at the output while retaining the digital state. This is important if you wish to drive multiple digital inputs from a single output. A normal digital gate has a limited **fan-out,** which defines the maximum number of similar digital inputs that can be driven by the gate's output (see Section 6.11.3). The buffer overcomes fan-out limitations by providing more output current.

Video Demo

5.1 Integrated circuits

Video Demo

5.2 Integrated circuit manufacturing process stages

5.3 Chip manufacturing process

5.4 How a CPU is made

All the gates in Table 6.3 are manufactured as **integrated circuits** where transistors, resistors, and diodes exist on a single chip of silicon. Video Demo 5.1 shows various types of IC packages, and Video Demos 5.2 through 5.4 describe and illustrate all of the steps of the IC manufacturing process.

There are two families of digital integrated circuits, called **TTL** for transistor-transistor logic and **CMOS** for complementary metal-oxide semiconductors. Voltage levels define **logic low** (0) and **logic high** (1) at the inputs and outputs. The ranges for the logic levels vary depending on the device family. A designer must be careful when mixing different types of digital integrated circuits because different digital devices have specified current source and sink capabilities that affect how much fan-out is allowed and how they may be mixed with each other. The TTL and CMOS families are described in detail in Section 6.11.

EXAMPLE 6.2   Combinational Logic

This example illustrates how to determine signal expressions and values in a logic diagram. Here is an example logic circuit:

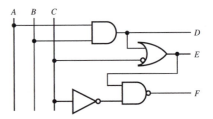

Signals *A, B,* and *C* are inputs and signals *D, E,* and *F* are outputs. Each of the signals can be high (1) or low (0). When analyzing a logic circuit, start by writing out logic expressions for all signals in the circuit. *D* is the most straightforward because it is simply the AND combination of signals *A* and *B:*

$$D = A \cdot B$$

Signal *E* depends on signal *D* and uses the circle inversion symbol on the *C* input to the OR gate, so

$$E = D + \overline{C}$$

This can also be written as

$$E = (A \cdot B) + \overline{C}$$

Finally, signal *F* is the result of a NAND combination of signal *E* and the inverse of *C:*

$$F = \overline{E \cdot \overline{C}}$$

Note that the inversion bar goes over the entire expression $E \cdot \overline{C}$ because the NAND gate inverts the output of the AND combination of *E* and $\overline{C}$.

Another way to express the functionality of the logic circuit is to summarize all the possible input and output combinations in a truth table as follows. Each output value can be determined from the logic expressions. For example, for input combination $A = 0$, $B = 0$, and $C = 0$, $D = 0 \cdot 0 = 0$, $E = 0 + \bar{0} = 0 + 1 = 1$, $F = \overline{1 \cdot \bar{0}} = \overline{1 \cdot 1} = \bar{1} = 0$.

A	B	C	D	E	F
0	0	0	0	1	0
0	0	1	0	0	1
0	1	0	0	1	0
0	1	1	0	0	1
1	0	0	0	1	0
1	0	1	0	0	1
1	1	0	1	1	0
1	1	1	1	1	1

When you become comfortable with logic expressions and truth tables, you will be able to construct the table very quickly using some short-cuts. For example, because D is the AND combination of A and B, D is 1 only if A is 1 *and* B is 1; otherwise, D is 0. Therefore, you can quickly fill in column D with all zeros except where A and B are both 1. Likewise, with column E, which is an OR combination, the output is 1 if D is 1 *or* if $\bar{C}$ is 1 (i.e., C is 0). Another way to interpret this is that E is 0 only if both D is 0 and $\bar{C}$ is 0 (i.e., C is 1). F is 0 (1 inverted) if E is 1 and $\bar{C}$ is 1 (i.e., C is 0).

6.4 TIMING DIAGRAMS

In order to analyze complex logic circuits it helps to sketch a timing diagram, which shows the simultaneous levels of the inputs and outputs in a circuit vs. time. The timing diagram can be used to illustrate every possible combination of input values and corresponding outputs, providing a graphical summary of the input/output relationships. Timing diagrams for the AND and OR gates are shown in Figures 6.2 and 6.3 as examples. Multiple-input digital oscilloscopes and logic analyzers have the capability to display timing diagrams for digital circuits.

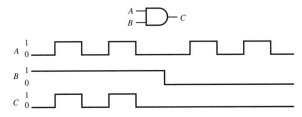

Figure 6.2 AND gate timing diagram.

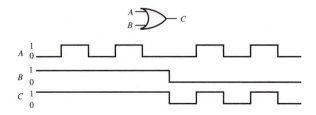

Figure 6.3 OR gate timing diagram.

6.5 BOOLEAN ALGEBRA

In formulating mathematical expressions for logic circuits, it is important to have knowledge of Boolean algebra, which defines the rules for expressing and simplifying binary logic statements. The basic Boolean laws and identities follow. A bar over a symbol indicates the Boolean operation NOT, which corresponds to inversion of a signal.

Boolean Algebra Laws and Identities

Fundamental Laws

OR	AND	NOT	
$A + 0 = A$	$A \cdot 0 = 0$		
$A + 1 = 1$	$A \cdot 1 = A$	$\bar{\bar{A}} = A$	(6.10)
$A + A = A$	$A \cdot A = A$	(double inversion)	
$A + \bar{A} = 1$	$A \cdot \bar{A} = 0$		

Commutative Laws

$$A + B = B + A \tag{6.11}$$

$$A \cdot B = B \cdot A \tag{6.12}$$

Associative Laws

$$(A + B) + C = A + (B + C) \tag{6.13}$$

$$(A \cdot B) \cdot C = A \cdot (B \cdot C) \tag{6.14}$$

Distributive Laws

$$A \cdot (B + C) = (A \cdot B) + (A \cdot C) \tag{6.15}$$

$$A + (B \cdot C) = (A + B) \cdot (A + C) \tag{6.16}$$

Other Useful Identities

$$A + (A \cdot B) = A \tag{6.17}$$

$$A \cdot (A + B) = A \tag{6.18}$$

$$A + (\bar{A} \cdot B) = A + B \qquad\qquad (6.19)$$

$$(A + B) \cdot (A + \bar{B}) = A \qquad\qquad (6.20)$$

$$(A + B) \cdot (A + C) = A + (B \cdot C) \qquad\qquad (6.21)$$

$$A + B + (A \cdot \bar{B}) = A + B \qquad\qquad (6.22)$$

$$(A \cdot B) + (B \cdot C) + (\bar{B} \cdot C) = (A \cdot B) + C \qquad\qquad (6.23)$$

$$(A \cdot B) + (A \cdot C) + (\bar{B} \cdot C) = (A \cdot B) + (\bar{B} \cdot C) \qquad\qquad (6.24)$$

De Morgan's laws are also useful in rearranging or simplifying longer Boolean expressions or in converting between AND and OR gates:

$$\overline{A + B + C + \cdots} = \bar{A} \cdot \bar{B} \cdot \bar{C} \cdots \qquad\qquad (6.25)$$

$$\overline{A \cdot B \cdot C \cdots} = \bar{A} + \bar{B} + \bar{C} + \cdots \qquad\qquad (6.26)$$

If we invert both sides of these equations and apply the double NOT law from Equation 6.10, we can write De Morgan's laws in the alternative form:

$$A + B + C + \cdots = \overline{\bar{A} \cdot \bar{B} \cdot \bar{C} \cdots} \qquad\qquad (6.27)$$

$$A \cdot B \cdot C \cdots = \overline{\bar{A} + \bar{B} + \bar{C} + \cdots} \qquad\qquad (6.28)$$

Truth tables can be very helpful in verifying an identity. For example, to show that Equation 6.19 is valid, we can construct the following truth table where each term in the identity is evaluated for all input combinations:

A	$\bar{A}$	B	$\bar{A} \cdot B$	$(A + \bar{A} \cdot B$	$A + B$
1	0	0	0	1	1
1	0	1	0	1	1
0	1	0	0	0	0
0	1	1	1	1	1

Because both sides of the identity are equal for every input combination, the identity is valid.

Simplifying a Boolean Expression   EXAMPLE 6.3

Simplify the following expression using Boolean laws and identities:

$$X = (A \cdot B \cdot C) + (B \cdot C) + (\bar{A} \cdot B)$$

First, we can rewrite this equation using the associative law and the fundamental law that $Z \cdot 1 = Z$:

$$X = A \cdot (B \cdot C) + 1 \cdot (B \cdot C) + (\bar{A} \cdot B) \qquad\qquad (continued)$$

(continued)

In this form, it is clear that we can use the distributive law to factor out the $(B \cdot C)$ term:

$$X = (A + 1) \cdot (B \cdot C) + (\bar{A} \cdot B)$$

Because $A + 1 = 1$ and $1 \cdot (B \cdot C) = B \cdot C,$

$$X = (B \cdot C) + (\bar{A} \cdot B)$$

Furthermore, using the associative and distributive laws we can factor out B:

$$X = B \cdot (C + \bar{A})$$

We have reduced the number of operators from seven in the original expression (including the inversion) to three in the final expression. This is important because it reduces the number of gates required to build the circuit.

6.6 DESIGN OF LOGIC NETWORKS

As an example illustrating the application of combinational logic to a real engineering problem, suppose you are asked to design a circuit for a simple security protection system for a home. The homeowner wants an alarm to sound if someone breaks into the house through a door or window or if something is moving around in the house while the occupants are away. Under certain conditions, the users may also want to disable portions of the alarm system. We assume that there are sensors to detect if windows or doors are disturbed and to detect motion. To accomplish the goals of this security system, we design a combinational logic circuit using two switches that can be set by the owner.

The following steps facilitate the design of a digital circuit to solve this type of problem:

1. Define the problem in words.
2. Write quasi-logic statements in English that can be translated into Boolean expressions.
3. Write the Boolean expressions.
4. Simplify and optimize the Boolean expressions, if possible.
5. Write an all-AND, all-NAND, all-OR, or all-NOR realization of the circuit to minimize the number of required logic gate IC components.
6. Draw the logic schematic for the electronic realization of the circuit.

Each of these steps is carried out for the security lock example in the following sections.

6.6.1 Define the Problem in Words

We begin our logic design by translating the problem into a series of word statements that reflect what should be happening in the system. We want the alarm system to create a high signal, sounding the alarm for certain combinations of the house sensors. Also, we want the user to be able to select one of three operating states:

1. An active state where the alarm will sound only if the windows or doors are disturbed. This state is useful when the occupants are sleeping.

2. An active state where the alarm will sound if the windows or doors are disturbed or if there is motion in the house. This state is useful when the occupants are away.

3. A disabled state where the alarm will not sound. This state is useful during normal household activity.

At this time, we must define Boolean variables that will represent the inputs and outputs of the circuit. The following Boolean variables are used to design the security system logic:

- *A:*   state of the door and window sensors
- *B:*   state of the motion detector
- *Y:*   output used to sound the alarm
- *C D:*   2-bit code set by the user to select the operating state defined by

$$CD = \begin{cases} 0\ 1 & \text{operating state 1} \\ 1\ 0 & \text{operating state 2} \\ 0\ 0 & \text{operating state 3} \end{cases}$$

The inputs to the system are *A, B, C,* and *D,* and the output is *Y.* We assume **positive logic** for *A, B,* and *Y,* where a 1 implies active or ON and a 0 implies inactive or OFF.

6.6.2 Write Quasi-Logic Statements

We further translate the word statements into logic-like statements. The quasi-logic statements for the security system are

> Activate the alarm ($Y = 1$) if *A* is high *and* the code *C D* is 0 1 *or* activate the alarm if *A or B* is high *and* the code is 1 0.

Note the italicized quasi-Boolean operators, which should aid in writing the Boolean expression.

6.6.3 Write the Boolean Expression

Now we write the Boolean expression based on the quasi-logic statement. To create a product of 1 for the active control code 0 1, we need to form the expression $\overline{C} \cdot D$; to create a product of 1 for the other active control code 1 0, we need to form the expression $C \cdot \overline{D}$. Based on this, the complete Boolean expression for the security system is

$$Y = A \cdot (\overline{C} \cdot D) + (A + B) \cdot (C \cdot \overline{D}) \tag{6.29}$$

The alarm will sound ($Y = 1$) if the expression $A \cdot (\overline{C} \cdot D)$ is 1 or if the expression $(A + B) \cdot (C \cdot \overline{D})$ is 1; otherwise, the alarm will not sound ($Y = 0$). The first expression will be 1 if and only if A is 1 *and* C is 0 *and* D is 1; the second expression will be 1 if and only if A *or* B is 1 *and* C is 1 *and* D is 0.

For this particular problem, we can simplify Equation 6.29 by looking at a truth table for the C and D terms for the different control code combinations.

C	D	$(\overline{C} \cdot D)$	$(C \cdot \overline{D})$
0	0	0	0
1	0	0	1
0	1	1	0

Note that $(\overline{C} \cdot D) = D$ and $(C \cdot \overline{D}) = C$ for the given control code combinations. If we disallow the state $C\,D = 1\,1$ (see Question 6.21), Equation 6.29 can be simplified as

$$Y = (A \cdot D) + (A + B) \cdot C \tag{6.30}$$

6.6.4 AND Realization

Once a Boolean expression is simplified, it may be desirable to manipulate the result further in order to convert all operations to a preferred type of gate (e.g., AND or OR). The reason for this is that logic gates come packaged on integrated circuits (logic ICs) in groups of four, six, or eight. Therefore, we may be able to reduce the total number of ICs required by using all of one type of gate. Converting from one gate type to another is easily accomplished with a repeated application of De Morgan's laws. For the security system example, an all-AND representation is achieved by applying Equation 6.27 to convert each OR gate to an AND gate. Starting with Equation 6.30:

$$Y = (A \cdot D) + (A + B) \cdot C \tag{6.31}$$

First, we convert the second OR gate expression $A + B$:

$$Y = (A \cdot D) + (\overline{\overline{A} \cdot \overline{B}}) \cdot C \tag{6.32}$$

Now we have one expression OR-ed into another. Again, Equation 6.30 states that to convert the OR gate to an AND gate, invert each expression, change the gate, and invert the whole expression:

$$Y = \overline{\overline{A \cdot D} \cdot \overline{(\overline{\overline{A} \cdot \overline{B}}) \cdot C}} \tag{6.33}$$

6.6.5 Draw the Circuit Diagram

Now it is relatively straightforward to draw the circuit using only AND gates and inverters from inspection of the final Boolean expression in Equation 6.33, by building the subexpressions one at a time and connecting them as shown in Figure 6.4. It is a good idea to label intermediate signals in a logic circuit with their corresponding Boolean expressions, as shown. This will help you prevent or find errors in your logic.

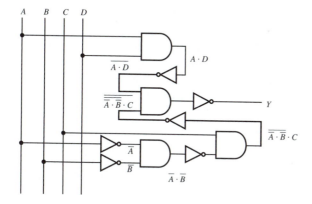

Figure 6.4 AND realization schematic of the security system.

Because there are a total of four AND gates and six inverters, the circuit can be constructed with two ICs: one quad AND gate IC (e.g., the 7408), which contains four AND gates, and one hex inverter IC (e.g., the 7404), which contains six inverters. Equation 6.30 could also be implemented using two ICs because only two OR gates and two AND gates are required. Therefore, for the security system, the all-AND realization did not reduce the number of ICs. However, for more complex Boolean expressions, a single-type gate realization will usually reduce the number of ICs.

The solution just presented is known as a hardware solution because it uses integrated circuit gates to provide the desired logic. An alternative is to implement the logic using a program running on a microcontroller. This solution, called a software solution, is presented in Example 7.5 in Section 7.5.2.

■ **CLASS DISCUSSION ITEM 6.3**
Everyday Logic

Make a list of devices that you interact with on a daily basis that use logic for control purposes. For each, describe what logic is being performed.

6.7 FINDING A BOOLEAN EXPRESSION GIVEN A TRUTH TABLE

As an alternative to the method presented in Sections 6.6.1 and 6.6.2, where we defined a logic problem in words and then wrote quasi-logic statements, sometimes it is more convenient to express the complete input/output combinations with a truth table. In these situations, there are two methods for directly obtaining the Boolean expression that performs the logic specified in the truth table. Both methods are described here, and an example is given to demonstrate their application.

The first method is known as the **sum-of-products method.** It is based on the fact that we can represent an output as a sum of products containing combinations of the inputs. For example, if we have three inputs A, B, and C and an output X, the sum of products would be a Boolean expression containing input terms AND-ed together to form product terms that are OR-ed together to define the output X as a Boolean sum. The following equation is an example of what a sum-of-products expression looks like:

$$X = (\bar{A} \cdot B \cdot C) + (\bar{A} \cdot \bar{B} \cdot C) + (A \cdot B \cdot \bar{C}) \tag{6.34}$$

If we form a product for every row in the truth table that results in an output of 1 and take the sum of the products, we can represent the complete logic of the table. For rows whose output values are 1, we must ensure that the product representing that row is 1. In order to do this, any input whose value is 0 in the row must be inverted in the product. By expressing a product for every input combination whose value is 1, we have completely modeled the logic of the truth table because every other combination will result in a 0.

The second method is known as the **product-of-sums method.** It is based on the fact that we can represent an output as a product of sums containing combinations of the inputs. For example, if we have three inputs A, B, and C and an output X, the product of sums would be a Boolean expression containing input terms OR-ed together to form sum terms that are AND-ed together to define the output X as a Boolean product. The following equation is an example of what a product-of-sums expression looks like:

$$X = (\bar{A} + B + C) \cdot (\bar{A} + \bar{B} + C) \cdot (A + B + \bar{C}) \tag{6.35}$$

If we form a sum for every row in the truth table that results in an output of 0 and take the product of the sums, we can represent the complete logic of the table. For rows whose output values are 0, we must ensure that the sum representing that row is 0. In order to do this, any input whose value is 1 in the row must be inverted in the sum. By expressing a sum for every input combination (row) whose value is 0, we have completely modeled the logic of the truth table because every other combination will result in a 1.

EXAMPLE 6.4   Sum of Products and Product of Sums

In performing binary arithmetic, the simplest operation is summing the two least significant bits resulting in a sum bit and a carry bit. The four possible combinations for adding two bits are shown below.

		1		C
0	0	1	1	A
+0	+1	+0	+1	+B
0	1	1	0	S

The last column shows the terminology used. The two input bits are labeled A and B, the sum of the two bits is labeled S, and the carry bit, if there is one, is labeled C. Only in the last case (1 + 1) is the carry bit 1; otherwise it is 0.

The truth table for this operation is

A	B	S	C
0	0	0	0
0	1	1	0
1	0	1	0
1	0	0	1

We apply the sum-of-products and product-of-sums methods to both of the these outputs to illustrate how the methods differ.

Using the procedure in the paragraph after Equation 6.34, the sum-of-products method applied to output S yields

$$S = (\bar{A} \cdot B) + (A \cdot \bar{B})$$

The product (AND) terms represent rows two and three where S is 1.

Using the procedure in the paragraph after Equation 6.35, the product-of-sums method applied to output S yields

$$S = (A + B) \cdot (\bar{A} + \bar{B})$$

The sum (OR) terms represent rows one and four where S is 0.

The sum-of-products method applied to output C yields

$$C = (A \cdot B)$$

The product (AND) term represents row four where C is 1.

The product-of-sums method applied to output C yields

$$C = (A + B) \cdot (A + \bar{B}) \cdot (\bar{A} + B)$$

The sum (OR) terms represent rows one, two, and three, where C is 0.

Note that the sum-of-products method was easier to apply to output C because only a single row has an output of 1. You can verify all of the derived expressions by testing each with a truth table, comparing the results to the desired truth table (Question 6.29).

If we use the product-of-sums result for S and the sum-of-products result for C, we obtain a circuit using the fewest number of gates:

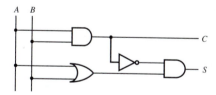

This circuit is known as a **half adder** because it applies only to the two least significant bits of a sum of two multiple-bit numbers. The higher-order bits require a lower-order carry bit as an additional input, and the circuit is called a **full adder** (see Question 6.31).

Draw the logic circuits for *S* and *C* in Example 6.4 using only the product-of-sums results and then do the same for the sum-of-products results. Compare your circuits to the one shown in the example. Also, show that the sum-of-products and product-of-sums results are equivalent.

Internet Link

6.3 Karnaugh
Mapping

An alternative to the sum-of-products and product-of-sums methods is Karnaugh mapping. This method results in a simplified Boolean expression through truth table manipulation. Internet Link 6.3 describes the method and provides examples.

6.8   SEQUENTIAL LOGIC

Combinational logic devices generate an output based on the input values, independent of the input timing. With **sequential logic** devices, however, the timing or sequencing of the input signals is important. Devices in this class include flip-flops, counters, monostables, latches, and more complex devices such as microprocessors. Sequential logic devices usually respond to inputs when a separate trigger signal transitions from one level to another. The trigger signal is usually referred to as the **clock** (CK) signal. The clock signal can be a periodic square wave or an aperiodic collection of pulses. Figure 6.5 illustrates edge terminology in relation to a clock pulse, where an arrow is used to indicate edges where state transitions occur. **Positive edge-triggered** devices respond to a low-to-high (0 to 1) transition, and **negative edge-triggered** devices respond to a high-to-low (1 to 0) transition. This topic is addressed again in Section 6.9.1 where it is applied to flip-flops.

6.9   FLIP-FLOPS

Because digital data is stored in the form of bits, digital memory devices such as computer random access memory (RAM) require a means for storing and switching between the two binary states. A **flip-flop** is a sequential logic device that can perform this function. The flip-flop is called a **bistable** device, because it has two and only two possible stable output states: 1 (high) and 0 (low). It has the capability to remain in a particular output state (i.e., storing a bit) until input signals cause it to change state. This is the basis of all semiconductor information storage and processing in digital computers; in fact, flip-flops perform many of the basic functions critical to the operation of almost all digital devices.

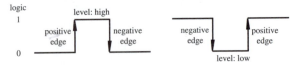

Figure 6.5 Clock pulse edges.

Figure 6.6 RS flip-flop.

A fundamental flip-flop, an **RS flip-flop,** is schematically shown in Figure 6.6. S is the **set** input, R is the **reset** input, and Q and $\overline{Q}$ are the **complementary outputs.** Most flip-flops include both outputs, where one output is the inverse (NOT) of the other. The RS flip-flop operates based on the following rules:

1. As long as the inputs S and R are both 0, the outputs of the flip-flop remain unchanged.
2. When S is 1 and R is 0, the flip-flop is *set* to $Q = 1$ and $\overline{Q} = 0$.
3. When S is 0 and R is 1, the flip-flop is *reset* to $Q = 0$ and $\overline{Q} = 1$.
4. It is "not allowed" (NA) to place a 1 on S and R simultaneously because the output will be unpredictable.

A truth table is a valuable tool for describing the functionality of a flip-flop. The truth table for a basic RS flip-flop is given in Table 6.4. The first row shows the memory state where the flip-flop retains the last value set or reset. Q_0 is the value of the output Q before the indicated input conditions were established; 1 is logic high and 0 is logic low. The NA in the last row indicates that the input condition for that row is not allowed. Because we are precluded from applying the $S = 1, R = 1$ input condition, the RS flip-flop is seldom used in actual designs. Other more versatile flip-flops that avoid the NA limitation are presented in subsequent sections.

To understand how flip-flops and other sequential logic circuits function, we will look at the internal design of an RS flip-flop illustrated in Figure 6.7a. It consists of combinational logic gates with internal feedback from the outputs to the inputs of the NAND gates. Figure 6.7b illustrates the timing of the various signals, which are affected by very short propagation delays through the NAND gates. Immediately after signal R transitions from 0 to 1, the inputs to the lower NAND gate are 0 and Q, which is still 1. This changes $\overline{Q}$ to 1 after a slight propagation delay Δt_1. Feedback of $\overline{Q}$ to the top NAND gate drives Q to 0 after a slight delay Δt_2. Now the flip-flop is

Table 6.4 Truth table for the RS flip-flop

Inputs		Outputs	
S	R	Q	$\overline{Q}$
0	0	Q_0	$\overline{Q}_0$
1	0	1	0
0	1	0	1
1	1	NA	

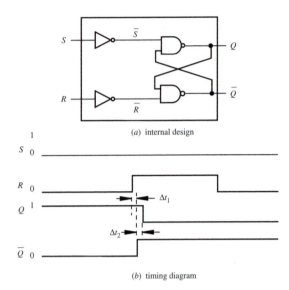

(a) internal design

(b) timing diagram

Figure 6.7 RS flip-flop internal design and timing.

reset, and it remains in this state even after R returns to 0. The set operation functions in a similar manner. The propagation delays Δt_1 and Δt_2 are usually in the nanosecond range. All sequential logic devices depend on feedback and propagation delays for their operation.

6.9.1 Triggering of Flip-Flops

Flip-flops are usually **clocked;** that is, a signal designated "clock" coordinates or synchronizes the changes of the output states of the device. This allows the design of complex circuits such as a microprocessor where all system changes are triggered by a common clock signal. This is called **synchronous** operation because changes in state are coordinated by the clock pulses. The outputs of different types of clocked flip-flops can change on either a positive edge or a negative edge of a clock pulse. These flip-flops are termed **edge-triggered** flip-flops. Positive edge triggering is indicated schematically by a small angle bracket on the clock input to the flip-flop (see Figure 6.8a). Negative edge triggering is indicated schematically by a small circle and angle bracket on the clock input (see Figure 6.8b).

The function of the edge-triggered RS flip-flop is defined by the following rules:

1. If S and R are both 0 when the clock edge is encountered, the output state remains unchanged.
2. If S is 1 and R is 0 when the clock edge is encountered, the flip-flop output is **set** to 1. If the output is at 1 already, there is no change.
3. If S is 0 and R is 1 when the clock edge is encountered, the flip-flop output is **reset** to 0. If the output is at 0 already, there is no change.
4. S and R should never both be 1 when the clock edge is encountered.

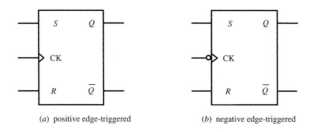

(a) positive edge-triggered   (b) negative edge-triggered

Figure 6.8 Edge-triggered RS flip-flops.

The truth table for a positive edge-triggered RS flip-flop (Figure 6.8a) is given in Table 6.5. The up-arrow ↑ in the clock (CK) column represents the positive edge transition from 0 to 1. The NA in the second to last row indicates that the input condition for that row is not allowed. As long as there is no positive edge transition, the values of S and R have no effect on the output as shown by the X symbols in the last row of the table. An example timing diagram is shown in Figure 6.9. The output is reset (Q = 0) at the first positive edge of the clock signal, where R = 1 and S = 0, and the output is set (Q = 1) at the second positive edge, where S = 1 and R = 0.

There are special devices that are not edge triggered in the way just described. An important example is called a **latch.** Its schematic symbol is shown in Figure 6.10. The output Q tracks the input D as long as CK is high. When a negative edge occurs (i.e., when CK goes low), the flip-flop will store (latch) the value that D had at the negative edge, and that value will be retained at the output. Because the output follows the input when the clock is high, we say the latch is **transparent** during this

Table 6.5 Positive edge-triggered RS flip-flop truth table

S	R	CK	Q	$\overline{Q}$
0	0	↑	Q_0	$\overline{Q_0}$
1	0	↑	1	0
0	1	↑	0	1
1	1	↑	NA	
X	X	0,1,↓	Q_0	$\overline{Q_0}$

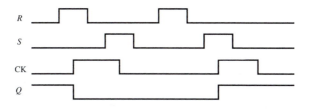

Figure 6.9 Positive edge-triggered RS flip-flop timing diagram.

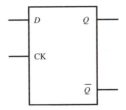

Figure 6.10 Latch.

Table 6.6 Latch truth table

D	CK	Q	$\overline{Q}$
0	1	0	1
0	1	1	0
X	0	Q_0	$\overline{Q_0}$

time. The latch can also be referred to as a positive-level-triggered device. The truth table for a latch is given in Table 6.6, and a timing diagram example is shown in Figure 6.11. Note how the output (Q) tracks the input (D) while the clock level is high (CK = 1). The X in the last row of the table indicates that the value of *D* has no effect on the output as long as CK is low.

6.9.2 Asynchronous Inputs

Flip-flops may have preset and clear functions that instantaneously override any other inputs. They are called **asynchronous inputs,** because their effect may be asserted at any time. They are not triggered by a clock signal. The **preset** input is used to set or initialize the output Q of the flip-flop to high (1). The **clear** input is used to clear or reset the output Q of the flip-flop to low (0). The small inversion symbols shown at the asynchronous inputs in Figure 6.12 are typical of most devices and imply that the function is asserted when the asynchronous input signal is low. Such an input is referred to as an **active low** input. Both preset and clear should not be asserted simultaneously. Either of these inputs can be used to define the state of a flip-flop after power-up; otherwise, at power-up the output of a flip-flop is uncertain.

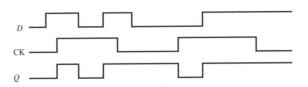

Figure 6.11 Latch timing diagram.

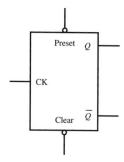

Figure 6.12 Preset and clear flip-flop functions.

6.9.3 D Flip-Flop

The **D flip-flop,** also called a data flip-flop, has a single input D whose value is stored and presented at the output Q at the edge of a clock pulse. A positive edge-triggered D flip-flop is illustrated in Figure 6.13, and its truth table is given in Table 6.7. Unlike a latch, a D flip-flop does not exhibit transparency. The output changes only when triggered by the appropriate clock edge (in this case, a positive edge).

Lab Exercise 7 explores latches and D flips-flops and shows how their functions differ. The exercise also deals with logic gates and shows how to wire switches for input into logic circuits.

Lab Exercise

Lab 7 Digital circuits—logic and latching

6.9.4 JK Flip-Flop

The **JK flip-flop** is similar to the RS flip-flop where the J is analogous to the S (set) input and the K is analogous to the R (reset) input. The major difference is that the J and K inputs may both be high simultaneously. This state causes the output to **toggle,** which means the output changes value (i.e., a 1 would become 0, and a 0 would become 1). The schematic representation and truth table for a negative edge-triggered JK flip-flop are shown in Figure 6.14 and Table 6.8. The first two rows of the table describe the preset or clear functions that can be used to initialize the output of the flip-flop. Recall that these features are active low and override the other inputs. The third row precludes presetting and clearing simultaneously. The symbol $\downarrow$ represents the negative edge of the clock signal, which causes the change in the

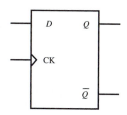

Figure 6.13 Positive edge-triggered D flip-flop.

Table 6.7 Positive edge-triggered D flip-flop truth table

D	CK	Q	$\overline{Q}$
0	↑	0	1
1	↑	1	0
X	0	Q_0	$\overline{Q_0}$
X	1	Q_0	$\overline{Q_0}$

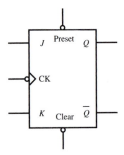

Figure 6.14 Negative edge-triggered JK flip-flop.

Table 6.8 Truth table for a negative edge-triggered JK flip-flop

Preset	Clear	CK	J	K	Q	$\overline{Q}$
0	1	X	X	X	1	0
1	0	X	X	X	0	1
0	0			NA		
1	1	↓	0	0	Q_0	$\overline{Q_0}$
1	1	↓	1	0	1	0
1	1	↓	0	1	0	1
1	1	↓	1	1	$\overline{Q_0}$	Q_0
1	1	0, 1	X	X	Q_0	$\overline{Q_0}$

output. The last row describes the memory feature of the flip-flop in the absence of a negative edge.

The JK flip-flop has a wide range of applications, and all flip-flops can easily be constructed from it with proper external wiring. The **T (toggle) flip-flop** serves as a good example of this. The symbol for a positive edge-triggered T flip-flop and the equivalent JK flip-flop implementation are shown in Figure 6.15. The T flip-flop simply toggles the output every time it is triggered. The preset and clear functions are necessary to provide direct control over the output because the T input alone provides no mechanism for initializing the output value. The truth table is given in Table 6.9.

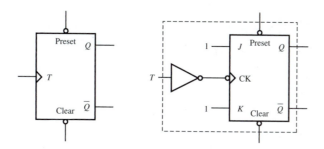

Figure 6.15 Positive edge-triggered T flip-flop.

Table 6.9 Positive edge-triggered T flip-flop truth table

T	$\overline{Preset}$	$\overline{Clear}$	Q	Q
↑	1	1	$\overline{Q}_0$	Q_0
0	1	1	Q_0	$\overline{Q}_0$
1	1	1	Q_0	$\overline{Q}_0$
X	0	1	1	0
X	1	0	0	1

Flip-Flop Circuit Timing Diagram   EXAMPLE 6.5

Given the following sequential logic circuit that includes RS, T, and JK flip-flops with the inputs as indicated in the timing diagram, the digital outputs D, E, and F will be as shown. The signals D, E, and F are assumed to be low at the beginning of the timing diagram. Observe how signal D updates (sets or resets) at positive edges of C, signal E updates (toggles) at positive edges of D, and signal F updates (sets, resets, or toggles) at negative edges of C.

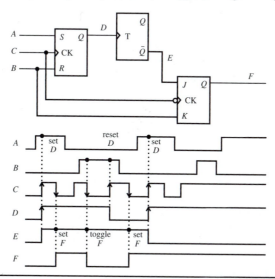

6.10 APPLICATIONS OF FLIP-FLOPS

We have just seen that there are a variety of flip-flops. In the subsections below, we illustrate a number of applications that use flip-flops as their functional units.

6.10.1 Switch Debouncing

When mechanical switches are opened or closed, there are brief current oscillations due to mechanical bouncing or electrical arcing. This phenomenon is called **switch bounce.** As illustrated in Figure 6.16, a single closing of a switch can result in multiple voltage transitions that usually occur within a few milliseconds. Video Demo 6.2 shows how switch bounce can be visualized on an oscilloscope. Video Demo 6.3 shows a very dramatic example of the arcing that occurs when a high-voltage switch is opened or closed. In this case, a power transmission line is being shut down by a large disconnect switch, and a very large arc forms in the process.

The sequential logic circuit shown in Figure 6.17 can provide an output that is free from multiple transitions associated with switch bounce. As the switch breaks contact with *B*, signal bounce occurs on the *B* line. There is a small delay as the switch moves from contact *B* to *A*, and then signal bounce occurs on the *A* line as contact is established with *A*. However, as a result of the feedback and logic, the

Video Demo

6.2 Switch bounce

6.3 High-voltage disconnect switch

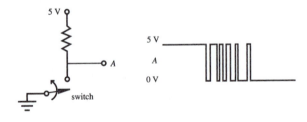

Figure 6.16 Switch bounce.

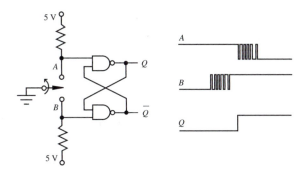

Figure 6.17 Switch debouncer circuit.

output signal (Q) experiences only a single transition from low to high (i.e., the output is bounce free). The circuit functions very much like a flip-flop (see Class Discussion Item 6.7).

■ **CLASS DISCUSSION ITEM 6.7**
Switch Debouncer Function

Track the inputs and outputs of the two NAND gates in Figure 6.17 as the switch moves from contact *B* to contact *A* and create a timing diagram. Also, draw an equivalent circuit for the debouncer using an RS flip-flop. Hint: Consider the gate switch delay as was done in Figure 6.7.

The switch shown in Figure 6.17 is called an SPDT switch, which is short for single-pole, double-throw (see Section 9.2.1 for more information). An SPDT switch has three leads. Section 6.12.2 shows a circuit that can be used to debounce SPST (single-pole, single-throw) switches, which only have two leads. Class Discussion Item 7.8 will also explore how microcontroller software can be used to debounce switch inputs directly. This is the most efficient solution if a design happens to include a microcontroller.

6.10.2 Data Register

Figure 6.18 shows a 4-bit **data register** that uses negative edge-triggered D flip-flops to transfer data from four data lines to the outputs of four AND gates. It does this in two distinct steps. First, the data values D_i are transferred to the outputs Q of the flip-flops on the negative edge of the load signal. Then a pulse on the read line presents the data at the register outputs R_i of the AND gates. Data registers are used in microprocessors to hold data for arithmetic calculations. Data registers can be cascaded to store as many bits as are required.

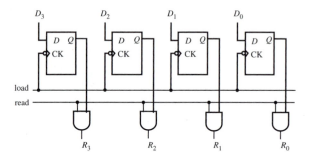

Figure 6.18 4-bit data register.

6.10.3   Binary Counter and Frequency Divider

Figure 6.19 shows a 4-bit **binary counter** consisting of four negative edge-triggered toggle flip-flops connected in sequence. The timing diagram is also shown for the first 10 input pulses. The four output bits B_i change according to the binary number counting sequence, counting from 0 ($B_3B_2B_1B_0 = 0000$), to 1 ($B_3B_2B_1B_0 = 0001$), etc., to 15 ($B_3B_2B_1B_0 = 1111$), and then returning back to 0. This circuit may also be used as a **frequency divider.** Output B_0 is a divide-by-2 output because its frequency is 1/2 the input pulse train frequency. B_1, B_2, and B_3 are divide-by -4, -8, and -16 outputs, respectively.

6.10.4   Serial and Parallel Interfaces

Figures 6.20 and 6.21 show flip-flop circuits that convert between serial and parallel data. **Serial data** consists of a sequence of bits, or train of pulses, that occurs on a single data line. **Parallel data** consist of a set of bits that occurs

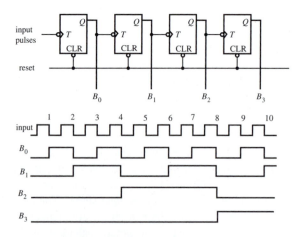

Figure 6.19 4-bit binary counter.

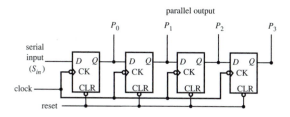

Figure 6.20 Serial-to-parallel converter.

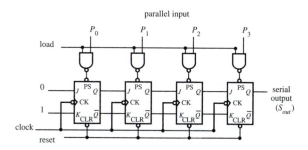

Figure 6.21 Parallel-to-serial converter.

simultaneously in parallel on a set of data lines. The serial-to-parallel converter utilizes negative edge-triggered D flip-flops, and the parallel-to-serial converter utilizes negative edge-triggered JK flipflops. In both circuits, the serial input or output is synchronized by a clock signal. A reset line is used to clear the flip-flops prior to loading a set of bits. The reset line is active low, meaning that, when the line goes low, the flip-flops are cleared, causing the outputs Q to go low (0). The load line for the parallel-to-serial converter passes the data through the NAND gates, storing the data line values in the flip-flops using the active low flip-flop preset. When the load line goes high and a parallel bit (P_i) is high (1), then the respective flip-flop is preset, resulting in a high (1) output Q. The figures show 4-bit converters, but the flip-flops can be cascaded for a larger number of bits.

■ **CLASS DISCUSSION ITEM 6.8**
Converting Between Serial and Parallel Data

Looking at the Figures 6.20 and 6.21, explain in detail the function of the circuits that convert between serial and parallel data. Also, how is the **baud rate** (bits per second) for serial transmission related to the converter's clock speed?

6.11 TTL AND CMOS INTEGRATED CIRCUITS

Now that we have discussed digital signals, Boolean algebra, the formulation of digital logic expressions, and logic devices, we are prepared to present the characteristics of the actual integrated circuits that perform the various digital functions. There are two families of logic devices called *TTL* and *CMOS*. TTL stands for **transistor-transistor logic** devices and CMOS for **complementary metal-oxide semiconductor** devices. In general, any combinational or sequential logic circuit can be constructed with either family or with a mix of the two families, but to do this correctly, we need to understand the differences in the electronic characteristics of each family.

In Chapter 3, we described bipolar junction transistors that are the building blocks for TTL logic, and MOSFETs that are the building blocks for CMOS logic. The two states of a digital device are defined by voltages occurring within specified acceptable ranges. The states and voltages for the two families are shown in Figure 6.22. For comparison purposes, we assume that both families are powered by a 5 V DC supply, although CMOS, unlike TTL, can be powered with a DC supply between 3 V and 18 V. For a TTL digital input, **logic zero** (0) or **low** (*L*) is defined as a value less than 0.8 V, and **logic one** (1) or **high** (*H*) is defined as a value greater

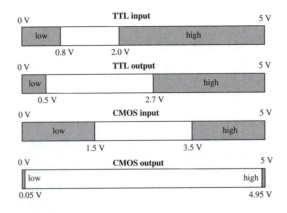

Figure 6.22 TTL and CMOS input and output levels.

than 2.0 V. The digital output of a TTL device typically ranges between 0 and 0.5 V for low and between 2.7 V and 5 V for high. The input voltage range 0.8 V to 2.0 V between the logic 0 and logic 1 states is a dead zone where the input state is undefined. For a CMOS digital input, logic zero (0) or low (L) is defined as a value less than 1.5 V, and logic high is defined as a value greater than 3.5 V. The digital output of a CMOS device typically ranges between 0 and 0.05 V for low and between 4.95 V and 5 V for high. The input voltage range 1.5 V to 3.5 V is a dead zone where the input state is undefined.

When interfacing digital devices, in addition to understanding the voltage levels, it is also important to know the input and output current characteristics of the devices. Important characteristics are the amount of current a device can **source** (produce) when the output is high and the amount of current the device can **sink** (draw) when the output voltage is low. In manufacturer data sheets for digital devices, these characteristics are usually labeled as I_{OH} or "high-level output current" for the sourcing capability and I_{OL} or "low-level output current" for the sinking capability. Example data sheets for TTL and CMOS devices are presented later in the section.

We now examine the equivalent output circuits for TTL and CMOS devices. Referring to Figure 6.23, TTL logic switches between states by forward biasing one of the two output transistors. This output circuit is called a **totem pole** configuration, where two bipolar junction transistors are stacked between power and ground. When the upper transistor is forward biased and the bottom transistor is off, the output is high. The resistor, transistor, and diode drop the actual output voltage to a value typically about 3.4 V. When the lower transistor is forward biased and the top transistor is off, the output is low. You can see that the TTL device sources current when there is a high output and sinks current when the output is low. The values of the sink and source currents depend on the TTL subfamily. When the output of a TTL device is connected to the input of another, the TTL device dissipates power continuously regardless of whether the output is high or low.

CMOS logic ICs employ complementary pairs of p-type and n-type enhancement-mode MOS transistors at their outputs, hence the name **complimentary MOS (CMOS).** Referring to the CMOS output circuit in Figure 6.23, if the input

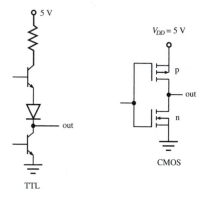

Figure 6.23 TTL and CMOS output circuits.

signal to this output stage is high, the p-type transistor (top) is off and the n-type transistor (bottom) is on, so the output is pulled low. When the input is low, the top transistor is on and the bottom transistor is off, so the output is pulled high. When the output is high, the device sources current; and when the output is low, the device sinks current if there is a load attached to the output. Because the MOSFET gates are insulated, CMOS devices consume power only when switching between states or when there is a load attached. Therefore, a major difference between CMOS and TTL is that TTL devices require power continuously (see Class Discussion Item 6.10).

CMOS is often recommended over TTL for the following reasons:

- When an output is unloaded or connected to other CMOS devices, CMOS requires power only when an output switches its logic state. Therefore, CMOS is useful in battery-operated applications where power is limited.

- The wide power supply range of CMOS (3–18 V) provides more design flexibility and allows use of less tightly regulated power supplies.

There are some disadvantages of CMOS:

- CMOS is sensitive to static discharge even with internal protective diodes. Protective packaging and static discharge during handling and assembly are necessary; otherwise, the devices are easily damaged.

- CMOS requires negligible input current, but its output current is also small compared to TTL. This limits the ability of CMOS to drive large TTL fan-out or other high current devices.

■ **CLASS DISCUSSION ITEM 6.10**
CMOS and TTL Power Consumption

Figure 6.23 shows the output circuits for TTL and CMOS devices. Study these circuits and convince yourself why TTL devices require power to maintain output levels when connected to other TTL devices and CMOS devices do not require power when connected to other CMOS devices.

6.11.1 Using Manufacturer IC Data Sheets

Manufacturers provide **data books** that contain **data sheets** for all the devices they manufacture. The data sheets contain all the information you might need to use the devices in your designs, including internal schematics, pin-out connections, maximum ratings, operating conditions, and electrical and switching characteristics. The labeling system used in TTL data books is usually in the form AAxxyzz, where AA is the manufacturer's prefix (SN for TI and others; DM for National Semiconductor); xx distinguishes between military (xx = 54) and industrial (xx = 74) quality; y distinguishes between different internal designs (no letter: standard TTL; L: low-power dissipation; H: high-power dissipation; S: Schottky type; AS: advanced Schottky,

LS: low-power Schottky; ALS: advanced low-power Schottky); and zz is the device number in the data book. Schottky devices have faster switching speeds and require less power. CMOS devices are available in the 40XXB series and the 74CXX series. The latter is pin compatible with the TTL 74XX series. There are also different varieties of the 74CXX family that provide different speed and power characteristics. They include the 74HCXX (high-speed CMOS), 74ACXX (advanced CMOS), and 74HCTXX and 74ACTXX (high-speed CMOS with TTL threshold).

Figures 6.24 through 6.26 illustrate some of the information available in a TTL data book. The 74LS00 QUAD NAND IC is the specific example shown here. Internet Link 6.4 points to the entire datasheet that contains a wealth of information, much of which you will find superfluous. Similar to other gates, the internal design is composed of transistors, resistors, and diodes, which are easily manufactured as an integrated circuit on a silicon chip. The IC is manufactured as a **dual in-line package (DIP)** or **surface mount package (SOP)** with the pin connections illustrated in Figure 6.25. This particular DIP has four NAND gates on a single silicon chip, hence the name QUAD NAND gate. As mentioned previously, two important parameters in Figure 6.26 are the current sourcing and sinking capacities. The TTL-LS NAND gate shows an output current sourcing limit (I_{OH}) of -0.4 mA and an output current sinking limit (I_{IH}) of 8 mA. The standard convention is to use a positive number for current entering the device. TTL devices can usually sink much more current than they can source.

Internet Link

6.4 74LS00 TTL QUAD NAND data sheet

■ **CLASS DISCUSSION ITEM 6.11**
NAND Magic

Even though the NAND gate circuit shown in Figure 6.24 is much more complicated than basic transistor circuits we've seen previously, convince yourself that it results in NAND logic.

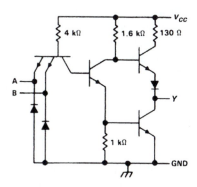

Figure 6.24 NAND gate internal design.
(Courtesy of Texas Instruments, Dallas, TX)

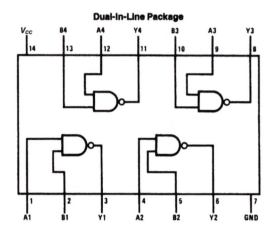

Figure 6.25 QUAD NAND gate IC pin-out.
(Courtesy of National Semiconductor, Santa Clara, CA)

Internet Link

6.5 4011B CMOS QUAD NAND data sheet

Figures 6.27 and 6.28 illustrate some of the information available in a CMOS databook. The 4011B QUAD NAND IC is the specific example shown. Internet Link 6.5 points to the entire data sheet. A CMOS output is composed of two complementary FETs, and the high output is at the supply voltage and the low output at ground. The positive supply voltage is denoted as V_{DD}, and the low side, usually ground, is denoted as V_{SS}. In Figure 6.28, the CMOS NAND operating with a supply voltage of 5 V can sink or source 1 mA.

6.11.2 Digital IC Output Configurations

Three different types of output circuits are used in TTL devices. The most common is the **totem pole** configuration, where two transistors are stacked between power and ground as shown in Figures 6.23 and 6.24. The second type of output circuit is known as an **open-collector output.** As shown in Figure 6.29, this configuration requires an external **pull-up resistor** connected to a power supply to produce the output states. When the output transistor is saturated (ON), V_{out} is low, and when it is in cutoff (OFF), V_{out} is high. Some devices that include this type of output include the 7401, 7403, 7405, and 7406. The third type of output circuit is known as the **tristate output,** where an additional input signal controls a third output state. When enabled, the third state produces a high-output impedance that effectively disconnects the output from any circuit to which it is attached. This permits attaching multiple devices to a single line where only one output would be enabled at a time (e.g., in the bus of a computer).

Some CMOS devices, instead of having a full CMOS output stage as shown in Figures 6.23 and 6.27, have an **open-drain** output, which is analogous to the TTL open-collector output.

Absolute Maximum Ratings (Note)

If Military/Aerospace specified devices are required, please contact the National Semiconductor Sales Office/Distributors for availability and specifications.

Supply Voltage	7V
Input Voltage	7V
Operating Free Air Temperature Range	
DM54LS and 54LS	$-55°C$ to $+125°C$
DM74LS	$0°C$ to $+70°C$
Storage Temperature Range	$-65°C$ to $+150°C$

Note: *The "Absolute Maximum Ratings" are those values beyond which the safety of the device cannot be guaranteed. The device should not be operated at these limits. The parametric values defined in the "Electrical Characteristics" table are not guaranteed at the absolute maximum ratings. The "Recommended Operating Conditions" table will define the conditions for actual device operation.*

Recommended Operating Conditions

Symbol	Parameter	DM54LS00			DM74LS00			Units
		Min	Nom	Max	Min	Nom	Max	
V_{CC}	Supply Voltage	4.5	5	5.5	4.75	5	5.25	V
V_{IH}	High Level Input Voltage	2			2			V
V_{IL}	Low Level Input Voltage			0.7			0.8	V
I_{OH}	High Level Output Current			-0.4			-0.4	mA
I_{OL}	Low Level Output Current			4			8	mA
T_A	Free Air Operating Temperature	-55		125	0		70	°C

Electrical Characteristics over recommended operating free air temperature range (unless otherwise noted)

Symbol	Parameter	Conditions		Min	Typ (Note 1)	Max	Units
V_I	Input Clamp Voltage	V_{CC} = Min, I_I = -18 mA				-1.5	V
V_{OH}	High Level Output Voltage	V_{CC} = Min, I_{OH} = Max, V_{IL} = Max	DM54	2.5	3.4		V
			DM74	2.7	3.4		
V_{OL}	Low Level Output Voltage	V_{CC} = Min, I_{OL} = Max, V_{IH} = Min	DM54		0.25	0.4	
			DM74		0.35	0.5	V
		I_{OL} = 4 mA, V_{CC} = Min	DM74		0.25	0.4	
I_I	Input Current @ Max Input Voltage	V_{CC} = Max, V_I = 7V				0.1	mA
I_{IH}	High Level Input Current	V_{CC} = Max, V_I = 2.7V				20	µA
I_{IL}	Low Level Input Current	V_{CC} = Max, V_I = 0.4V				-0.36	mA
I_{OS}	Short Circuit Output Current	V_{CC} = Max (Note 2)	DM54	-20		-100	mA
			DM74	-20		-100	
I_{CCH}	Supply Current with Outputs High	V_{CC} = Max			0.8	1.6	mA
I_{CCL}	Supply Current with Outputs Low	V_{CC} = Max			2.4	4.4	mA

Switching Characteristics at V_{CC} = 5V and T_A = 25°C (See Section 1 for Test Waveforms and Output Load)

Symbol	Parameter	R_L = 2 kΩ				Units
		C_L = 15 pF		C_L = 50 pF		
		Min	Max	Min	Max	
t_{PLH}	Propagation Delay Time Low to High Level Output	3	10	4	15	ns
t_{PHL}	Propagation Delay Time High to Low Level Output	3	10	4	15	ns

Figure 6.26 DM74LS00 NAND gate IC data sheet.
(Courtesy of National Semiconductor, Santa Clara, CA)

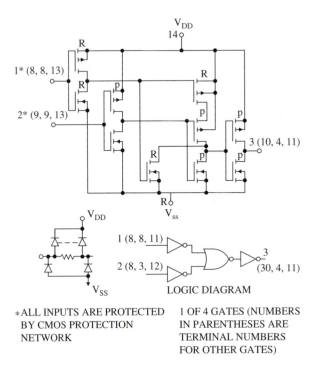

Figure 6.27 CMOS 4011B NAND gate internal design.
(Courtesy of Texas Instruments, Dallas, TX)

■ CLASS DISCUSSION ITEM 6.12
Driving an LED

There are three different ways to drive an LED with a TTL digital device, depending on whether the output type is totem pole or open collector and whether or not the device sinks or sources current through the LED. Sketch the three possible output circuits. Which of these three circuits exhibits positive logic (i.e., when the output of the device is HI, the LED is ON and when the output is LOW, the LED is OFF)? For each of the three circuits, indicate whether it is sourcing or sinking current. In which circuit would you expect the LED to be the brightest? Why?

6.11.3 Interfacing TTL and CMOS Devices

There may be situations where we need to connect devices from the same and/or different families. We consider connecting a TTL device to other TTL and CMOS devices and connecting a CMOS device to other CMOS and TTL devices. When designing digital systems, we recommend using only one family of devices (TTL or CMOS), but sometimes you may be required to interface between families.

STATIC ELECTRICAL CHARACTERISTICS

CHARACTER- ISTIC	CONDITIONS			LIMITS AT INDICATED TEMPERATURES (°C)							UNITS
	V_O (V)	V_{IN} (V)	V_{DD} (V)	−55	−40	+85	+125	+25			
								Min.	Typ.	Max.	
Quiescent Device Current, I_{DD} Max.	−	0.5	5	0.25	0.25	7.5	7.5	−	0.01	0.25	μA
	−	0.10	10	0.5	0.5	15	15	−	0.01	0.5	
	−	0.15	15	1	1	30	30	−	0.01	1	
	−	0.20	20	5	5	150	150	−	0.02	5	
Output Low (Sink) Current I_{OL} Min.	0.4	0.5	5	0.64	0.61	0.42	0.36	0.51	1	−	mA
	0.5	0.10	10	1.6	1.5	1.1	0.9	1.3	2.6	−	
	1.5	0.15	15	4.2	4	2.8	2.4	3.4	6.8	−	
Output High (Source) Current, I_{OH} Min.	4.6	0.5	5	−0.64	−0.61	−0.42	−0.36	−0.51	−1	−	
	2.5	0.5	5	−2	−1.8	−1.3	−1.15	−1.6	−3.2	−	
	9.5	0.10	10	−1.6	−1.5	−1.1	−0.9	−1.3	−2.6	−	
	13.5	0.15	15	−4.2	−4	−2.8	−2.4	−3.4	−6.8	−	
Output Voltage: Low-Level, V_{OL} Max.	−	0.5	5	0.05				−	0	0.05	V
	−	0.10	10	0.05				−	0	0.05	
	−	0.15	15	0.05				−	0	0.05	
Output Voltage: High-Level, V_{OH} Min.	−	0.5	5	4.95				4.95	5	−	
	−	0.10	10	9.95				9.95	10	−	
	−	0.15	15	14.95				14.95	15	−	
Input Low Voltage, V_{IL} Max.	4.5	−	5	1.5				−	−	1.5	V
	9	−	10	3				−	−	3	
	13.5	−	15	4				−	−	4	
Input High Voltage, V_{IH} Min.	0.5, 4.5	−	5	3.5				3.5	−	−	
	1.9	−	10	7				7	−	−	
	1.5, 13.5	−	15	11				11	−	−	
Input Current I_{IN} Max.		0.18	18	±0.1	±0.1	±1	±1	−	$\pm 10^{-5}$	±0.1	μA

Figure 6.28 CMOS 4011B NAND gate IC data sheet. *(Courtesy of Texas Instruments, Dallas, TX)*

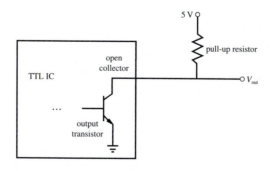

Figure 6.29 Open-collector output with pull-up resistor.

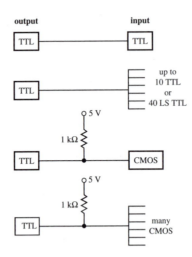

Figure 6.30 Interfacing TTL to digital devices.

Figure 6.30 shows how to interface TTL to different combinations of digital ICs. The output of a TTL device sinks current when it is low and sources current when it is high. The TTL low sink current (I_{OL}) is the limiting factor when interfacing to multiple TTL inputs. A TTL output can drive up to 10 standard TTL inputs or up to 40 Low-power Schottky (LS) TTL inputs. TTL outputs are easy to interface to CMOS due to the insulating gate input, which draws no steady state current. It is necessary only to ensure voltages match when connecting TTL outputs to CMOS inputs. Referring back to Figure 6.22, a TTL low output will match a CMOS low input just fine. However, a TTL high may be as low as 2.7 V, which is not enough to match the 3.5 V input required by CMOS. As shown in Figure 6.30, using a pull-up resistor on the TTL output will raise the output voltage above the 3.5 V required by a CMOS input. The pull-up resistor must be large enough (e.g., 1 kΩ) that the TTL low sink current (I_{OL}) for the device is not exceeded. If power consumption is a concern, the pull-up resistor can be even larger.

Figure 6.31 shows how to interface CMOS to different combinations of digital ICs. A CMOS device neither sources nor sinks current at its input due to the insulating gate at the input. Therefore, CMOS can interface to multiple CMOS inputs. If CMOS is powered by a 5 V supply voltage (V_{DD}) and is used to drive TTL devices, a CMOS high is no problem because the CMOS device can source enough current to drive the TTL input. However, a CMOS low can sink only enough current to drive one LS TTL input. A CMOS 4049 buffer can be used to provide adequate fan-out for up to two standard TTL inputs or approximately 10 LS TTL inputs.

Section 7.8 provides additional information on how to interface TTL and CMOS inputs and outputs to various types of devices. Although the material is specific to the PIC microcontroller, it applies to any TTL and CMOS device. Included are interfaces to sensors, switches, keyboards, Schmitt triggers, power transistors, and relays.

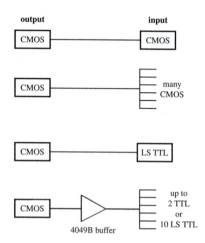

Figure 6.31 Interfacing CMOS to digital devices.

In summary, when using ICs of one logic family exclusively, you need not be concerned with voltage levels and current drives as long as the fan-out is less than 10 for TTL (CMOS can be higher). CMOS is better for general use because it draws no current unless switching, and the output swings nearly from ground to the positive supply value. However, at high frequency, CMOS can dissipate nearly the power required by an equivalent TTL circuit.

6.12 SPECIAL PURPOSE DIGITAL INTEGRATED CIRCUITS

This section presents some important special purpose ICs useful for digital design. They include the decade counter, the Schmitt trigger, and the 555 timer ICs.

6.12.1 Decade Counter

Section 6.10.3 presented a flip-flop circuit that can be used to perform binary counting. Another common counter, called a **decade counter,** can be constructed using a 7490 IC (Question 6.49). It is a negative edge-triggered counter, and the output is in binary coded decimal (BCD) consisting of 4 bits, making it useful for decimal counting applications. Table 6.10 shows the output sequence for the 4 bits as the counter increments from 0 to 9. The timing diagram for the input and four outputs of the decade counter are shown in Figure 6.32. Note that the counter output cycles back to 0000 (0_{10}) after 1001 (9_{10}).

As illustrated in Figure 6.33, BCD counters can be cascaded in order to count in powers of 10. Output D can be used as the clock input for a second 7490, thus cascading the two together to raise the range for counting from 0 to 99. Further cascading allows counting of higher powers of 10. The 7490 is a versatile IC and can

Table 6.10 7490 decade counter BCD coding

Decimal Count	BCD Output			
	D	C	B	A
0	0	0	0	0
1	0	0	0	1
2	0	0	1	0
3	0	0	1	1
4	0	1	0	0
5	0	1	0	1
6	0	1	1	0
7	0	1	1	1
8	1	0	0	0
9	1	0	0	1
0	0	0	0	0

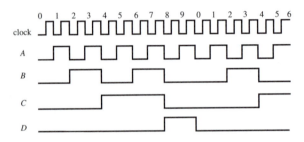

Figure 6.32 Decade counter timing.

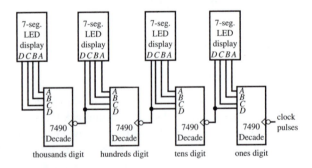

Figure 6.33 Cascaded decade counters.

Internet Link

6.6 7490 decade counter data sheet

be wired in a variety of useful ways. Examples include binary counters, divide-by-2 counters, and divide-by-4 counters. Internet Link 6.6 points to the entire data sheet for the 7490. The data sheet provides instructions for wiring the IC for the various configurations it supports.

A convenient device for viewing a BCD output is a seven-segment LED display (see Figure 6.34) driven by a 7447 BCD-to-seven-segment decoder. Figure 6.35 shows a two-segment LED display that can display numbers ranging from 0 to 99. Also shown in the photograph is the 7447 display decoder IC. The 7447 converts the 4 BCD bits at its inputs into a 7-bit code to drive the LED segments properly. The function table describing the input (BCD) to output (negative logic seven-segment LED code) relationship for the 7447 is shown in Table 6.11. Verify some of the rows in this table by drawing the digits with the appropriate segments. Figure 6.36 illustrates the internal design of the 7447, which consists of combinational logic circuits. Internet Link 6.7 points to the full data sheet for the 7447.

Internet Link

6.7 7447 seven-segment display decoder data sheet

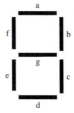

Figure 6.34 Seven-segment LED display.

2-digit LED display

7447 display decoder

Figure 6.35 Two-digit LED display and a 7447 display decoder.

Table 6.11 7447 BCD to seven-segment decoder

Decimal digit	Input				Output						
	D	C	B	A	$\bar{a}$	$\bar{b}$	$\bar{c}$	$\bar{d}$	$\bar{e}$	$\bar{f}$	$\bar{g}$
0	0	0	0	0	0	0	0	0	0	0	1
1	0	0	0	1	1	0	0	1	1	1	1
2	0	0	1	0	0	0	1	0	0	1	0
3	0	0	1	1	0	0	0	0	1	1	0
4	0	1	0	0	1	0	0	1	1	0	0
5	0	1	0	1	0	1	0	0	1	0	0
6	0	1	1	0	1	1	0	0	0	0	0
7	0	1	1	1	0	0	0	1	1	1	1
8	1	0	0	0	0	0	0	0	0	0	0
9	1	0	0	1	0	0	0	0	1	0	0

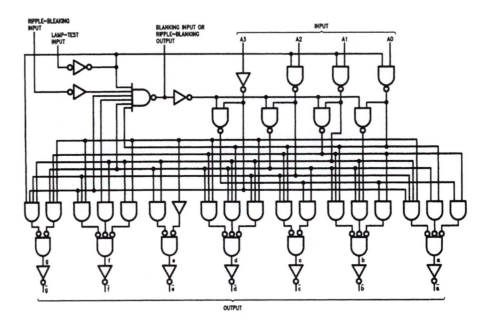

Figure 6.36 7447 internal design. *(Courtesy of National Semiconductor, Santa Clara, CA)*

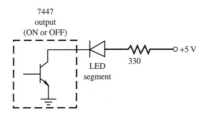

Figure 6.37 7447 output circuit.

Video Demo

6.4 Decade
counter and
display circuit
driven by a 555
timer

If the 7447 decoder driver is properly connected to a seven-segment LED display, the output from a counter can be displayed in numerical digit form. Note that
the decoder driver does not actually drive the segment LEDs by supplying current
to them; rather, it sinks current from them as illustrated in Figure 6.37. The outputs
of the 7447 are open-collector outputs. The resistor and LED segment complete the
circuit for each 7447 output. Therefore, the LED is ON (i.e., illuminated) when the
7447 output is low (0), allowing current to flow to ground. 330 Ω resistors are used
in series with the segments to limit the current into the decoder driver, thus preventing damage to the segments.

Lab Exercise 8 involves building a counter circuit with an LED digit display.
Video Demo 6.4 shows the final functioning circuit.

■ **CLASS DISCUSSION ITEM 6.13**
Up-Down Counters

Using a TTL logic handbook or an online search, look up the digital IC known as an **up-down counter.** Note that the device can be used as a binary, hexadecimal, or decimal counter. Discuss the purpose of counting both up and down and give examples of devices that might use the different configurations.

Lab Exercise

Lab 8 Digital circuits—counter and LED display

6.12.2 Schmitt Trigger

In some applications, digital pulses may not exhibit sharp edges; instead, the signal may ramp from 0 to 5 V over a finite time period, and it may do so in a "noisy" (jumpy) fashion as illustrated in Figure 6.38. The Schmitt trigger is a device that can convert such a signal into a sharp pulse using the threshold hysteresis effect illustrated in the figure. The output goes high when the input exceeds the high threshold and remains high until the input falls below the low threshold. The hysteresis between the low and high thresholds results in the distinct edges in the output. Six Schmitt triggers are usually packaged on a single IC (e.g., the 7414 Hex Schmitt Trigger Inverter).

Section 6.10.1 presented a circuit that can be used to debounce an SPDT switch, which has three leads. That circuit cannot be used to debounce signals from an SPST switch, which only has two leads. Common pushbutton switches are normally of this variety (see Section 9.2.1 for more information). Figure 6.39 shows a Schmitt trigger *RC* circuit that can be used to debounce an SPST switch. As the inversion circle on the output of the 7414 indicates, the 7414 is a Schmitt trigger inverter, meaning the output will be the inverse of the input. With the switch open, as shown, the capacitor is fully charged to 5 V, and the output of the 7414 is low. When the switch is closed, the capacitor is shorted to ground, and the 7414's output goes high. Depending on the capacitor value, the capacitor voltage might discharge below the low threshold of the 7414 on first contact. If not, the capacitor will eventually discharge as any bouncing occurs and continual contact

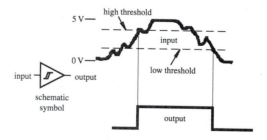

Figure 6.38 Input and output of a Schmitt trigger.

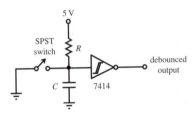

Figure 6.39 Schmitt trigger SPST debounce circuit.

is established. When the switch is opened, the capacitor again begins to charge. Any switch bounce occurring when the switch is opened just momentarily interrupts the charging. When the charge on the capacitor exceeds the high threshold of the 7414, the output again goes low. The result is a clean, bounce-free, output pulse.

6.12.3   555 Timer

The 555 integrated circuit is known as the "time machine" because it performs a wide variety of timing tasks. It is a combination of digital and analog circuits. The block diagram for the 555, including pin terminology and numbering, is shown in Figure 6.40. Two packages and pin configurations for the 555 are shown in Figure 6.41. Manufacturers usually list 555 ICs (e.g., TI's NE555) in the special function section of their *Linear* data books. Internet Link 6.8 points to the complete data sheet for the 555 IC. Applications for the 555 include bounce-free switches, cascaded timers, frequency dividers, voltage-controlled oscillators, pulse generators, LED flashers, and many other useful circuits.

A circuit easily built with a 555 IC is the **monostable multivibrator** shown in Figure 6.42. It is constructed by adding an external capacitor and resistor to a 555.

Internet Link

6.8 NE555 timer data sheet

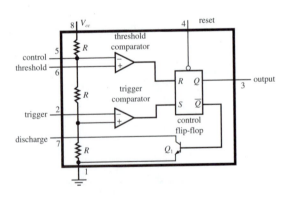

Figure 6.40 Block diagram of the 555 IC.

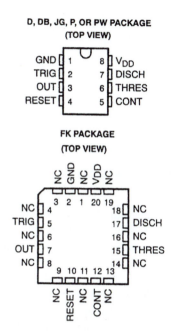

D, DB, JG, P, OR PW PACKAGE
(TOP VIEW)

FK PACKAGE
(TOP VIEW)

NC – No internal connection

Figure 6.41 555 pin-out. *(Courtesy of Texas Instruments, Dallas, TX)*

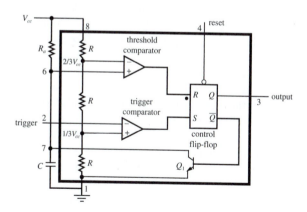

Figure 6.42 Monostable multivibrator (one-shot).

The circuit generates a single pulse of desired duration when it receives a trigger signal, hence it is also called a **one-shot.** The time constant of the resistor-capacitor combination determines the length of the pulse. In Figure 6.43, the sequence of operation is as follows: When the circuit is powered up or after the reset is activated,

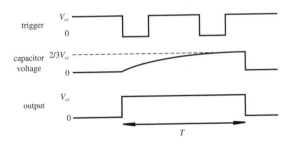

Figure 6.43 One-shot timing.

the output will be low ($Q = 0$), transistor Q_1 is saturated (ON) shorting capacitor C, and the outputs of both comparators are low. When the trigger pulse goes below $(1/3)V_{cc}$, the trigger comparator goes high, setting the flip-flop. Now the output is high ($Q = 1$), transistor Q_1 cuts off, and the capacitor begins to charge with time constant $\tau = R_a C$. The time constant is the time required for the capacitor voltage to reach 63.2% of its full charge value V_{cc} (see Section 4.9). When the capacitor voltage reaches $(2/3)V_{cc}$, the threshold comparator resets the flip-flop, which resets the output low ($Q = 0$) and discharges the capacitor again. If the trigger is pulsed while the output is high, it has no effect. The length of the pulse is given approximately (Question 6.52) by

$$\Delta T \approx 1.1 R_a C \qquad (6.36)$$

As alternatives to the 555 one-shot circuit, the TTL logic family also includes devices that can be used as one-shots. Examples are the 74121 and the 74123. They also require an external resistor and capacitor to control the pulse width.

Another important circuit that can be constructed with a 555 timer is an **astable multivibrator,** also called a square-wave generator or an oscillator. The schematic for this circuit is shown in Figure 6.44. When the circuit is powered, the capacitor C charges through the series resistors R_1 and R_2 with a time constant $(R_1 + R_2)C$. When the voltage on C reaches $(2/3)V_{cc}$, the flip-flop resets, turning on the discharge transistor that discharges the capacitor through resistor R_2 with time constant $R_2 C$. When the voltage on C drops to $(1/3)V_{cc}$, the flip-flop sets causing the cycle to repeat. The result is a signal consisting of repeating and regular pulses.

Figure 6.45 shows how the capacitor voltage (V_C) and the output signal (Q) vary over time as the astable multivibrator cycle repeats. From Section 4.9, an RC circuit, which is a first-order system with time constant $\tau = RC$, charges in response to a step input V_{cc} as follows:

$$V_C(t) = V_{cc}\left(1 - e^{-\frac{t}{RC}}\right) \qquad (6.37)$$

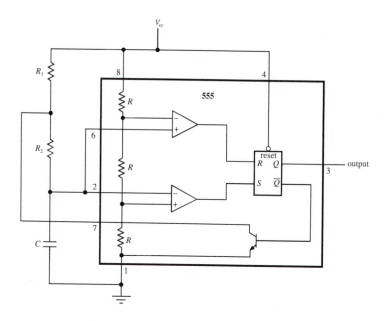

Figure 6.44 Astable multivibrator.

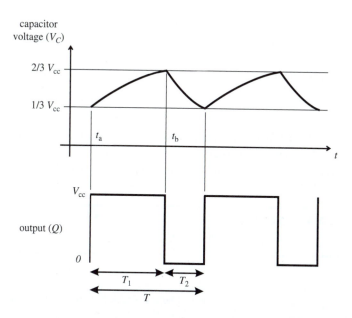

Figure 6.45 Astable multivibrator capacitor voltage and output signal.

The time it would take the capacitor to charge from 0 V to $2/3V_{cc}$, through series resistors R_1 and R_2, can be found from:

$$\frac{2}{3}V_{cc} = V_{cc}\left(1 - e^{\frac{t_b}{(R_1 + R_2)C}}\right) \tag{6.38}$$

Solving this equation gives:

$$t_b = -(R_1 + R_2)C\ln\left(\frac{1}{3}\right) \tag{6.39}$$

Similarly, the time to charge from 0 V to $1/3V_{cc}$ would be:

$$t_a = -(R_1 + R_2)C\ln\left(\frac{2}{3}\right) \tag{6.40}$$

Therefore, the period of charging through series resistors R_1 and R_2 from $1/3V_{cc}$ to $2/3V_{cc}$, during which time the output is high, is:

$$T_1 = t_b - t_a = (R_1 + R_2)\,C\ln(2) \tag{6.41}$$

Similarly, the period of discharge through resistor R_2 from $2/3V_{cc}$ to $1/3V_{cc}$, during which time the output is low, can be found to be (Question 6.54):

$$T_2 = R_2C\ln(2) \tag{6.42}$$

Therefore the total period for the pulse cycle is:

$$T = T_1 + T_2 = \ln(2)(R_1 + 2R_2)C \approx 0.693\,(R_1 + 2R_2)C \tag{6.43}$$

and the cycle frequency is:

$$f = \frac{1}{T} = \frac{1}{\ln(2)(R_1 + 2R_2)C} \approx \frac{1.443}{(R_1 + 2R_2)C} \tag{6.44}$$

To obtain a near-symmetrical square wave, the ON and OFF times for the pulse widths, T_1 and T_2, need to be as close to equal as possible. This can be achieved by making resistance value R_2 much larger than resistance value R_1 (see Class Discussion Item 6.14).

■ CLASS DISCUSSION ITEM 6.14
Astable Square-Wave Generator

How could you attempt to produce a perfectly symmetric square wave with the circuit in Figure 6.44? Are there any potential problems with this approach? Also, if the capacitor has a partial charge on it before power is applied to the circuit, what effect would this have on the circuit output?

6.13 INTEGRATED CIRCUIT SYSTEM DESIGN

For most digital design applications, integrated circuits can be used as building blocks to create the desired functionality. A myriad of ICs on the market provide almost every conceivable digital function. Listings of ICs along with data sheets describing their functionality can be found online or in manufacturer TTL or logic data books. The digital tachometer example in Design Example 6.1 illustrates a digital design solution employing several commercially available ICs. Internet Link 6.9 is an excellent resource providing a review of digital electronics fundamentals, logic devices, circuit analysis and design, and application examples.

Internet Link

6.9 All about circuits — Vol. IV — digital

Digital Tachometer

DESIGN EXAMPLE 6.1

Our objective is to design a system to measure and display the rotational speed of a shaft. A simple method to measure rotational speed is to count the number of shaft rotations during a given period of time. The resulting count will be directly proportional to the shaft speed.

A number of different sensors can detect shaft rotation. A proximity sensor that uses magnetic, optical, or mechanical principles to detect some feature on the shaft is one example. We can use an LED-phototransistor pair as an optical sensor and place a small piece of reflective tape on the shaft. Each time the tape passes by the photo-optic pair, the 7414 Schmitt trigger inverter provides a single pulse to a counting circuit. The following figure illustrates the sensor and signal conditioning circuit.

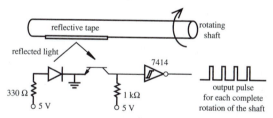

Now we need a circuit to count and display the pulses over a given interval of time T. The following figure illustrates all the required components. The 7490 decade counter counts the pulses and is reset by a negative edge on signal R after the time period T. The period T is set by a resistor-capacitor combination using a 555 oscillator circuit. If the count can exceed nine during the period T, additional 7490s must be cascaded to provide the full count. Just prior to counter reset, the output is stored by 7475 data latches that are enabled by a brief pulse on signal L. The latches are necessary to hold the previous count for display while the counter begins a new count cycle. One of the two 74123 one-shots is positive edge-triggered by the clock signal CK to generate a latch pulse L of length Δt. Note that the latch and reset pulse widths must be small ($\Delta t \ll T$) to maintain count accuracy (see Class Discussion Item 6.15). The trailing edge of the latch pulse triggers the second one-shot, which is negative edge-triggered, to produce a delayed reset pulse R for the counter. The 7447 LED decoder and driver converts

(continued)

(continued) the latched BCD count into the seven signals required to drive the LED display. The display reports the number of pulses that have occurred during the counting period T.

The shaft speed in revolutions per minute is related to the displayed pulse count by

$$\text{rpm} = \frac{\text{pulse count/ppr}}{T} \, 60$$

where ppr is the number of pulses per revolution generated by the sensor (e.g., 1 for a single piece of reflective tape).

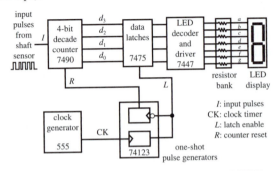

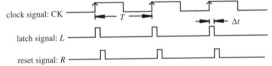

■ CLASS DISCUSSION ITEM 6.15
Digital Tachometer Accuracy

What effect does the choice of Δt have on the accuracy of the counts displayed by the digital tachometer presented in Design Example 6.1?

■ CLASS DISCUSSION ITEM 6.16
Digital Tachometer Latch Timing

One potential problem with the digital tachometer circuit in Design Example 6.1 could occur if latching happens at the exact instant when the counter is incrementing in response to an input pulse. Explain why. This problem can be resolved by blocking the input pulses during the latch with a logic gate. What would you need to add to the circuit schematic to accomplish this?

■ **CLASS DISCUSSION ITEM 6.17**
Using Storage and Bypass Capacitors in Digital Design

It is standard practice to include one or more large tantalum or electrolytic **storage capacitors** (e.g., 100 μf) across the power supply lines within a digital system and to include a small ceramic **bypass** or **decoupling capacitor** (e.g., 0.1 μf) between the V_{cc} and ground leads supplying power to each IC. Why?

Digital Control of Power to a Load Using Specialized ICs

DESIGN EXAMPLE 6.2

In Design Example 3.4, we introduced a discrete component power driver for mechatronic system peripheral devices. The large semiconductor manufacturers provide specialized integrated circuits that vastly aid the mechatronic designer in many applications. Example ICs include relay drivers, lamp drivers, motor drivers, and solenoid drivers. In this chapter, we learned how to design logic circuits to provide digital control outputs. These outputs can then be interfaced to peripheral power devices to control mechanical power. Peripheral power requirements are quite varied, creating a need for interface ICs that have a large degree of adaptability. Specialized interface ICs offer some benefits that are difficult for the designer to include using discrete components. Example features include short circuit protection at outputs, glitch-free power-up, inductive flyback protection, and negative transient protection.

Consider a situation where digital outputs are provided via an 8-bit bus that could be used, for example, to control eight separate devices. A good choice for a peripheral driver is the National Semiconductor DP7311 octal latched peripheral driver. This device can provide up to 100 mA DC at each output at a maximum operating voltage of 30 V to drive LEDs, motors, sensors, solenoids, or relays. It has open-collector outputs that require external pull-up resistors when interfaced to high-current external discrete drivers capable of controlling much higher currents than the IC itself. Let us consider two possible configurations.

The first configuration shows a typical output (one of eight) for a high-side (load between supply and collector), large-current bipolar power transistor driver. The *data in* line is active high, and the clear (CLR) and strobe (STR) lines are active low. The latches are cleared (the outputs are turned off) when CLR goes low, and the *data in* values are latched when STR goes low.

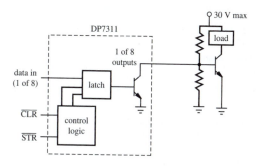

(continued)

(continued)　　　　An alternative high-current n-channel MOSFET design is shown next:

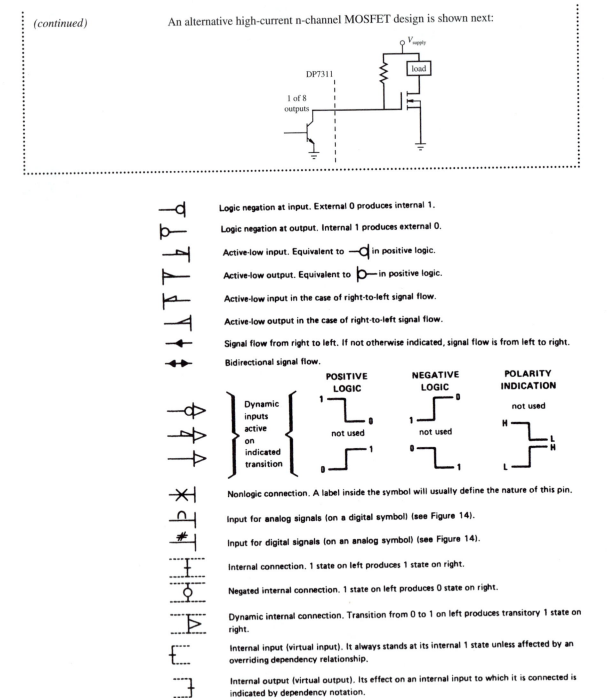

Figure 6.46 IEEE standard symbols for digital ICs. *(Courtesy of Texas Instruments, Dallas, TX)*

6.13.1 IEEE Standard Digital Symbols

The IEEE standard symbols for illustrating digital input and output conditions are shown in Figure 6.46. When drawing or reading digital schematics, it is important to recognize these conventions. You will find these symbols used in some manufacturers' TTL data books.

QUESTIONS AND EXERCISES

Section 6.2   Digital Representations

6.1. In computers, integers are sometimes represented by 16 bits. What is the largest positive base 10 integer that can be represented by a 16 bit binary number?

6.2. Convert each of the following base 10 integers to their equivalent binary representations. Show your work.

   a. 128

   b. 127

6.3. Convert each of the following base 10 integers to their equivalent hexadecimal representations. Show your work.

   a. 128

   b. 127

6.4. Perform each of the following binary arithmetic operations and check your results with decimal equivalents. Show your work.

   a. $1101 + 1001$

   b. $1101 - 1001$

   c. 1101×1001

   d. $111 + 111$

   e. 111×111

Section 6.3   Combinational Logic and Logic Classes

6.5. Draw the schematics of logic circuits that produce the following logic expressions:

   a. $\overline{A} + \overline{B}$

   b. $\overline{A} \cdot \overline{B}$

6.6. Create an inverter using a single NAND gate. Draw the schematic.

6.7. Write out a simplified Boolean expression and construct a truth table for each of the following circuits. Assume 0 = low = 0 V and 1 = high = 5 V.

(a)

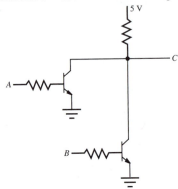

Hint: Try to write out a quasi-logic statements (see Sections 6.6.1 and 6.6.2) or a truth table first, based on how the transistors function. Assume ideal transistors operating either in cutoff or saturation.

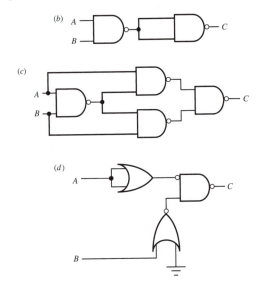

Section 6.4 Timing Diagrams

6.8. Construct a timing diagram showing the complete functionality of a NOR gate.

6.9. Construct a timing diagram showing the complete functionality of an XOR gate.

6.10. Construct a timing diagram showing the results of the truth table in Example 6.2.

Section 6.5 Boolean Algebra

6.11. Find the result of each of the following expressions:

a. $\overline{1 \cdot \overline{0}} + 1 \cdot (0 + 1) + \overline{\overline{0}} \cdot (1 + \overline{0})$

b. $A \cdot \overline{B} + A \cdot (A + B)$

6.12. Determine a simplified Boolean expression for X in the combinational logic circuit that follows. Also, complete the timing diagram.

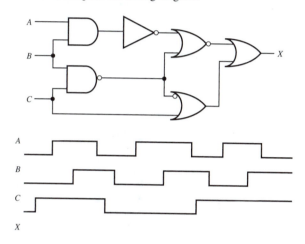

6.13. Prove Equation 6.20 using the basic Boolean algebra laws.

6.14. Prove Equation 6.21 using the basic Boolean algebra laws.

6.15. Verify that Equation 6.22 is correct using a truth table.

6.16. Prove Equation 6.23 using a truth table.

6.17. Prove that the following Boolean identity is correct:

$$AB + AC + \overline{B}C = AB + \overline{B}C$$

6.18. Prove whether the following Boolean equations are valid or not:

a. $(A \cdot B) + (B \cdot C) + (\overline{B} \cdot C) = (A \cdot B) + \overline{C}$

b. $A \cdot B \cdot C = \overline{\overline{A} + \overline{B} + \overline{C}}$

c. $(A \cdot B) + (B \cdot C) + (\overline{B} \cdot C) = (A \cdot B) + C$

6.19. Use a truth table to prove the validity of De Morgan's laws (Equations 6.25 and 6.26) for two signals (A and B).

6.20. The following circuit is called a **multiplexer.** Construct a truth table and write out a Boolean expression for X. Also, explain why the circuit is termed a multiplexer.

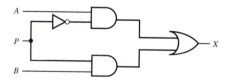

6.21. For the security system simplified Boolean expression derived in Section 6.6.3, if the input state $CD = 11$ resulted from a malfunction, explain how the alarm system would respond in different situations?

Section 6.6 Design of Logic Networks

6.22. Create an all-AND and an all-OR representations for the simplified Boolean expression in Example 6.3. Given that AND and OR gates are available on ICs in groups of four per IC and inverters are available on ICs in groups of six per IC, how many ICs would be required to implement the original and each of the two alternative representations?

6.23. Create an all-OR realization for Equation 6.30 and draw the resulting circuit.

6.24. Design and draw a logic circuit that will drive segment c of a seven-segment LED display given the 4-bit BCD input DCBA representing decimal numbers from 0 to 9. Note that the logic circuit will be used in the driver circuit that follows.

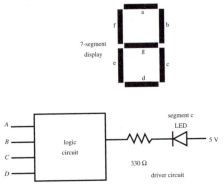

6.25. Draw a logic circuit for the following Boolean expression using NAND gates and inverters only:

$$X = \overline{A} \cdot B \cdot C + (A + B) \cdot \overline{C}$$

The NAND gates may only have two inputs each.

6.26. Determine a simplified Boolean expression for the following circuit and draw an equivalent circuit using NOR gates and inverters only:

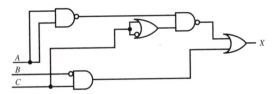

The NOR gates may only have two inputs each.

6.27. Design a logic circuit for a simple automobile door and seat belt buzzer system. Assume that sensors are available to provide digital signals representing the door and seat belt states. Signal A is from the door, where high implies the door is closed; signal B is from the seat belt, where high implies the seat belt is fastened; and signal C indicates whether the ignition switch is turned on. Your circuit should output a signal X that can be used to turn a buzzer on or off, where high implies on. The buzzer should be on if the ignition is on and if the door is open or the seat belt is not fastened. Draw your logic circuit and construct a complete truth table.

Section 6.7   Finding a Boolean Expression Given a Truth Table

6.28. Derive the simplified Boolean expression for X in Question 6.20 using the sum-of-products method.

6.29. Verify the four expressions for S and C (two each) from Example 6.4 by testing each with a truth table.

6.30. Document a complete and thorough answer to Class Discussion Item 6.4.

6.31. Design and draw a **full-adder** circuit that has two sum bits A_i and B_i and a lower-order carry bit C_{i-1} as inputs (see Example 6.4). Include a complete truth table and include sum-of-products Boolean expressions for the output sum bit S_i and carry bit C_i.

Section 6.9   Flip-Flops

6.32. Answer CDI 6.5, demonstrating the results of every row in Table 6.8.

6.33. Construct a complete truth table for a negative edge-triggered T flip-flop including preset and clear inputs.

6.34. Construct a complete truth table and timing diagram for a negative edge-triggered D flip-flop including preset and clear inputs.

6.35. Complete the timing diagram for the following circuit.

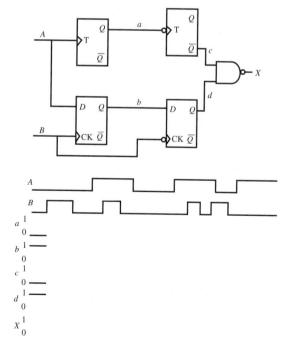

6.36. Complete the timing diagram (see below) for the following circuit.

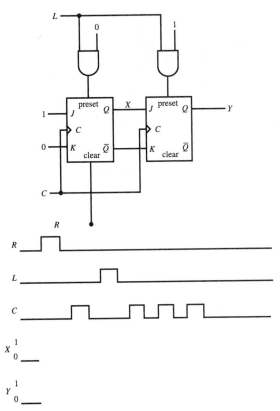

6.37. Complete the timing diagram (see below) for the following circuit.

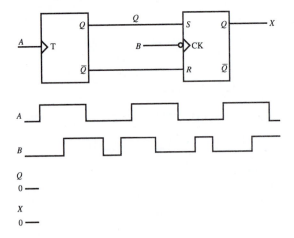

Section 6.10 Applications of Flip-Flops

6.38. Document a complete and thorough answer to Class Discussion Item 6.7.

6.39. Design a circuit that will store data from four binary sensors (presenting either a high or low value) when an enable pulse from a digital source first goes high. Assume that the enable input is normally low and the pulse is high for 1 sec.

6.40. Draw a timing diagram for all of the signals in Figure 6.20, showing all steps in the transmission of the nibble (4-bit string) 1011, starting with the MSB first.

6.41. Draw a timing diagram for all of the signals in Figure 6.21, showing all steps in the transmission of the nibble (4-bit string) 1011, starting with the MSB first.

6.42. Using JK flip-flops, design a circuit that will produce a clock signal of half the frequency of an input clock signal. If the input clock signal is asymmetric, will the output clock also be asymmetric?

6.43. You have a digital sensor such as a photo-interrupter that is normally low, but when activated will go high for a short but unspecified period of time. Design a circuit that will capture the output from the sensor and hold it until you apply a reset signal.

6.44. Design a sequential logic circuit to store the state of an SPDT switch when a bounce-free NO button (see Section 9.2.1 for details) is first pressed down and increment a counter when the button is released if the switch was down when the button was first pressed. Assume the bounce-free button and counter are purchased off the shelf and need not be designed. Draw a complete circuit diagram of your solution and show a timing diagram, assuming the SPDT switch exhibits switch bounce.

Section 6.11 TTL and CMOS Integrated Circuits

6.45. Document a complete and thorough answer to Class Discussion Item 6.12.

6.46. You are constrained in a digital design to interface a 74LS00 NAND gate to a 4011B NAND gate. Draw the schematic of a circuit necessary to do this properly.

6.47. Why can you not have large TTL fan-out from the output of a CMOS device?

Section 6.12 Special Purpose Digital Integrated Circuits

6.48. Derive a Boolean expression for the output columns $\bar{c}$ and $\bar{e}$ of Table 6.11 using either the product-of-sums or sum-of-products techniques presented in Section 6.7.

6.49. Using the information in a TTL data book, draw a complete schematic for a 7490 decade counter with BCD output. Include a reset feature using the four R lines in the timing diagram that follows. Also, complete the timing diagram.

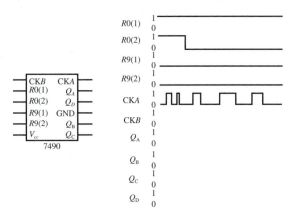

6.50. How does the decade counter differ from the 4-bit binary counter presented in Figure 6.19?

6.51. Given a Schmitt trigger operating between 0 and 5 V, assume that the lower threshold is 1 V and the upper threshold is 4 V. Given the following input signals applied to the Schmitt trigger, sketch the corresponding output from the Schmitt trigger for one cycle of each signal:

   a. $2.5 + 1.0 \sin (2\pi t)$ V

   b. $2.5 + 2.0 \sin (2\pi t)$ V

   c. $1.5 + 1.5 \sin (2\pi t)$ V

   d. $3.0 + 1.5 \sin (2\pi t)$ V

6.52. Determine the exact pulse width as a function of the resistance and capacitance in a one-shot circuit (see Equation 6.36).

6.53. Design and draw the schematic for a 555 oscillator that generates a 1-Hz clock signal.

6.54. Show all of the detailed algebraic steps necessary to derive Equation 6.42.

6.55. Document a complete and thorough answer to Class Discussion Item 6.14.

Section 6.13   Integrated Circuit System Design

6.56. Document a complete and thorough answer to Class Discussion Item 6.16.

6.57. Look up the 74LS90 decade counter in a digital handbook. Using the information found there, design a circuit that will turn on an LED after 100 occurrences of a digital event. The design should be a schematic. Include details of input codes to the IC necessary for decade counting.

6.58. For the circuit you designed in Question 6.44, or for a design provided by your instructor, verify whether or not switch bounce (from the SPDT switch) has an effect. Also, would the circuit still work if the NO button were not bounce-free? If not, how could the design be modified to deal with this?

6.59. You are provided with a coin slot through which people push coins. The bottom of the coin always slides along the bottom of the slot. People feed dimes, nickels, and quarters into the slot. Photo sensors are available and can be placed at any height to detect when portions of coins pass through the slot. The output of a sensor is high when its beam is interrupted; otherwise, it is low. Design a system that will light a red LED when a dime passes through, a yellow LED when a nickel passes through, and a green LED when a quarter passes through.

6.60. Assuming you have been successful in Question 6.59, extend the design to also count the number of dimes, nickels, and quarters passing through the slot.

BIBLIOGRAPHY

Horowitz, P. and Hill, W., *The Art of Electronics,* 2nd Edition, Cambridge University Press, New York, 1989.

Mano, M., *Digital Logic and Computer Design,* Prentice-Hall, Englewood Cliffs, NJ, 1979.

McWhorter, G. and Evans, A., *Basic Electronics,* Master Publishing, Richardson, TX, 1994.

Mims, F., *Getting Started in Electronics,* Radio Shack Archer Catalog No. 276-5003A, 1991.

Mims, F., *Engineers Mini-Notebook: 555 Circuits,* Radio Shack Archer Catalog No. 276-5010, 1984.

Mims, F., *Engineer's Mini-Notebook: Digital Logic Circuits,* Radio Shack Archer Catalog No. 276-5014, 1986.

Stiffler, A., *Design with Microprocessors for Mechanical Engineers,* McGraw-Hill, New York, 1992.

Texas Instruments, *TTL Logic Data Book,* Dallas, TX, 1988.

Texas Instruments, *Operational Amplifiers and Comparators,* Volume B, Dallas, TX, 1995.

Texas Instruments, *TTL Linear Circuits Data Book,* Volume 3, Dallas, TX, 1992.

7

CHAPTER

Microcontroller Programming and Interfacing

This chapter describes how to program and interface a microcontroller. Various input and output devices are also presented. ■

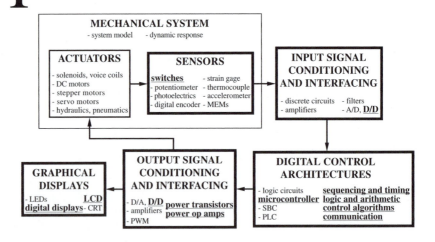

CHAPTER OBJECTIVES

After you read, discuss, study, and apply ideas in this chapter, you will:

1. Understand the differences among microprocessors, microcomputers, and microcontrollers

2. Know the terminology associated with a microcomputer and microcontroller

3. Understand the architecture and principles of operation of a microcontroller

4. Understand the basic concepts of assembly language programming

5. Understand the basics of higher level programming languages such as PicBasic Pro

6. Be able to write programs to control PIC microcontrollers

7. Be able to interface microcontrollers to input and output devices

8. Be able to design microcontroller-based mechatronic systems

9. Be aware of several practical considerations that will help you prototype, program, and debug microcontroller-based systems

10. Be able to select an appropriate source of power for a microcontroller-based system

7.1 MICROPROCESSORS AND MICROCOMPUTERS

The digital circuits presented in Chapter 6 allow the implementation of combinational and sequential logic operations by interconnecting ICs containing gates and flip-flops. This is considered a **hardware** solution because it consists of a selection of specific ICs, which when hardwired on a circuit board, carry out predefined functions. To make a change in functionality, the hardware circuitry must be modified and may require a redesign. This is a satisfactory approach for simple design tasks (e.g., the security system presented in Section 6.6 and the digital tachometer presented in Design Example 6.1). However, in many mechatronic systems, the control tasks may involve complex relationships among many inputs and outputs, making a strictly hardware solution impractical. A more satisfactory approach in complex digital design involves the use of a microprocessor-based system to implement a **software** solution. Software is a procedural program consisting of a set of instructions to execute logic and arithmetic functions and to access input signals and control output signals. An advantage of a software solution is that, without making changes in hardware, the program can be easily modified to alter a mechatronic system's functionality.

A **microprocessor** is a single, very-large-scale-integration (VLSI) chip that contains many digital circuits that perform arithmetic, logic, communication, and control functions. When a microprocessor is packaged on a printed circuit board with other components, such as interface and memory chips, the resulting assembly is referred to as a **microcomputer** or **single-board computer.** The overall architecture of a typical microcomputer system using a microprocessor is illustrated in Figure 7.1.

The microprocessor, also called the **central processing unit** (CPU) or **microprocessor unit** (MPU), is where the primary computation and system control operations occur. The **arithmetic logic unit** (ALU) within the CPU executes mathematical functions on data structured as binary words. A **word** is an ordered set of bits, usually 8, 16, 32, or 64 bits long. The instruction decoder interprets instructions

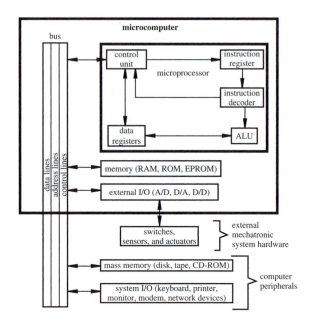

Figure 7.1 Microcomputer architecture.

fetched sequentially from memory by the control unit and stored in the instruction register. Each instruction is a set of coded bits that commands the ALU to perform bit manipulation, such as binary addition and logic functions, on words stored in the CPU data registers. The ALU results are also stored in data registers and then transferred to memory by the control unit.

The **bus** is a set of shared communication lines that serves as the central nervous system of the computer. Data, address, and control signals are shared by all system components via the bus. Each component connected to the bus communicates information to and from the bus via its own bus controller. The data lines, address lines, and control lines allow a specific component to access data addressed to that component. The **data lines** are used to communicate words to and from data registers in the various system components such as memory, CPU, and input/output (I/O) peripherals. The **address lines** are used to select devices on the bus or specific data locations within memory. Devices usually have a combinational logic address decoder circuit that identifies the address code and activates the device. The **control lines** transmit read and write signals, the system clock signal, and other control signals such as system interrupts, which are described in subsequent sections.

A key to a CPU's operation is the storage and retrieval of data from a memory device. Different types of memory include **read-only memory** (ROM), **random-access memory** (RAM), and **erasable-programmable ROM** (EPROM). ROM is used for permanent storage of data that the CPU can read, but the CPU cannot write data to ROM. ROM does not require a power supply to retain its data and therefore is called nonvolatile memory. RAM can be read from or written to at any time, provided power is maintained. The data in RAM is considered volatile because it is lost when power is removed. There are two main types of RAM: **static RAM** (SRAM), which retains its data in flip-flops as long as the memory is powered, and **dynamic RAM** (DRAM), which consists of capacitor storage of data that must be refreshed (rewritten) periodically because of charge leakage. Data stored in an EPROM can be erased with ultraviolet light applied through a transparent quartz window on top of the EPROM IC. Then new data can be stored on the EPROM. Another type of EPROM is **electrically erasable (EEPROM).** Data in EEPROM can be erased electrically and rewritten through its data lines without the need for ultraviolet light. Because data in RAM are volatile, ROM, EPROM, EEPROM, and peripheral mass memory storage devices such as magnetic, optical, and solid-state drives are sometimes needed to provide permanent data storage.

Communication to and from the microprocessor occurs through I/O devices connected to the bus. External computer peripheral I/O devices include keyboards, printers, displays, and network devices. For mechatronic applications, analog-to-digital (A/D), digital-to-analog (D/A), and digital I/O (D/D) devices provide interfaces to switches, sensors, and actuators.

The instructions that can be executed by the CPU are defined by a binary code called **machine code.** The instructions and corresponding codes are microprocessor dependent. Each instruction is represented by a unique binary string that causes the microprocessor to perform a low-level function (e.g., add a number to a register or move a register's value to a memory location). Microprocessors can be programmed using **assembly language,** which has a mnemonic command corresponding to each instruction (e.g., *ADD* to add a number to a register and *MOV* to move a register's value to a memory location). However, assembly language must be converted to machine code, using software called an **assembler,** before it can be executed on the microprocessor. When the set of instructions is small, the microprocessor is known as a **RISC** (reduced instruction-set computer) microprocessor. RISC microprocessors are cheaper to design and manufacture and are usually faster. However, more programming steps may be required for complex algorithms, due to the limited set of instructions.

Programs can also be written in a higher level language such as BASIC or C, provided that a compiler is available that can generate machine code for the specific microprocessor being used. The advantages of using a high-level language are that it is easier to learn and use, programs are easier to **debug** (the process of finding and removing errors), and programs are easier to comprehend. A disadvantage is that the resulting machine code may be less efficient (i.e., slower and require more memory) than a corresponding well-written assembly language program.

7.2 MICROCONTROLLERS

There are two branches in the ongoing evolution of the microprocessor. One branch supports CPUs for the personal computer and workstation industry, where the main constraints are high speed and large word size (32 and 64 bits). The other branch includes development of the **microcontroller,** which is a single IC containing specialized circuits and functions that are applicable to mechatronic system design. It contains a microprocessor, memory, I/O capabilities, and other on-chip resources. It is basically a microcomputer on a single IC. Examples of microcontrollers are Microchip's PIC, Motorola's 68HC11, and Intel's 8096.

Internet Link

7.1 Microcontroller online resources and manufacturers

Internet Link 7.1 points to various microcontroller online resources and manufacturers. A wealth of information is available online, and it is constantly changing as manufacturers continually release products with faster speeds, larger memories, and more functionality. Factors that have driven development of the microcontroller are low cost, versatility, ease of programming, and small size. Microcontrollers are attractive in mechatronic system design because their small size and broad functionality allow them to be physically embedded in a system to perform all of the necessary control functions.

Microcontrollers are used in a wide array of applications including home appliances, entertainment equipment, telecommunication equipment, automobiles, trucks, airplanes, toys, and office equipment. All these products involve devices that require some sort of intelligent control based on various inputs. For example, the microcontroller in a microwave oven monitors the control panel for user input, updates the graphical displays when necessary, and controls the timing and cooking functions. In an automobile, there are many microcontrollers to control various subsystems, including cruise control, antilock braking, ignition control, keyless entry, environmental control, and air and fuel flow. An office copy machine controls actuators to feed paper, uses photo sensors to scan a page, sends or receives data via a network connection, and provides a user interface complete with menu-driven controls. A toy robot dog has various sensors to detect inputs from its environment (e.g., bumping into obstacles, being patted on the head, light and dark, voice commands), and an onboard microcontroller actuates motors to mimic actual dog behavior (e.g., bark, sit, and walk) based on this input. All of these powerful and interesting devices are controlled by microcontrollers and the software running on them.

Figure 7.2 is a block diagram for a typical full-featured microcontroller. Also included in the figure are lists of typical external devices that might interface to the microcontroller. The components of a microcontroller include the CPU, RAM, ROM, digital I/O ports, a serial communication interface, timers, analog-to-digital (A/D) converters, and digital-to-analog (D/A) converters. The CPU executes the software stored in ROM and controls all the microcontroller components. The RAM is used to store settings and values used by an executing program. The ROM is used to store the program and any permanent data. A designer can have a program and data permanently stored in ROM by the chip manufacturer, or the ROM can be in the form of EPROM or EEPROM, which can be reprogrammed by the user. Software permanently stored in ROM is referred to as **firmware.** Microcontroller manufacturers offer programming devices that can download compiled machine code from a PC directly to the EEPROM of the micro-controller, usually via the PC serial port and special-purpose pins on the microcontroller. These pins can usually be used for other purposes once the device is programmed. Additional EEPROM may also be available and used by the program to store settings and parameters generated or modified during execution. The data in EEPROM is nonvolatile, which means the program can access the data when the microcontroller power is turned off and back on again.

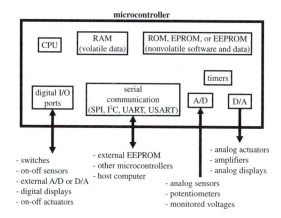

Figure 7.2 Components of a typical full-featured microcontroller.

The digital I/O **ports** allow binary data to be transferred to and from the microcontroller using external pins on the IC. These pins can be used to read the state of switches and on-off sensors, to interface to external A/D and D/A converters, to control digital displays, and to control on-off actuators. The I/O ports can also be used to transmit signals to and from other microcontrollers to coordinate various functions. The microcontroller can also use a serial port to transmit data to and from external devices, provided these devices support the same serial communication protocol. Examples of such devices include external EEPROM memory ICs that might store a large block of data for the microcontroller, other microcontrollers that need to share data and coordinate control, and a host computer that might download a program into the microcontroller's onboard EEPROM. There are various standards or protocols for serial communication including SPI (serial peripheral interface), I^2C (interintegrated circuit), UART (universal asynchronous receiver-transmitter), and USART (universal synchronous-asynchronous receiver-transmitter).

The A/D converter allows the microcontroller to convert an external analog voltage (e.g., from a sensor) to a digital value that can be processed or stored by the CPU. The D/A converter allows the microcontroller to output an analog voltage to a nondigital device (e.g., a motor amplifier). A/D and D/A converters and their applications are discussed in Chapter 8. Onboard timers are usually provided to help create delays or ensure events occur at precise time intervals (e.g., reading the value of a sensor).

Microcontrollers typically have less than 1 kilobyte to several tens of kilobytes of program memory, compared with microcomputers whose RAM memory is measured in megabytes or gigabytes. Also, microcontroller clock speeds are slower than those used for microcomputers. For some applications, a selected microcontroller may not have enough speed or memory to satisfy the needs of the application. Fortunately, microcontroller manufacturers usually provide a wide range of products to accommodate different applications. Also, when more memory or I/O capability is required, the functionality of the microcontroller can be expanded with additional external components (e.g., RAM or EEPROM chips, external A/D and D/A converters, and other microcontrollers).

In the remainder of this chapter, we focus on the Microchip PIC microcontroller due to its popularity, abundant information resources and support products, low cost, and ease of use. **PIC** is an acronym for peripheral interface controller, the phrase Microchip uses to refer to its line of microcontrollers. Microchip offers a large and diverse family of low-cost PIC products. They vary in footprint (physical size), the number of I/O pins available, the size of the EEPROM and RAM space for storing programs and data, and the availability of A/D and D/A converters. Obviously, the more features and capacity a microcontroller has, the higher the cost. Information for Microchip's entire line of products can be found on its website at *www.microchip.com* (see Internet Link 7.2). We focus specifically on the **PIC16F84,** which is a low-cost 8-bit microcontroller with EEPROM flash memory for program and data storage. It has no built-in A/D, D/A, or serial communication capability, but it supports 13 digital I/O lines and serves as a good learning platform because it is low cost and easy to program. Once you know how to interface and program one microcontroller, it is easy to extend that knowledge to other microcontrollers with different features and programming options.

Another good reason to focus on the PIC16F84 is that many other PIC microcontrollers are upward compatible and pin compatible with the PIC16F84. Examples include the PIC16F84A, PIC16F88, and PIC16F819. These three (and other) microcontrollers can be configured to mimic the PIC16F84, and everything learned on the PIC16F84 can be directly applied to these and many other Microchip microcontrollers. So what are the differences that distinguish the 84A, 88, and 819 from the 84? First, all three can be run at faster clock speeds (up to 20 MHz). The 88 and 819 have internal oscillators (that can provide the clock signal without external components), allowing for more I/O lines, and they have more memory and additional software-configurable functionality (e.g., onboard comparators and pulse-width-modulation generators). The 88 and 819 also includes software-configurable A/D converters, which allow one to interface the PIC to analog sensors that output a continuously varying voltage rather than a digital signal.

Internet Link

7.2 Microchip, Inc.

■ CLASS DISCUSSION ITEM 7.1
Car Microcontrollers

List various automobile subsystems that you think are controlled by microcontrollers. In each case, identify all of the inputs to and outputs from the microcontroller and describe the function of the software.

7.3 THE PIC16F84 MICROCONTROLLER

The block diagram for the PIC16F84 microcontroller is shown in Figure 7.3. This diagram, along with complete documentation of all of the microcontroller's features and capabilities, can be found in the manufacturer's data sheets. The PIC16F8X

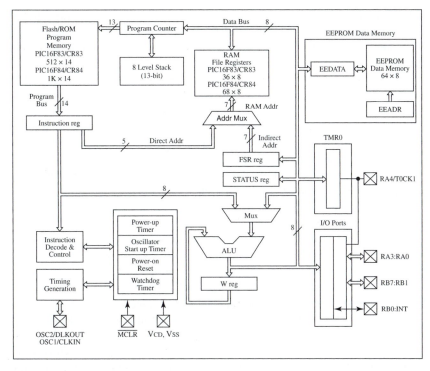

Figure 7.3 PIC16F84 block diagram. *(Courtesy of Microchip Technology, Inc., Chandler, AZ)*

data sheets are contained in a book available from Microchip and as a PDF file on its website (see Internet Link 7.3). The PIC16F84 is an 8-bit CMOS microcontroller with 1792 bytes of flash EEPROM program memory, 68 bytes of RAM data memory, and 64 bytes of nonvolatile EEPROM data memory. The 1792 bytes of program memory are subdivided into 14-bit words, because machine code instructions are 14 bits wide. Therefore, the EEPROM can hold up to 1024 (1 k) instructions. The PIC16F84 can be driven at a clock speed up to 10 MHz but is typically driven at 4 MHz.

When a program is compiled and downloaded to a PIC, it is stored as a set of binary machine code instructions in the flash program memory. These instructions are sequentially fetched from memory, placed in the instruction register, and executed. Each instruction corresponds to a low-level function implemented with logic circuits on the chip. For example, one instruction might load a number stored in RAM or EEPROM into the **working register,** which is also called the **W register** or **accumulator;** the next instruction might command the ALU to add a different number to the value in this register; and the next instruction might return this summed value to memory. Because an instruction is executed every four clock cycles, the PIC16F84 can do calculations, read input values, store and retrieve information from memory, and perform other functions very quickly. With a clock speed

Internet Link

7.3 Microchip PIC16F84 data sheet

of 4 MHz, an instruction is executed every microsecond and 1 million instructions can be executed every second. The microcontroller is referred to as 8-bit, because the data bus is 8 bits wide, and all data processing and storage and retrieval occur using bytes.

A useful special purpose timer, called a **watch-dog timer,** is included on PIC microcontrollers. This is a count-down timer that, when activated, needs to be continually reset by the running program. If the program fails to reset the watch-dog timer before it counts down to 0, the PIC will automatically reset itself. In a critical application, you might use this feature to have the microcontroller reset if the software gets caught in an unintentional endless loop.

The RAM, in addition to providing space for storing data, maintains a set of special purpose byte-wide locations called **file registers.** The bits in these registers are used to control the function and indicate the status of the microcontroller. Several of these registers are described below.

The PIC16F84 is packaged on an 18-pin DIP IC that has the pin schematic (pinout) shown in Figure 7.4. The figure also shows the minimum set of external components recommended for the PIC to function properly. Figure 7.5 shows what the schematic shown in Figure 7.4 looks like when assembled on a breadboard with actual components. Table 7.1 lists the pin identifiers in natural groupings, along with their descriptions. The five pins RA0 through RA4 are digital I/O pins collectively referred to as **PORTA,** and the eight pins RB0 through RB7 are digital I/O pins collectively referred to as **PORTB.** In total, there are 13 I/O lines, called **bidirectional** lines because each can be individually configured in software as an input or output. PORTA and PORTB are special purpose file registers on the PIC that provide the interface to the I/O pins. Although all PIC registers contain 8 bits, only the 5 least significant bits (LSBs) of PORTA are used.

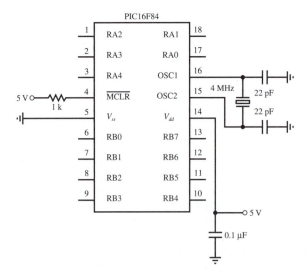

Figure 7.4 PIC16F84 pin-out and required external components.

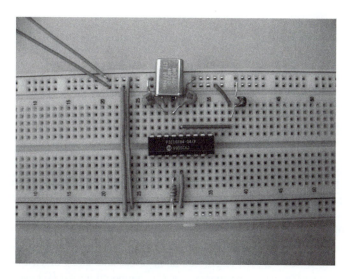

Figure 7.5 Required PIC16F84 components on a breadboard.

Table 7.1 PIC16F84 pin name descriptions

Pin identifier	Description
RA[0–4]	5 bits of bidirectional I/O (PORTA)
RB[0–7]	8 bits of bidirectional I/O (PORTB)
V_{ss}, V_{dd}	Power supply ground reference (*ss*: source) and positive supply (*dd*: drain)
OSC1, OSC2	Oscillator crystal inputs
$\overline{\text{MCLR}}$	Master clear (active low)

An important feature of the PIC, available with most microcontrollers, is its ability to process interrupts. An **interrupt** occurs when a specially designated input changes state. When this happens, normal program execution is suspended while a special interrupt handling portion of the program is executed. This is discussed further in Section 7.6. On the PIC16F84, pins RB0 and RB4 through RB7 can be configured as interrupt inputs.

Power and ground are connected to the PIC through pins V_{dd} and V_{ss}. The *dd* and *ss* subscripts refer to the drain and source notation used for MOS transistors, since a PIC is a CMOS device. The voltage levels (e.g., $V_{dd} = 5$ V and $V_{ss} = 0$ V) can be provided using a DC power supply or batteries (e.g., four AA batteries in series or a 9 V battery connected through a voltage regulator). The master clear pin ($\overline{\text{MCLR}}$) is active low and provides a reset feature. Grounding this pin causes the PIC to reset and restart the program stored in EEPROM. This pin must be held high during normal program execution. This is accomplished with the pull-up resistor shown in Figure 7.4. If this pin were left unconnected (floating), the chip might reset

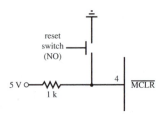

Figure 7.6 Reset switch circuit.

itself sporadically. To provide a manual reset feature to a PIC design, you can add a normally open (NO) pushbutton switch as shown in Figure 7.6. Closing the switch grounds the pin and causes the PIC to reset.

The PIC **clock** frequency can be controlled using different methods, including an external RC circuit, an external clock source, or a clock crystal. In Figure 7.4, we show the use of a clock crystal to provide an accurate and stable clock frequency at relatively low cost. The clock frequency is set by connecting a 4-MHz crystal across the OSC1 and OSC2 pins with the 22 pF capacitors grounded as shown in Figure 7.4.

7.4 PROGRAMMING A PIC

To use a microcontroller in mechatronic system design, software must be written, tested, and stored in the ROM of the microcontroller. Usually, the software is written and compiled using a personal computer (PC) and then downloaded to the microcontroller ROM as machine code. If the program is written in assembly language, the PC must have software called a **cross-assembler** that generates machine code for the microcontroller. An assembler is software that generates machine code for the microprocessor in the PC, whereas a cross-assembler generates machine code for a different microprocessor, in this case the microcontroller.

Various software development tools can assist in testing and debugging assembly language programs written for a microcontroller. One such tool is a **simulator,** which is software that runs on a PC and allows the microcontroller code to be simulated (run) on the PC. Most programming errors can be identified and corrected during simulation. Another tool is an **emulator,** which is hardware that connects a PC to the microcontroller in a prototype mechatronic system. It usually consists of a printed circuit board connected to the mechatronic system through ribbon cables. The emulator can be used to load and run a program on the actual microcontroller attached to the mechatronic system hardware (containing sensors, actuators, and control circuits). The emulator allows the PC to monitor and control the operation of the microcontroller while it is embedded in the mechatronic system.

The assembly language used to program a PIC16F84 consists of 35 commands that control all functions of the PIC. This set of commands is called the **instruction set** for the microcontroller. Every microcontroller brand and family has its own specific instruction set that provides access to the resources available on the chip. The complete instruction set and brief command descriptions for the

PIC16F84 are listed in Table 7.2. Each command consists of a name called the **mnemonic** and, where appropriate, a list of operands. Values must be provided for each of these operands. The letters f, d, b, and k correspond, respectively, to a file register address (a valid RAM address), result destination (0: W register, 1: file register), bit number (0 through 7), and literal constant (a number between 0 and 255). Note that many of the commands refer to the working register W, also called the accumulator. This is a special CPU register used to temporarily store values (e.g., from memory) for calculations or comparisons. At first, the mnemonics and

Table 7.2 PIC16F84 instruction set

Mnemonic and operands	Description
ADDLW k	Add literal and W
ADDWF f, d	Add W and f
ANDLW k	AND literal with W
ANDWF f, d	AND W with f
BCF f, b	Bit clear f
BSF f, b	Bit set f
BTFSC f, b	Bit test f, skip if clear
BTFSS f, b	Bit test f, skip if set
CALL k	Call subroutine
CLRF f	Clear f
CLRW	Clear W
CLRWDT	Clear watch-dog timer
COMF f, d	Complement f
DECF f, d	Decrement f
DECFSZ f, d	Decrement f, skip if 0
GOTO k	Go to address
INCF f, d	Increment f
INCFSZ f, d	Increment f, skip if 0
IORLW k	Inclusive OR literal with W
IORWF f, d	Inclusive OR W with f
MOVF f, d	Move f
MOVLW k	Move literal to W
MOVWF f	Move W to f
NOP	No operation
RETFIE	Return from interrupt
RETLW k	Return with literal in W
RETURN	Return from subroutine
RLF f, d	Rotate f left 1 bit
RRF f, d	Rotate f right 1 bit
SLEEP	Go into standby mode
SUBLW k	Subtract W from literal
SUBWF f, d	Subtract W from f
SWAPF f, d	Swap nibbles in f
XORLW k	Exclusive OR literal with W
XORWF f, d	Exclusive OR W with f

Internet Link

7.3 Microchip
PIC16F84
datasheet

descriptions in the table may seem cryptic, but after you compare functionality with the terminology and naming conventions, it becomes much more understandable. In Example 7.1, we introduce a few of the statements and provide some examples. We illustrate how to write a complete assembly language program in Example 7.2.

For more information (e.g., detailed descriptions and examples of each assembly statement), refer to the PIC16F8X data sheet available on Microchip's website (see Internet Link 7.3).

EXAMPLE 7.1	## Assembly Language Instruction Details

Here, we provide more detailed descriptions and examples of a few of the assembly language instructions to help you better understand the terminology and the naming conventions.

BCF f, b
(read *BCF* as "bit clear f")
clears bit *b* in file register *f* to 0, where the bits are
numbered from 0 (LSB) to 7 (MSB)

For example, *BCF PORTB, 1* makes bit 1 in PORTB go low (where PORTB is a constant containing the address of the PORTB file register). If PORTB contained the hexadecimal (hex) value FF (binary 11111111) originally, the final value would be hex FC (binary 11111101). If PORTB contained the hex value A8 (binary 10101000) originally, the value would remain unchanged.

MOVLW k
(read *MOVLW* as "move literal to W")
stores the literal constant *k* in the accumulator (the W register)

For example, *MOVLW 0xA8* would store the hex value A8 in the W register. In assembly language, hexadecimal constants are identified with the *0x* prefix.

RLF f, d
(read *RLF* as "rotate f left")
shifts the bits in file register *f* to the left 1 bit, and stores the result in *f* if *d* is 1
or in the accumulator (the W register) if *d* is 0. The value of the LSB will become 0,
and the original value of the MSB is lost.

For example, if the current value in PORTB is hex 1F (binary 00011111), then *RLF PORTB, 1* would change the value to hex 3E (binary 00111110).

SWAPF f, d
(read *SWAPF* as "swap nibbles in f")
exchanges the upper and lower nibbles (a nibble is 4 bits or half a byte)
of file register *f* and stores the result in *f* if *d* is 1
or in the accumulator (the W register) if *d* is 0

For example, if the memory location at address hex 10 contains the value hex AB, then *SWAPF 0x10, 0* would store the value hex BA in the W register. *SWAPF 0x10, 1* would change the value at address hex 10 from hex AB to hex BA.

Assembly Language Programming Example EXAMPLE 7.2

The purpose of this example is to write an assembly language program that will turn on an LED when the user presses a pushbutton switch. When the switch is released, the LED is to turn off. After the switch is pressed and released a specified number of times, a second LED is to turn on and stay lit. The hardware required for this example is shown in the following figure:

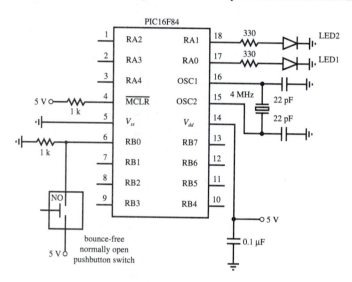

The pushbutton switch is assumed to be bounce-free, implying that when it is pressed and then released, a single pulse is produced (the signal goes high when it is pressed and goes low when it is released).

Assembly language code that will accomplish the desired task follows the text below. A remark or comment can be inserted anywhere in a program by preceding it with a semicolon (;). Comments are used to clarify the associated code. The assembler ignores comments when generating the hex machine code. The first four active lines (*list . . . target*) are assembler directives that designate the processor and define constants that can be used in the remaining code. Defining constants (with the *equ* directive) at the beginning of the program is a good idea because the names, rather than hex numbers, are easier to read and understand in the code and because the numbers can be conveniently located and edited later. Assembly language constants such as addresses and values are written in hexadecimal, denoted with a *0x* prefix.

The next two lines of code, starting with *movlw,* move the literal constant *target* into the W register and then from the W register into the *count* address location in memory. The target value (0x05) will be decremented until it reaches 0x00. The next section of code initializes the special function registers PORTA and TRISA to allow output to pins RA0 and RA1, which drive the LEDs. These registers are located in different banks of memory, hence the need for the *bsf* and *bcf* statements in the program. All capitalized words in the program are constant addresses or values predefined in the processor-dependent include

(continued)

(continued) file (p16f84.inc). The function of the TRISA register is discussed more later; but by clearing the bits in the register, the PORTA pins are configured as outputs.

The main loop uses the *btfss* (bit test in file register; skip the next instruction if the bit is set) and *btfsc* (bit test in file register; skip the next instruction if the bit is clear) statements to test the state of the signal on pin RB0. The tests are done continually within loops created by the *goto* statements. The words *begin* and *wait* are statement labels used as targets for the *goto* loops. When the switch is pressed, the state goes high and the statement *btfss* skips the *goto begin* instruction; then LED1 turns on. When the switch is released, pin RB0 goes low and the statement btfsc skips the *goto wait* instruction; then LED1 turns off.

After the switch is released and LED1 turns off, the statement *decfsz* (decrement file register; skip the next instruction if the count is 0) executes. The *decfsz* decrements the *count* value by 1. If the *count* value is not yet 0, *goto begin* executes and control shifts back to the label *begin*. This resumes execution at the beginning of the main loop, waiting for the next switch press. However, when the *count* value reaches 0, *decfsz* skips the *goto begin* statement and LED2 is turned on. The last *goto begin* statement causes the program to again jump back to the beginning of the main loop to wait for another button press.

```
; bcount.asm (program file name)

; Program to turn on an LED every time a pushbutton switch is pressed and turn on
; a second LED once it has been pressed a specified number of times

; I/O:
; RB0: bounce-free pushbutton switch (1:pressed, 0:not pressed)
; RA0: count LED (first LED)
; RA1: target LED (second LED)

; Define the processor being used
list p=16f84
include <p16F84.inc>

; Define the count variable location and the initial count-down value
count equ 0x0c                          ; address of count-down variable
target equ 0x05                         ; number of presses required

; Initialize the counter to the target number of presses
movlw target                            ; move the count-down value into the
                                        ;  W register
movwf count                             ; move the W register into the count memory
                                        ;  location

; Initialize PORTA for output and make sure the LEDs are off
bcf        STATUS, RP0                  ; select bank 0
clrf       PORTA                        ; initialize all pin values to 0
bsf        STATUS, RP0                  ; select bank 1
clrf       TRISA                        ; designate all PORTA pins as outputs
bcf        STATUS, RP0                  ; select bank 0
```

```
; Main program loop
begin

        ; Wait for the pushbutton switch to be pressed
        btfss   PORTB, 0
        goto    begin

        ; Turn on the count LED1
        bsf     PORTA, 0

wait

        ; Wait for the pushbutton switch to be released
        btfsc   PORTB, 0
        goto    wait

        ; Turn off the count LED1
        bcf     PORTA, 0

        ; Decrement the press counter and check for 0
        decfsz count, 1
        goto begin      ; continue if count-down is still > 0

        ; Turn on the target LED2
        bsf             PORTA, 1

        goto begin      ; return to the beginning of the main loop

        end             ; end of instructions
```

■ **CLASS DISCUSSION ITEM 7.2**
Decrement Past 0

In Example 7.2, the *decfsz* statement is used to count down from 5 to 0. When 0 (0x00) is decremented, the resulting value at address *count* will be 255 (0xFF), then 254 (0xFE), and so forth. What effect, if any, does this have on the operation of the code and the LED display? To have the program continue its normal operation after the decrement to zero, what code would you need to add?

Learning to program in assembly language can be very difficult at first and may result in errors that are difficult to debug. Fortunately, high-level language compilers are available that allow us to program a PIC at a more user-friendly level. The particular programming language we discuss in the remainder of the chapter is **PicBasic Pro.** The compiler for PicBasic Pro is available from microEngineering Labs, Inc. (see Internet Link 7.4, which points to *www.melabs.com*). PicBasic Pro is much easier to learn and use than assembly language. It provides easy access to all of the PIC capabilities and provides a rich set of advanced functions and features to support various applications. PicBasic Pro is also closely compatible with the BASIC language

Internet Link

7.4 micro Engineering Labs

Internet Link

7.5 Microchip
PIC microcontroller
resources

used to control Basic Stamp minicontrollers (Parallax, Inc., Rocklin, CA), which are popular development boards that utilize the PIC microcontroller.

For additional information on the PIC and related products, see Internet Link 7.5. This site contains many useful links to manufacturers and other useful web pages that provide resources and information on support literature, useful accessories, and PicBasic Pro.

7.5 PICBASIC PRO

PIC programs can be written in a form of BASIC called **PicBasic Pro.** The PicBasic Pro complier can compile these programs, producing their assembly language equivalents, and this assembly code can then be converted to hexadecimal machine code (hex code) that can be downloaded directly to the PIC flash EEPROM through a programming device attached to a PC. Once loaded, the program begins to execute when power is applied to the PIC if the necessary additional components, such as those shown in Figure 7.4, are connected properly.

Internet Link

7.6 PicBasic
Pro manual

We do not intend to cover all aspects of PicBasic Pro programming. Instead, we present an introduction to some of the basic programming principles, provide a brief summary of the statements, and then provide some examples. The PicBasic Pro Complier manual available online (see Internet Link 7.6) is a necessary supplement to this chapter if you need to solve problems requiring more functionality than the examples we present here. If you have not used a programming language such as BASIC, C, C++, or FORTRAN, then Section 7.5.1 may be challenging for you. Even if this the case, as you read through the examples that follow, the concepts should become clearer.

7.5.1 PicBasic Pro Programming Fundamentals

To illustrate the fundamentals of PicBasic Pro, we start with a very simple example. The goal is to write a program to turn on an LED for a second, then turn it off for a second, repeating for as long as power is applied to the circuit. The code for this program, called *flash.bas,* follows. The hardware required is shown in Figure 7.7. Pin RA2 is used as an output to source current to an LED through a current-limiting resistor. The first two lines in the program are comments that identify the program and its purpose. **Comment lines** must begin with an apostrophe. On any line, information on the right side of an apostrophe is treated as a comment and ignored by the compiler. The label *loop* allows the program to return control to this label at a later time using the *Goto* command. The statement *High PORTA.2* causes pin RA2 to go high and the LED turns on. The *Pause* command delays execution of the next line of code a given number of milliseconds (in this case, 1000, which corresponds to 1000 milliseconds or 1 second). The statement *Low PORTA.2* causes pin RA2 to go low, turning the LED off. The next *Pause* causes a 1 sec delay before executing the next line. The *Goto loop* statement returns control to the program line labeled *loop,* and the program continues indefinitely. The *End* statement on the last line of the program terminates execution. In this example, the loop continues until power is removed,

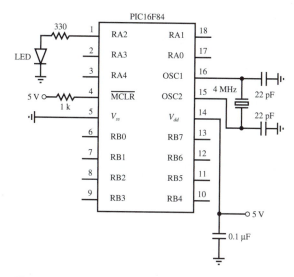

Figure 7.7 Circuit schematic for the flash.bas example.

and the *End* statement is never reached. However, to be safe you should always terminate a program with an *End* statement.

```
' flash.bas
' Example program to flash an LED once every two seconds

loop: High PORTA.2          ' turn on LED connected to pin RA2
      Pause 1000            ' delay for one second (1000 ms)

      Low PORTA.2           ' turn off LED connected to pin RA2
      Pause 1000            ' delay for one second (1000 ms)

      Goto loop             ' go back to label "loop" and repeat
                            '  indefinitely

      End
```

Lab Exercise 9 explores how to implement simple programs like *flash.bas* above. The exercise includes steps for wiring circuits to a PIC, entering and compiling PicBasic Pro programs, and downloading executable code from a PC to the flash memory on the PIC.

As illustrated in the simple *flash.bas* example, PicBasic Pro programs consist of a sequence of program statements that are executed one after another. The programmer must be familiar with the syntax of PicBasic Pro, but it will be easier to learn and debug than assembly language programs. **Comments,** any text preceded by an apostrophe, can be placed anywhere in the program to help explain the purpose of specific lines of the code. Any user-defined labels, variable names, or constant names are called **identifiers** (e.g., *loop* in the preceding example). You can use any combination of characters for these identifiers, provided they do not start with a number. Also, identifiers must be different from all the words reserved by PicBasic Pro

Lab Exercise

Lab 9 Programming a PIC microcontroller— part I

(e.g., keywords like *High* and *Low*). Identifiers may be any length, but PicBasic Pro ignores all the characters after the first 32. PicBasic Pro is not case sensitive so it does not matter whether or not letters are capitalized. Therefore, any combination of lower or upper case letters can be used for identifiers, including labels, variables, statements, and register or bit references. For example, to PicBasic Pro, *High* is equivalent to *HIGH* or *high*. However, when writing code, it is best to use a consistent pattern that helps make the program more readable. In the examples presented in this chapter, all variables and labels are written in lower case, all keywords in statements are written with an initial capital, and all registers and constants are written in upper case.

In some applications, you need to store a value for later use in the program (e.g., a counter that is incremented each time a pushbutton switch is pressed). PicBasic Pro lets you create variables for this purpose. The syntax for creating a **variable** is

$$\text{name Var type} \tag{7.1}$$

where *name* is the identifier to be used to refer to the variable and *type* describes the type and corresponding data storage size of the variable. The type can be **BIT** to store a single bit of information (0 or 1), **BYTE** to store an 8-bit positive integer that can range from 0 to 255 ($2^8 - 1$), or **WORD** to store a 2-byte (16-bit) positive integer that can range from 0 to 65,535 ($2^{16} - 1$). The following lines are examples of variable declarations and assignment statements that store values in variables:

```
my_bit Var BIT
my_byte Var BYTE

my_bit = 0
my_byte = 187
```

The *Var* keyword can also be used to give identifier names to I/O pins or to bits within a byte variable using the following syntax:

$$\text{name Var byte.bit} \tag{7.2}$$

For example,

```
led Var PORTB.0
lsb Var my_byte.0
```

would designate *led* as the state of pin RB0 and *lsb* as bit 0 of byte variable *my_byte*.

Another type of variable is an **array,** which can be used to store a set or vector of numbers. The syntax for declaring an array is

$$\text{name Var type[size]} \tag{7.3}$$

where *type* defines the storage type (BIT, BYTE, or WORD) and *size* indicates the number of elements in the array. A particular **element** in an array can be accessed or referenced with the following syntax:

$$\text{name[i]} \tag{7.4}$$

where *i* is the index of the element being referenced. The elements are numbered from 0 to size–1. For example, *values Var byte[5]* would define an array of 5 bytes, and the elements of the array would be *values[0], values[1], values[2], values[3],* and *values[4].*

Constants can be given names in a program using the same syntax as that used for variables (Equation 7.1) by replacing the *Var* keyword with *Con* and by replacing the *type* keyword by a constant value. When specifying values in a program, the prefix *$* denotes a hexadecimal value and the prefix *%* denotes a binary value. If there is no prefix, the number is assumed to be a decimal value. For example, with the following variable and constant definitions, all of the assignment statements that follow are equivalent:

```
number Var BYTE
CONSTANT Con 23

number = 23
number = CONSTANT
number = %10111
number = $17
```

Normally, constants and results of calculations are assumed to be unsigned (i.e., zero or positive), but certain functions, such as *Sin* and *Cos,* use a different byte format, where the MSB is used to represent the sign of the number. In this case, the byte can take on values between -127 and 127. Some of the fundamental expressions using **mathematical operators** and functions available in PicBasic Pro are listed in Table 7.3. See other operators and functions, more details, and examples in the PicBasic Pro Compiler manual.

Table 7.3 Selected PicBasic Pro math operators and functions

Math operator or function	Description
A + B	Add A and B
A − B	Subtract B from A
A * B	Multiply A and B
A / B	Divide A by B
A // B	Return the remainder (modulo) of the division of B into A
A << n	Shift A n bits to the left
A >> n	Shift A n bits to the right
COS A	Return the cosine of A
A MAX B	Return the maximum of A and B
A MIN B	Return the minimum of A and B
SIN A	Return the sine of A
SQR A	Return the square root of A
A & B	Return the bitwise AND of A and B
A	B	Return the bitwise OR of A and B
A ^ B	Return the bitwise Exclusive OR of A and B
~A	Return the bitwise NOT of A

Lab Exercise

Lab 11 Pulse-width-modulation motor speed control with a PIC

When doing integer arithmetic with fixed-bit-length variables (e.g., BYTE and WORD), one must check for truncation and overflow errors. As pointed out above, each variable type can only store numbers within a certain range (e.g., 0 to 255 for a BYTE variable). If you try to assign an expression to a variable, and the expression's value exceeds the maximum value allowed for the variable, an error will result. This is called **overflow. Truncation** occurs with integer division. If an integer division calculation results in a fraction, the remainder of the division (the decimal portion) is discarded. Sometimes the effects of truncation can be minimized or avoided by rearranging terms in an expression so divisions are performed at strategic points. These principles are presented in detail, with examples, in Lab Exercise 11.

There is a collection of PicBasic Pro statements that allow you to read, write, and process inputs from and outputs to the I/O port pins. To refer to an I/O pin, you use the following syntax:

$$\text{port\_name.bit} \tag{7.5}$$

where *port_name* is the name of the port (PORTA or PORTB) and *bit* is the bit location specified as a number between 0 and 7. For example, to refer to pin RB1, you would use the expression PORTB.1. When a bit is configured as an output, the output value (0 or 1) on the pin can be set with a simple assignment statement (e.g., *PORTB.1 = 1*). When a bit is configured as an input, the value on the pin (0 or 1) can be read by referencing the bit directly (e.g., *value = PORTA.2*). All of the bits within a port can be set at one time using an assignment statement of the following form:

$$\text{port\_name} = \text{constant} \tag{7.6}$$

where *constant* is a number between 0 and 255 expressed in binary, hexadecimal, or decimal. For example, *PORTA = %00010001* sets the PORTA.0 and PORTA.4 bits to 1, and sets all other bits to 0. Because the three most significant bits in PORTA are not used, they need not be specified (i.e., *PORTA = %10001* is equivalent and more appropriate).

The I/O status of the PORTA and PORTB bits are configured in two special registers called **TRISA** and **TRISB.** The prefix *TRIS* is used to indicate that tristate gates control whether or not a particular pin provides an input or an output. The input and output circuits for PORTA and PROTB on the PIC16F84 are presented in Section 7.8, where we deal with interfacing. When a TRIS register bit is set high (1), the corresponding PORT bit is considered an input, and when the TRIS bit is low (0), the corresponding PORT bit is considered an output. For example, TRISB = %01110000 would designate pins RB4, RB5, and RB6 as inputs and the other PORTB pins as outputs. At power-up, all TRIS register bits are set to 1 (i.e., TRISA and TRISB are both set to $FF or %11111111), so all pins in PORTA and PORTB are treated as inputs by default. You must redefine them if necessary for your application, in the initialization statements in your program.

The port bit access syntax described by Equation 7.5 can also be used to access individual bits in byte variables. For example, given the following declarations,

```
my_byte Var byte
my_array Var byte[10]
```

my_byte.3 = 1 would set bit 3 in *my_byte* to 1 and *my_array[9]*.7 = 0 would set the MSB of the last element of *my_array* to 0. All bits within a variable can be set by assigning a value or expression to the variable with an **assignment statement:**

$$\text{variable} = \text{expression} \qquad (7.7)$$

For example,

```
my_byte = 231
my_array[2] = my_byte - 12
```

Two important features in any programming language are statements to perform logical comparisons and statements to branch, loop, and iterate. In PicBasic Pro, logic is executed within an *If . . . Then . . . Else . . .* statement construct, where a logical comparison is made and if the result of the comparison is true, then the statements after *Then* are executed; otherwise, the statements after *Else* are executed. PicBasic Pro supports the **logical comparison operators** listed in Table 7.4. The keywords **And, Or, Xor** (exclusive Or), and **Not** can also be used in conjunction with parentheses to create general Boolean expressions for use in logical comparisons. Example 7.3 and other examples to follow illustrate use of logical expressions.

Table 7.4 PicBasic Pro logical comparison operators

Operator	Description
= or ==	equal
<> or !=	not equal
<	less than
>	greater than
<=	less than or equal to
>=	greater than or equal to

A PicBasic Pro Boolean Expression   EXAMPLE 7.3

The following PicBasic Pro statement turns on a motor controlled by a transistor connected to pin RA0 when the state of a switch connected to pin RB0 is high or when the state of a switch connected to pin RB1 is low, and when a byte variable *count* has a value less than or equal to 10:

```
If (((PORTB.0 == 1) OR (PORTB.1 == 0)) And (count <= 10)) Then
    High PORTA.0
```

The simplest form of looping is to use a statement label with a *goto* statement as illustrated earlier in the *flash.bas* example. PicBasic Pro also provides *For . . . Next* and *While . . . Wend* statement structures to perform looping and iteration. These constructs are demonstrated in examples through the remainder of the chapter.

Internet Link

7.6 PicBasic Pro
manual (online,
PDF file)

Table 7.5 lists all of the PicBasic Pro statements with corresponding descriptions. Complete descriptions and examples of the statements and their associated parameters and variables can be found in the PicBasic Pro Compiler manual available online at microEngineering, Inc.'s website (see Internet Link 7.6). In Table 7.5, the keywords are capitalized, and the parameters or variables that follow the keywords are shown in lower case. Also, any statement parameters enclosed within curly brackets ({...}) are optional. All the features and operators just presented and all the statements listed in the table are built from the limited set of assembly language instructions given in Table 7.2. PicBasic Pro eliminates the cryptic assembly language details for you and provides a high-level, more user-friendly language.

Table 7.5 PicBasic Pro statement summary

Statement	Description
@ assembly statement	Insert one line of assembly language code
ADCIN channel, var	Read the on-chip analog to digital converter (if there is one)
ASM . . . ENDASM	Insert an assembly language code section consisting of one or more statements
BRANCH index, [label1{, label2, . . .}]	Computed goto that jumps to a label based on index
BRANCHL index, [label1{, label2, . . .}]	Branch to a label that can be outside of the current page of code memory (for PICs with more than 2 k of program ROM)
BUTTON pin, down_state, auto_repeat_delay, auto_repeat_rate, countdown_variable, action_state, label	Read the state of a pin and perform debounce (by use of a delay) and autorepeat (if used within a loop)
CALL assembly_label	Call an assembly language subroutine
CLEAR	Zero all variables
CLEARWDT	Clear the watch-dog timer
COUNT pin, period, var	Count the number of pulses occurring on a pin during a period
DATA {@ location,} constant1 {, constant2, . . .}	Define initial contents of the on-chip EEPROM (same as the EEPROM statement)
DEBUG item1{, item2, . . .}	Asynchronous serial output to a pin at a fixed baud rate
DEBUGIN {timeout, label,} [item1 {, item2, . . .}]	Asynchronous serial input from a pin at a fixed baud rate
DISABLE	Disable ON INTERRUPT and ON DEBUG processing
DISABLE DEBUG	Disable ON DEBUG processing
DISABLE INTERRUPT	Disable ON INTERRUPT processing
DTMFOUT pin, {on_ms, off_ms,} [tone1 {, tone2, . . .}]	Produce touch tones on a pin
{EEPROM {@ location,} constant1 {, constant2, . . .}}	Define initial contents of on-chip EEPROM (same as the DATA statement)
ENABLE	Enable ON INTERRUPT and ON DEBUG processing
ENABLE DEBUG	Enable ON DEBUG processing
ENABLE INTERRUPT	Enable ON INTERRUPT processing
END	Stop execution and enter low power mode

Table 7.5 PicBasic Pro statement summary

Statement	Description
FOR count = start TO end {STEP {−} inc} {body statements} NEXT {count}	Repeatedly execute statements as count goes from start to end in fixed increment
FREQOUT pin, on_ms, freq1{, freq2}	Produce up to two frequencies on a pin
GOSUB label	Call a PicBasic subroutine at the specified label
GOTO label	Continue execution at the specified label
HIGH pin	Make pin output high
HSERIN {parity_label,} {time_out, label,} [item1{, item2, . . .}]	Hardware asynchronous serial input (if there is a hardware serial port)
HSEROUT [item1{, item2, . . .}]	Hardware asynchronous serial output (if there is a hardware serial port)
I2CREAD data_pin, clock_pin, control, { address,} [var1{, var2, . . .}]{, label}	Read bytes from an external I^2C serial EEPROM device
I2CWRITE data_pin, clock_pin, control, { address,} [var1{, var2, . . .}]{, label}	Write bytes to an external I^2C serial EEPROM device
IF log_comp THEN label	Conditionally jump to a label
IF log_comp THEN true_statements ELSE false_statements ENDIF	Conditional execution of statements
INPUT pin	Make pin an input
LCDIN {address,} [var1{, var2, . . .}]	Read RAM on a liquid crystal display (LCD)
LCDOUT item1{, item2, . . .}	Display characters on LCD
{LET} var = value	Assignment statement (assigns a value to a variable)
LOOKDOWN value, [const1 {, const2, . . .}], var	Search constant table for a value
LOOKDOWN2 value, {test} [value1 {, value2, . . .}], var	Search constant/variable table for a value
LOOKUP index, [const1{, const2, . . .}], var	Fetch constant value from a table
LOOKUP2 index, [value1{, value2, . . .}], var	Fetch constant/variable value from a table
LOW pin	Make pin output low
NAP period	Power down processor for a selected period of time
ON DEBUG GOTO label	Execute PicBasic debug subroutine at label after every statement if debug is enabled
ON INTERRUPT GOTO label	Execute PicBasic subroutine at label when an interrupt is detected
OUTPUT pin	Make pin an output
PAUSE period	Delay a given number of milliseconds
PAUSEUS period	Delay a given number of microseconds
{PEEK address, var}	Read byte from a register
{POKE address, var}	Write byte to a register
POT pin, scale, var	Read resistance of a potentiometer, or other variable resistance device, connected to a pin with a series capacitor to ground
PULSIN pin, state, var	Measure the width of a pulse on a pin
PULSOUT pin, period	Generate a pulse on a pin

(continued)

Table 7.5 PicBasic Pro statement summary *(Continued)*

Statement	Description
PWM pin, duty, cycles	Output a pulse width modulated (PWM) pulse train to pin
RANDOM var	Generate a pseudo-random number
RCTIME pin, state, var	Measure pulse width on a pin
READ address, var	Read a byte from on-chip EEPROM
READCODE address, var	Read a word from code memory
RESUME {label}	Continue execution after interrupt handling
RETURN	Continue execution at the statement following last executed GOSUB
REVERSE pin	Make output pin an input or an input pin an output
SERIN pin, mode,{ timeout, label,} {[qual1, qual2, . . .],} { item1{, item2, . . .}}	Asynchronous serial input (Basic Stamp 1 style)
SERIN2 data_pin{\flow_pin}, mode, {parity_ label,} {timeout, label,} [item1 {, item2, . . .}]	Asynchronous serial input (Basic Stamp 2 style)
SEROUT pin, mode, [item1{, item2, . . .}]	Asynchronous serial output (Basic Stamp 1 style)
SEROUT2 data_pin{\flow_pin}, mode, {pace,} {timeout, label,} [item1{, item2, . . .}]	Asynchronous serial output (Basic Stamp 2 style)
SHIFTIN data_pin, clock_pin, mode, [var1{\ bits1} {, var2{\bits2}, . . .}]	Synchronous serial input
SHIFTOUT data_pin, clock_pin, mode, [var1 {\bits1} {, var2{\bits2}, . . .}]	Synchronous serial output
SLEEP period	Power down the processor for a given number of seconds
SOUND pin, [note1, duration1{, note2, dura- tion2, . . .}]	Generate a tone or white noise on a specified pin
STOP	Stop program execution
SWAP var1, var2	Exchange the values of two variables
TOGGLE pin	Change the state of an output pin
WHILE logical_comp statements	Execute code while condition is true
WEND	
WRITE address, value	Write a byte to on-chip EEPROM
WRITECODE address, value	Write a word to code memory
XIN data_pin, zero_pin, {timeout, label,} [var1{, var2, . . .}]	Receive data from an external X-10 type device
XOUT data_pin, zero_pin, [house_code1\key_ code1{\repeat1}{, house_code2\key_code2{\ repeat2, . . .}}]	Send data to an external X-10 type device

7.5.2 PicBasic Pro Programming Examples

This section presents a series of problems that can be solved with a PIC16F84. The examples illustrate the application of PicBasic Pro. In Sections 7.7 and 7.8, we present more details on how to interface the PIC to a variety of input and output devices. In Section 7.9, we present a methodical design procedure that will help you create software and associated hardware when challenged to design a new microcontroller-based mechatronic system.

PicBasic Pro Alternative to the Assembly Language Program in Example 7.2

EXAMPLE 7.4

As in Example 7.2, the objective is to turn on an LED when the user presses a switch, and turn it off when the switch is released. After the switch is pressed and released a specified number of times, a second LED is to turn on and stay lit. Example 7.2 presented an assembly language solution to this problem. A corresponding PicBasic Pro solution follows. The comments included throughout the code help explain the function of the various parts of the program.

The *While . . . Wend* construct allows the program to wait for the first switch to be pressed or released. The *While* loop continually cycles and does nothing while the switch remains at a particular state. In most applications, there would be statements between the *While* and *Wend* lines that are executed each time through the loop, but none are required here.

The prefix *my_* is included as part of the identifiers *my_count* and *my_button* because the words *count* and *button* are **reserved words.** Reserved words are those used by PicBasic Pro as keywords in statements, predefined constants, and mathematical and logical functions. These words cannot be used as identifiers.

The only fundamental difference in the assembly language and PicBasic Pro solutions is the way the count is handled. Here, we are able to count up and detect when the count reaches the target value. In assembly language, this is not easily done, and we chose to count down instead. Another difference is that PicBasic Pro simplifies how memory is handled. In PicBasic Pro, you do not need to identify addresses for variables, specify memory banks, or move values through the accumulator. PicBasic Pro does all of this for you implicitly.

```
' bcount.bas
' Program to turn on an LED every time a pushbutton switch is pressed, and
' turn on a second LED once it has been pressed a specified number of times

' Define variables and constants
my_count      Var      BYTE      ' number of times switch has been pressed
TARGET        Con      5         ' number of switch presses required

' Define variable names for the I/O pins
my_button     Var      PORTB.0
led_count     Var      PORTA.0
led_target    Var      PORTA.1

' Initialize the counter and guarantee the LEDs are off
my_count = 0
Low led_count
Low led_target

begin:
      ' Wait for the switch to be pressed
      While (my_button == 0)      ' wait as long as switch is not pressed (0)
      Wend

      ' Turn on the count LED now that the switch has been pressed
      High led_count
```

(continued)

(continued)

```
      ' Wait for the switch to be released
      While (my_button == 1)      ' wait as long as switch is pressed (1)
      Wend

      ' Turn off the count LED and increment the counter now that the switch
      ' has been released
      Low led_count
      my_count = my_count + 1

      ' Check if the target has been reached; and if so, turn on the target LED
      If (my_count >= TARGET) Then
            High led_target
      Endif

Goto begin
End
```

If you compare the two solutions, you will see that PicBasic Pro code is easier to write and comprehend. This would be even more evident if the problem were more complex. Things that are easy to do using PicBasic Pro but very difficult using assembly language include variable and array management, assignment statements with complex calculations, logical comparison expressions, iteration, interrupts, pauses, and special purpose functions.

One disadvantage to using PicBasic Pro, however, is that it consumes more program EEPROM space. For this example, the assembly language version requires 17 words of program memory and the PicBasic Pro version requires 39 words, even though the function is identical. This is a consequence of using a high-level language such as PicBasic Pro. Fortunately, the inexpensive PIC16F84 allows up to 1024 words, which is adequate for rather complex programs. Also, many other PIC chips have larger memory capacities for longer programs. Furthermore, the cost of microcontrollers continues to fall and memory capacities continue to rise.

■ **CLASS DISCUSSION ITEM 7.3**
PicBasic Pro and Assembly Language Comparison

Compare the assembly language code in Example 7.2 to the PicBasic Pro code in Example 7.4. Comment on any differences in the way each program functions.

■ **CLASS DISCUSSION ITEM 7.4**
PicBasic Pro Equivalents of Assembly Language Statements

For each of the assembly language statement examples presented in Example 7.1, write corresponding PicBasic Pro code.

PicBasic Pro Program for Security System Example

EXAMPLE 7.5

This example presents a PicBasic Pro program used to control the security system described in Section 6.6. Please refer back to Section 6.6 to review the problem statement. The program comments that follow will help you understand how the code functions. Note how tabs, spaces, blank lines, parentheses, comments, and variable definitions all help make the program more readable. If all these formatting features and definitions were left out, the program would still run, but it would not be as easy for you to understand. The hardware required for PIC implementation is shown in the figure below the code.

The door and window sensors are assumed to be normally open switches that are closed when the door and window are closed. They are wired in series and connected to 5 V through a pull-up resistor; therefore, if either switch is open, then signal A will be high. Both the door and window must be closed for signal A to be low. This is called a **wired-AND** configuration because it is a hardwired solution providing the functionality of an AND gate.

The motion detector produces a high on line B when it detects motion. Single-pole, double-throw switches are used to set the 2-bit code C D. In the figure, the switches are both in the normally closed position; therefore, code C D is 0 0. The alarm buzzer sounds when signal Y goes high, forward biasing the transistor. When Y is high, the 1 k base resistor limits the output current to approximately 5 mA (5 V / 1 kΩ), which is well within the output current specification for a PORTA pin (20 mA as listed in Section 7.8.2).

```
' security.bas

' PicBasic Pro program to perform the control functions of the security
' system presented in Section 6.6

' Define variables for I/O port pins
door_or_window      Var       PORTB.0      ' signal A
motion              Var       PORTB.1      ' signal B
c                   Var       PORTB.2      ' signal C
d                   Var       PORTB.3      ' signal D
alarm               Var       PORTA.0      ' signal Y

' Define constants for use in IF comparisons
OPEN                Con       1        ' to indicate that a door OR window is open
DETECTED            Con       1        ' to indicate that motion is detected

' Make sure the alarm is off to begin with
Low alarm

' Main polling loop
always:
        If ((c == 0) And (d == 1)) Then       ' operating state 1 (occupants
                                              ' sleeping)
                If (door_or_window == OPEN) Then
                        High alarm
                Else
                        Low alarm
                Endif
```

(continued)

(continued)

```
        Else
        If ((c == 1) And (d == 0)) Then    ' operating state 2 (occupants away)
                If((door_or_window == OPEN) Or (motion == DETECTED)) Then
                        High alarm
                Else
                        Low alarm
                Endif
        Else                                ' operating state 3 or NA (alarm
                                            ' disabled)
                Low alarm
        Endif
    Endif
Goto always                                 ' continue to poll the inputs

    End
```

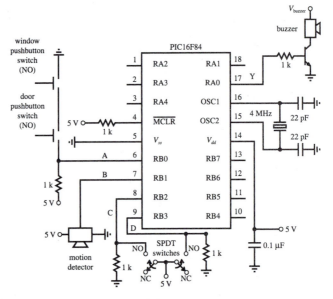

If all the variable and constant definitions, formatting, and comments were left out, the program would still run properly but the resulting code would be much more difficult to comprehend. This is what the resulting code would look like:

```
Low PORTA.0
always: If PORTB.2==0 And PORTB.3==1 Then
If PORTB.0==1 Then
High PORTA.0
Else
Low PORTA.0
Endif
Else
If PORTB.2==1 And PORTB.3==0 Then
```

```
If PORTB.0==1 Or PORTB.1==1 Then
High PORTA.0
Else
Low PORTA.0
Endif
Else
Low PORTA.0
Endif
Endif
Goto always
End
```

There is no advantage to leaving out the formatting, because it is ignored by the compiler. Also, the variable and constant declarations make no difference in the size of the compiled machine code.

■ CLASS DISCUSSION ITEM 7.5
Multiple Door and Window Security System

If you had more than one door and one window, how would modify the hardware design? Would you have to modify the software?

■ CLASS DISCUSSION ITEM 7.6
PIC vs. Logic Gates

In Example 7.5, a PIC solution to the security system problem was presented as an alternative to the logic gate solution presented in Section 6.6. What are the pros and cons of each approach? Which implementation do you think is the best choice in general and in this problem specifically?

Graphically Displaying the Value of a Potentiometer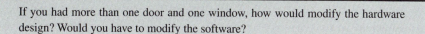

This example presents code designed to sample the resistance of a potentiometer and display a scaled value in binary form using a set of LEDs. The code uses the PicBasic Pro statement *Pot,* which can indirectly sample the resistance of a potentiometer or other variable resistance. The code for the program, *pot.bas,* and the necessary hardware are included below.

The wiper of a potentiometer is connected to pin RA3, and one end of the potentiometer is in series with a capacitor to ground. Note that the third lead of the potentiometer is unconnected.

Each of pins RB0 through RB7 in PORTB is connected to an LED in series with a current-limiting resistor to ground. When any of these pins goes high, the corresponding LED is on. The eight LEDs display a binary number corresponding to the current position

(continued)

(continued)

of the potentiometer. The value displayed can range from 0 to 255. This program uses the assignment statement *PORTB = value* to update the display, where *value* is a byte variable (8 bits) that contains the current sample from the potentiometer. The assignment statement drives the outputs of PORTB such that RB0 represents the least significant bit and RB7 represents the most significant bit of the scaled resistance sample.

The test LED attached to pin RA2 is used to indicate that the program is running. When the program is running, the test LED blinks. It is good practice to include some sort of program execution indicator, especially in the debugging stages of a project. The blinking LED signals that the PIC has power and the necessary support components and that the program is loaded, running, and looping properly. This is a simple example, and not much can go wrong with the program logic and sequencing. However, complicated programs containing complex logic, branching, looping, and interrupts may hang up unexpectedly or terminate prematurely, especially before they are fully debugged. A nonblinking LED would indicate a problem with the program.

The syntax for the *Pot* statement is

```
Pot pin, scale, var
```

where *pin* is the input pin identifier, *scale* is a number between 1 and 255 to adjust for the maximum time constant (RC) of the potentiometer and series capacitor, and *var* is the name of a byte variable used to store the value returned by the *Pot* statement. Refer to the PicBasic Pro manual for details on how to choose an appropriate value for *scale*. When the potentiometer is at minimum resistance, the value of *var* is minimum (0 for 0 Ω), and when the resistance is maximum the value is maximum (255 if *scale* is selected appropriately). For a 5 kΩ potentiometer and a 0.1 μF series capacitor, an appropriate value for scale is 200.

In this example, the *TRISB = %00000000* assignment statement designates all PORTB pins as outputs and is required because the scaled potentiometer value is written directly to PORTB with the *PORTB = value* assignment statement. In previous examples, when statements like *High* and *Low* were used, the TRIS registers did not need to be set explicitly because the statements themselves automatically designate the pins as outputs. *PORTB = value* sets all the PORTB outputs to the corresponding bit values in the byte variable *value*.

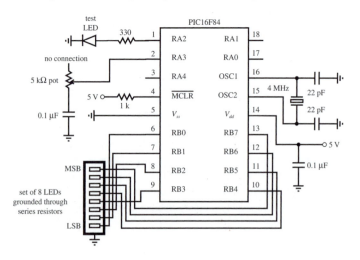

```
' pot.bas

' Graphically displays the scaled resistance of a potentiometer using a set of
'  LEDs corresponding to a binary number ranging from 0 to 255.

' Define variables, pin assignments, and constants
value     Var     BYTE              ' define an 8 bit (byte) variable capable of
                                    '  storing numbers between 0 and 255
test_led  Var     PORTA.2           ' pin to which a test LED is attached (RA2)
pot_pin   Var     PORTA.3           ' pin to which the potentiometer and series
                                    '  capacitor are attached (RA3)
SCALE     Con     200               ' value for Pot statement scale factor

' Define the input/output status of the I/O pins
TRISB = %00000000              ' designate all PORTB pins as outputs

loop:
      High test_led                 ' turn on the test LED

      Pot pot_pin, SCALE, value     ' read the potentiometer value
      PORTB = value                 ' display the binary value graphically
                                    ' with the 8 PORTB LEDs (RB0 through RB7)

      Pause 100                     ' wait one tenth of a second
      Low test_led                  ' turn off the test LED as an indication
                                    ' that the program and loop are running
      Pause 100                     ' wait one tenth of a second

Goto loop     ' continue to sample and display the potentiometer value and blink
              ' the test LED
End
```

This example illustrates how to sample the value of a resistance using the special Pic-Basic Pro statement *Pot*. PicBasic Pro provides an assortment of other high-level statements (e.g., *Button, Freqout, Lcdout, Lookdown, Lookup, Pwm, Serin, Serout,* and *Sound*) that help you create sophisticated functionality with only a few lines of code. Video Demo 7.1 shows a live demonstration of the hardware and software from this example. Eight LEDs provide a visual display of the angular position of a knob (a potentiometer) as it is turned.

Video Demo

7.1 Potentio-meter input and binary display

■ **CLASS DISCUSSION ITEM 7.7**
How Does Pot Work?

The PicBasic Pro statement *Pot* applied in Example 7.6 uses a digital I/O pin to measure the resistance of a potentiometer. *Pot* effectively converts the analog resistance value into a digital number, appearing to function as an A/D converter. How do you think PicBasic Pro accomplishes this? Hint: Consider the step response of an RC circuit and the use of a single pin as an output and then an input.

■ **CLASS DISCUSSION ITEM 7.8**
Software Debounce

Section 6.10.1 presented how to debounce a single-pole, double-throw switch using a NAND gate or flip-flop circuit. This is called a *hardware solution,* because it requires extra components wired together. If a switch is used to input data to a PIC design, debounce can be done in software instead. Assuming that a switch is connected to a PIC by only a single line, write PicBasic Pro code to perform the debounce. Note that the PicBasic Pro statement *Button* can be used for this purpose, but here we want you to think about how you would do it using more fundamental statements.

DESIGN EXAMPLE 7.1

Option for Driving a Seven-Segment Digital Display with a PIC

There will be PIC applications where you need to display a decimal digit using a seven-segment LED display. The display could represent some calculated or counted value (e.g., the number of times a switch was pressed). One approach is to drive the seven LED segments directly from seven output pins of a PIC. This would involve decoding in software to determine which segments need to be on or off to display the digit properly. If we label the segments as shown in the figure in the margin and if the PORTB pins are wired to the segments of the LED display, where the segments are connected to 5 V through a set of current-limiting resistors, the following initialization code must appear at the top of your program:

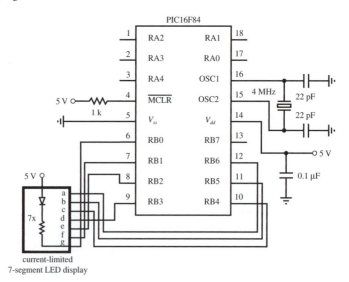

```
' Declare variables
number    var    BYTE    ' digit to be displayed (value assumed to be from 0
                        ' to 9)
pins  var  BYTE[10]   ' an array of 10 bytes used to store the 7-segment
                        ' display codes for each digit

' Initialize I/O pins
TRISB = %00000000      ' designate all PORTB pins as outputs (although, pin 7 is
                        ' not used)

' Segment codes for each digit where a 0 implies the segment is on and a 1 implies
' it is off, because the PIC sinks current from the LED display
'           %gfedcba              display
pins[0]  = %1000000              ' 0
pins[1]  = %1111001              ' 1
pins[2]  = %0100100              ' 2
pins[3]  = %0110000              ' 3
pins[4]  = %0011001              ' 4
pins[5]  = %0010010              ' 5
pins[6]  = %0000011              ' 6
pins[7]  = %1111000              ' 7
pins[8]  = %0000000              ' 8
pins[9]  = %0011000              ' 9
```

The remainder of your code might consist of a polling loop that needs to update the digital display periodically. A subroutine can be used to accomplish this. **Subroutines** are blocks of code to perform specialized functions that may need to be executed in numerous places within your program. Using the byte variable *number* declared in the code, the following subroutine could be used to display the value stored in the variable:

```
' Subroutine to display a digit on a 7-segment LED display. The value of the
' digit must be stored in a byte variable called "number." The value is assumed
' to be less than 10; otherwise, all segments are turned off to indicate an error.
display_digit:
        If (number < 10) Then
              PORTB = pins[number]      ' display the digit
        Else
              PORTB = %1111111          ' turn off all 7 segments
        Endif
Return
```

A number can be displayed at any point in your program by assigning the value to the variable *number* and calling the subroutine. For example, the following statements would display the digit 8:

```
number = 8
Gosub display_digit
```

The solution just presented above requires seven output pins. Because the PIC16F84 has a total of only 13 I/O pins, this could limit the addition of other I/O functions in your design. *(continued)*

(continued)

An alternative design that requires fewer output pins uses a seven-segment decoder IC (e.g., 7447). Here, only four I/O pins are required as shown next:

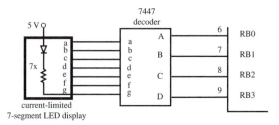

For this case, the *pins* array is not required, and only pins RB0 through RB3 require initialization as outputs. The subroutine would change to

```
display_digit:
      If (number < 10) Then
            PORTB = (PORTB & $F0) | number      ' display the digit
      Else
            PORTB = (PORTB & $F0) | $F           ' turn off all segments
      Endif
Return
```

The assignment statement for PORTB uses a **logic mask** to retain the four MSBs of PORTB, which may have been independently set by other program statements for other functions, and to assign the binary equivalent of *number* to the four LSBs that are output to the seven-segment display driver. A logic mask is a bit string used to protect selected bits in a binary number from change while allowing others to change. The bitwise AND (&) and OR (|) operators are used to help accomplish the mask operation. The term *PORTB & $F0* retains the four MSBs while clearing the four LSBs with zeros. For example, if PORTB's current value is %11011001, then *PORTB & $F0* yields the following result:

$$\begin{array}{ll} \%11011001 & (PORTB) \\ \& & \\ \%11110000 & (\$F0) \\ = & \\ \mathbf{\%11010000} & \mathbf{(PORTB \& \$F0)} \end{array}$$

By OR-ing this result with *number,* the four LSBs are replaced by the corresponding binary value of *number.* For example, if the current value of *number* is 7 (%0111), then *(PORTB & $F0) | number* would yield the following result:

$$\begin{array}{ll} \%11010000 & (PORTB \& \$F0) \\ | & \\ \%00000111 & (number) \\ = & \\ \mathbf{\%11010111} & \mathbf{((PORTB \& \$F0) \mid number)} \end{array}$$

As you can see, the four MSBs of PORTB remain unchanged, and the four LSBs have changed to the binary value of *number.*

The *$F* (15) in the *Else* clause is the input value required by the decoder IC to blank all seven segments.

If you lack the luxury of even four I/O pins in your design but still wish to display a digit, another alternative is to use a 7490 decade counter IC with reset and count inputs. The reset input is assumed to have positive logic, so when the line goes high, the counter is reset to 0. The count input is edge triggered, and in this example it does not matter if it is positive or negative edge-triggered. Only two PIC I/O pins are required to drive this alternative as shown next:

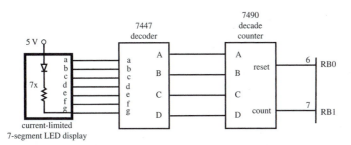

In this case, only pins RB0 and RB1 need be initialized as outputs, and a new counter variable (*i*) has to be declared for use in the subroutine. Also, two pin assignments can be made as follows:

```
i          Var        BYTE          ' counter variable used in FOR loop
reset      Var        PORTB.0       ' signal to reset the counter to 0
count      Var        PORTB.1       ' signal to increment the counter by 1
```

The subroutine would change to

```
display_digit:
        Pulsout reset, 1            ' send a full pulse to reset the counter
                                    ' to zero
        If (number < 10) Then
        ' Increment the counter "number" times to display the appropriate
        ' digit
        For i = 1 To number
                Pulsout count, 1    ' send a full pulse to increment the
                                    '  counter
        Next i
    Else
        ' Increment the counter 15 times to clear the display (all segments
        ' off)
        For i = 1 To 15
                Pulsout count, 1    ' send a full pulse to increment the
                                    '  counter
        Next i
    Endif
Return
```

(continued)

(concluded)

Here, the PicBasic Pro statement *Pulsout* is used to send a pulse to each of the control pins. The syntax of this command is

```
Pulsout pin, period
```

where *pin* is the pin identifier (e.g., PORTB.0) and *period* is the length of the pulse in tens of microseconds. The pulses generated by the code above are 10 microseconds wide.

In this example, the last two alternatives require additional components (the decoder and counter ICs). If the physical size of the design is no constraint and an objective is to minimize cost by not having to use an additional PIC16F84 or an alternative PIC with more I/O pins, then one of the latter alternatives might be attractive.

■ **CLASS DISCUSSION ITEM 7.9**
Fast Counting

In the third option presented in Design Example 7.1, the counter was incremented the appropriate number of times to display the desired decimal value on the display. Do you think that this counting will be detectable on the display? Why or why not?

Lab Exercise

Lab 10
Programming a PIC microcontroller—part II

Lab Exercise 10 shows how to wire components and write software to create some of the functionality presented in Design Example 7.1. The exercise expands upon the design example functionality by providing a display of all hexadecimal digits from 0 to F. Video Demos 7.2 and 7.3 show two different designs explored in Lab Exercise 10. The first solution uses a 555 timer circuit with data latches to debounce pushbutton switch input signals, and the second solution deals with switch bounce in software using delays.

Video Demo

7.2 Hexadecimal counter using a 555 timer and data flip-flops for switch debounce

7.3 Hexadecimal counter with software debounce

7.6 USING INTERRUPTS

The program to solve the security system problem presented in Example 7.9 uses a method called **polling,** where the program includes a check of the sensor inputs within a loop to update the output accordingly. Program flow is easy to understand, because all processing takes place within the main program loop. The loop repeats as long as the microcontroller is powered. For more complex applications, polling may not be suitable, because the loop may take too long to execute. In a long loop, the inputs may not be checked often enough. An alternative approach is to use an interrupt. In an interrupt-driven program, some inputs are connected to special input lines, designated as interrupts. When one or more of these lines changes level, the microcontroller temporarily suspends normal program execution while the change is acted on by a subprogram or function called an **interrupt service routine.** At the end of the service routine, control is returned to the main program at the point where the interrupt occurred. Because polling is easier to implement, it is preferred over interrupts, as long as the polling loop can run fast enough.

To detect interrupts, two specific registers on the PIC must be initialized correctly. These are the option register (**OPTION_REG**) and the interrupt control register (**INTCON**). The definition for each bit in the first register (OPTION_REG) follows. Recall that the least significant bit is on the right and designated as bit $0(b_0)$, while the most significant bit is on the left and designated as bit 7 (b_7):

$$OPTION\_REG = \%b_7b_6b_5b_4b_3b_2b_1b_0$$

bit 7: $\overline{\text{RBPU}}$: PORTB pull-up enable bit
   1 = PORTB pull-ups are disabled
   0 = PORTB pull-ups are enabled
bit 6: Interrupt edge select bit
   1 = Interrupt on rising edge of signal on pin RB0
   0 = Interrupt on falling edge of signal on pin RB0
bit 5: T0CS: TMR0 clock source select bit
   1 = External signal on pin RA4
   0 = Internal instruction cycle clock (CLKOUT)
bit 4: T0SE: TMR0 source edge select bit
   1 = Increment on high-to-low transition of signal on pin RA4
   0 = Increment on low-to-high transition of signal on pin RA4
bit 3: PSA: prescaler assignment bit
   1 = Prescaler assigned to the Watchdog timer (WDT)
   0 = Prescaler assigned to TMR0
bits 2, 1, and 0:
   3-bit value used to define the prescaler rate for the timer features

Value	TMR0 Rate	WDT Rate
000	1 : 2	1 : 1
001	1 : 4	1 : 2
010	1 : 8	1 : 4
011	1 : 16	1 : 8
100	1 : 32	1 : 16
101	1 : 64	1 : 32
110	1 : 128	1 : 64
111	1 : 256	1 : 128

In the *onint.bas* example presented below, OPTION_REG is set to $7F, which is %01111111. Setting bit 7 low enables PORTB pull-ups and setting bit 6 high causes interrupts to occur on the positive edge of a signal on pin RB0. When pull-ups are enabled, the PORTB inputs are held high until they are pulled low by the external input circuit (e.g., a switch to ground wired to pin RB0). Bits 0 through 5 are important only when using special purpose timers.

The definition for each bit in the second register (INTCON) follows:

bit 7: GIE: global interrupt enable bit
   1 = Enables all unmasked interrupts
   0 = Disables all interrupts

bit 6: EEIE: EE write complete interrupt enable bit
           1 = Enables the EE write complete interrupt
           0 = Disables the EE write complete interrupt
bit 5: T0IE: TMR0 overflow interrupt enable bit
           1 = Enables the TMR0 interrupt
           0 = Disables the TMR0 interrupt
bit 4: INTE: RB0 interrupt enable bit
           1 = Enables the RB0/INT interrupt
           0 = Disables the RB0/INT interrupt
bit 3: RBIE: RB port change interrupt enable bit (for pins RB4 through RB7)
           1 = Enables the RB port change interrupt
           0 = Disables the RB port change interrupt
bit 2: T0IF: TMR0 overflow interrupt flag bit
           1 = TMR0 has overflowed (must be cleared in software)
           0 = TMR0 did not overflow
bit 1: INTF: RB0 interrupt flag bit
           1 = The RB0 interrupt occurred
           0 = The RB0 interrupt did not occur
bit 0: RBIF: RB port change interrupt flag bit
           1 = At least one of the signals on pins RB4 through RB7 has changed
               state (must be cleared in software)
           0 = None of the signals on pins RB4 through RB7 has changed state

In the *onint.bas* example that follows, INTCON is set to $90, which is %10010000. For interrupts to be enabled, bit 7 must be set to 1. Bit 4 is set to 1 to check for interrupts on pin RB0. Bits 0 and 1 are used to indicate interrupt status during program execution. If more than one interrupt signal were required, bit 3 would be set to 1, which would enable interrupts on pins RB4 through RB7. In that case, INTCON would be set to $88 (%10001000). To check for interrupts on RB0 and RB4–7, INTCON would be set to $98 (%10011000). PORTA has no interrupt capability, and PORTB has interrupt capability only on pin RB0 and pins RB4 through RB7. Bits 6, 5, and 2 are for advanced features not used in this example.

A simple example, called *onint.bas,* that illustrates the use of interrupts follows. The corresponding schematic is shown in Figure 7.8. The program's function is described in detail in the following paragraphs.

```
' onint.bas

' This program turns on an LED and waits for an interrupt on PORTB.0. When RB0
' changes state, the program turns the LED off for 0.5 seconds and then resumes
' normal execution.

led    var    PORTB.7              ' designate pin RB7 as "led"

       OPTION_REG = $7F            ' enable PORTB pull-ups and detect positive
                                   '  edges on interrupt
       On Interrupt Goto myint     ' define interrupt service routine location
       INTCON = $90                ' enable interrupts on pin RB0
```

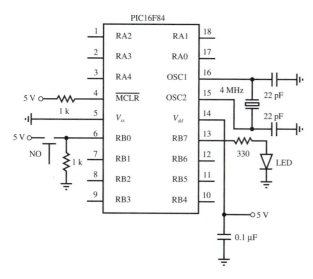

Figure 7.8 Interrupt example schematic.

```
' Turn LED on and loop until there is an interrupt
loop:   High led
        Goto loop

' Interrupt handling routine
Disable                         ' disable interrupts until the Enable
                                ' statement appears
myint:  Low led                 ' turn LED off
        Pause 500               ' wait 0.5 seconds
        INTCON.1 = 0            ' clear interrupt flag
        Resume                  ' return to main program
Enable                          ' allow interrupts again

End                             ' end of program
```

The *onint.bas* program turns on an LED connected to pin RB7 until an external interrupt occurs. A normally open pushbutton switch connected to pin RB0 provides the source for the interrupt signal. When the signal transitions from low to high, the interrupt routine executes, causing the LED to turn off for half a second. Control then returns to the main loop, causing the LED to turn back on again.

The first active line creates the variable name *led* to denote the pin identifier PORTB.7. In the next line, the OPTION_REG is set to $7F (or %01111111) to enable PORTB pull-ups and configure the interrupt to be triggered when the positive edge of a signal occurs on pin RB0. The PicBasic Pro statement *On Interrupt Goto* designates the label *myint* as the location to which program control jumps when an interrupt occurs. The value of the INTCON register is set to $90 (or %10010000) to properly enable RB0 interrupts.

The two lines starting with the label *loop* cause the program to loop continually, maintaining program execution while waiting for an interrupt. The LED connected to pin RB7 remains on during this loop. The loop is called an **infinite loop,** because it cycles as long as no interrupt occurs. Note that an active statement (such as *High led*) must exist between the label and the *Goto* of the loop for the interrupt to function, because PicBasic Pro checks for interrupts only after a statement is completed.

The final section of the program contains the interrupt service routine. *Disable* must precede the label, and *Enable* must follow the *Resume* to prevent checking for interrupts until control is returned to the main program. The interrupt service routine executes when control of the program is directed to the beginning of this routine, labeled by *myint,* when an interrupt occurs on pin RB0. At the identifier label *myint,* the statement *Low led* clears pin RB7, turning off the LED. The *Pause* statement causes a 500 millisecond delay, during which time the LED remains off. The next line sets the INTCON.1 bit to 0 to clear the interrupt flag. The interrupt flag was set internally to 1 when the interrupt signal was received on pin RB0, and this bit must be reset to 0 before exiting the interrupt routine, so that subsequent interrupts can be serviced. At the end of the *myint* routine, control returns to the main program loop where the interrupt occurred. Lab Exercise 9 explores the concepts and example presented in this section.

Lab Exercise

Lab 9
Program-
ming a PIC
microcontroller—
part I

Internet Link

7.7 PIC I/O
interface devices
and useful
accessories

7.7 INTERFACING COMMON PIC PERIPHERALS

This section introduces interfacing a PIC to two common peripheral devices. The first is a 12-button keypad that can be used to input numeric data. The second is a liquid crystal display that can be used to output messages and numeric information to the user. More information on these and other useful peripheral devices can be found online (see Internet Link 7.7).

7.7.1 Numeric Keypad

Figure 7.9 illustrates a common three-row, four-column 12-button **keypad.** Figure 7.10 shows a photographs of two common keypads. One has 12 keys with a ribbon cable interface, and the other has 16 keys with solder holes to which individual wires can be attached. Each key in a keypad is attached to a normally open pushbutton switch. When a key is pressed, the switch closes. Figure 7.11 illustrates the electrical schematic of the keypad with a recommended interface to the PIC16F84. A standard keypad has a seven-pin header for connection to a ribbon cable socket. There is one pin for each row and one pin for each column as numbered in Figure 7.11.

The four rows (row 1, row 2, row 3, row 4) are connected to pins RB7 through RB4, which are configured as inputs. Internal pull-up resistance is available as a software option on these pins, so external pull-up resistors are not required (see details in Section 7.8.1). The three columns (col 1, col 2, col 3) are connected to pins RB0 through RB2, which are configured as outputs. The following PicBasic Pro code contains initializations and a framework for a polling loop that can be used to process input from the keypad. The column outputs are cleared low one at a time, and each row input is polled to determine if the key switch in that column is closed.

Figure 7.9 Numeric keypad.

Figure 7.10 Photograph of 12-key and 16-key numeric keypads.

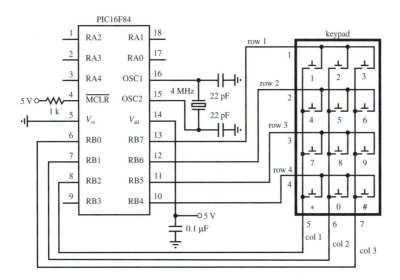

Figure 7.11 Numeric keypad schematic and PIC interface.

For example, if col 1 is low while col 2 and col 3 are high, and only the 1 key is being held down, then row 1 will be low while the remaining row lines will be high, indicating that only the 1 key is down. Statements could be added to the *If* statement blocks, in place of the comments, to process the input.

```
' Pin assignments
row1 Var PORTB.7
row2 Var PORTB.6
row3 Var PORTB.5
row4 Var PORTB.4
col1 Var PORTB.2
col2 Var PORTB.1
col3 Var PORTB.0

' Enable PORTB pull-ups
OPTION_REG = $7f

' Initialize the I/O pins (RB7:RB4 and RB3 as inputs and RB2:RB0 as outputs)
TRISB = %11111000

' Keypad polling loop
loop:
        ' Check column 1
        Low col1 : High col2 : High col3
        If (row1 == 0) Then
                ' key 1 is down
        Endif
        If (row2 == 0) Then
                ' key 4 is down
        Endif
        If (row3 == 0) Then
                ' key 7 is down
        Endif
        If (row4 == 0) Then
                ' key * is down
        Endif

        ' Check column 2
        High col1 : Low col2 : High col3
        If (row1 == 0) Then
                ' key 2 is down
        Endif
        If (row2 == 0) Then
                ' key 5 is down
        Endif
        If (row3 == 0) Then
                ' key 8 is down
        Endif
        If (row4 == 0) Then
                ' key 0 is down
        Endif
```

```
' Check column 3
High col1 : High col2 : Low col3
If (row1 == 0) Then
        ' key 3 is down
Endif
If (row2 == 0) Then
        ' key 6 is down
Endif
If (row3 == 0) Then
        ' key 9 is down
Endif
If (row4 == 0) Then
        ' key # is down
Endif
' Continue polling

Goto loop

End
```

Lab Exercise 11 explores how to wire and accept input from a numeric keypad. In this case, three keys of a keypad are used to control the motion of a DC motor. Video Demo 7.4 shows a demonstration of the example in action.

Lab Exercise

Lab 11 Pulse-width-modulation motor speed control with a PIC

Video Demo

7.4 Pulse-width-modulation speed control of a DC motor, with keypad input

7.7.2 LCD Display

The other common peripheral device we want to highlight is a standard Hitachi 44780-based **liquid crystal display (LCD).** LCDs come in different shapes and sizes that can support different numbers of rows of text and different numbers of characters per row. The standard choices for the number of characters and rows are 8×2, 16×1, 16×2, 16×4, 20×2, 24×2, 40×2, and 40×4. An example of a common 20×2 LCD is shown in Figure 7.12. It is illustrated schematically in the top of Figure 7.13. Applications of LCDs include displaying messages or information to the user (e.g., a home thermostat display, a microwave oven display, or a digital clock) and displaying a hierarchical input menu for changing settings and making selections (e.g., a copy machine or printer display).

For an LCD display with 80 characters or less (all but the 40×4 just listed), the display is controlled via 14 pins. The names and descriptions of these pins are listed in Table 7.6. PicBasic Pro offers a simple statement called *Lcdout* to control an LCD display. LCD displays with more than 80 characters (40×4) use a 16-pin header with different pin assignments not compatible with *Lcdout*. A 14-pin LCD can be controlled via four or eight data lines. PicBasic Pro supports both, but it is recommended that you use four lines to minimize the number of I/O pins required. Figure 7.13 shows the recommended interface to the PIC using a four-line data bus. Commands and data are sent to the display via lines DB4 through DB7, and lines DB0 through DB3 (pins 7 through 10) are not used. Pin RA4 is connected to 5 V through a pull-up resistor because it is an open drain output (see details in Section 7.8). The potentiometer connected to V_{ee} is used to adjust the contrast between the

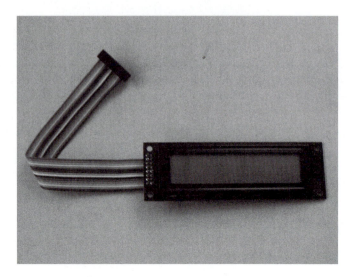

Figure 7.12 Photograph of an LCD.

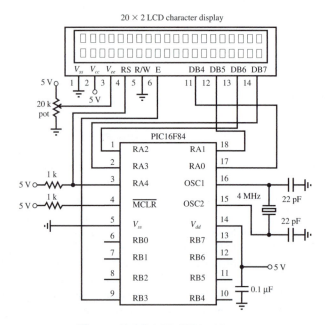

Figure 7.13 LCD PIC interface.

Internet Link

7.7 PIC I/O interface devices and useful accessories

foreground and background shades of the display. The RS, R/W, and E lines are controlled automatically by PicBasic Pro when communicating with the display. Detailed information about LCD displays and how to write your own interface can be found online (see Internet Link 7.7).

Table 7.6 Liquid crystal display pin descriptions

Pin	Symbol	Description
1	V_{ss}	Ground reference
2	V_{cc}	Power supply (5 V)
3	V_{ee}	Contrast adjustment voltage
4	RS	Register select (0: instruction input; 1: data input)
5	R/W	Read/write status (0: write to LCD; 1: read from LCD RAM)
6	E	Enable signal
7–14	DB0–DB7	Data bus lines

With the hardware interface shown in Figure 7.13, the display can be controlled with the PicBasic Pro statement *Lcdout*. The simplest form of this statement is *Lcdout text* (e.g., *Lcdout "Hello world"*) where *text* is a string constant. The statement also supports various commands for controlling the display and cursor and for outputting numbers and data in different formats. Refer to the description of the *Lcdout* statement in the PicBasic Pro compiler manual for more information on these options. Here is a simple example to illustrate the use of the commands and format controls. If *x* is defined as a byte variable and currently contains the value 123, the following statement,

```
Lcdout $FE, 1, "Current value for x:", $FE, $C0, " ", DEC x
```

would clear the display and output the following two-line message:

```
Current value for x:
123
```

The code word *$FE* indicates to the display that the next item is a command. In the example above, command *1* clears the display and command *$C0* moves the cursor to the beginning of the next line. The prefix *DEC* is used to instruct the display to output the following number in its decimal digit form rather than its corresponding ASCII character. Video Demo 7.5 shows a demonstration of the example *Lcdout* statement above, where the displayed number is incremented by a *For* loop.

Video Demo

7.5 LCD display

THREADED DESIGN EXAMPLE

DC motor position and speed controller—Keypad and LCD interfaces   **C.2**

The figure that follows shows the functional diagram for Threaded Design Example C (see Section 1.3 and Video Demo 1.8), with the portion described here highlighted.

The schematic on the next page shows the components and connections for this part of the design. A special integrated circuit available from E-Lab (*see* Internet Link 7.8) called the EDE1144 keypad decoder is used to monitor keypresses on the keypad and transmit them to the PIC via a serial interface. Detailed information about this device can be found in the data sheet (*see* Internet Link 7.9). The EDE1144, in addition to monitoring and transmitting keypress information, provides audio feedback to the user when a buzzer is connected as shown. The Beep signal switches the transistor on and off, causing the buzzer to oscillate. The LED in

(continued)

Video Demo

1.8 DC motor
position and
speed controller

parallel with the buzzer provides a visual cue that a keypad button is being pressed. The LCD is wired in the standard way shown in Figure 7.13, allowing convenient use of the PicBasic Pro statement: *Lcdout.*

Presented below is a portion of the PicBasic Pro code designed to accept input from the keypad and display a menu-driven user interface on the LCD. The remainder of the code will be presented in Threaded Design Example C.3. The *Serin* command in the first line of the "main" loop waits for keypress data to be transmitted from the EDE1144. The set of *If* statements then dispatches the appropriate subroutine based on the user selection. Again, more details will be shown in Threaded Design Example C.3.

```
' Define I/O pin name
key_serial Var PORTB.0          ' keypad serial interface input

' Declare Variables
key_value Var BYTE              ' code byte from the keypad

' Define constants
key_mode Con 0                  ' 2400 baud mode for serial connection to
                                  keypad.
key_1 Con $30                   ' hex code for the 1-key on the keypad
key_2 Con $31                   ' hex code for the 2-key on the keypad
key_3 Con $32                   ' hex code for the 3-key on the keypad

' Wait for a keypad button to be pressed (i.e., polling loop)
Gosub main_menu                 ' display the main menu on the LCD

main:
    Serin key_serial, key_mode, key_value
    If (key_value = key_1) Then
       Gosub position
    Else
    If (key_value = key_2) Then
       Gosub speed
```

Internet Link

7.8 E-Lab, Inc.
7.9 EDE1144
keypad decoder

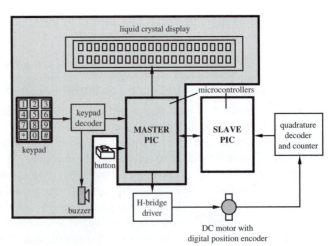

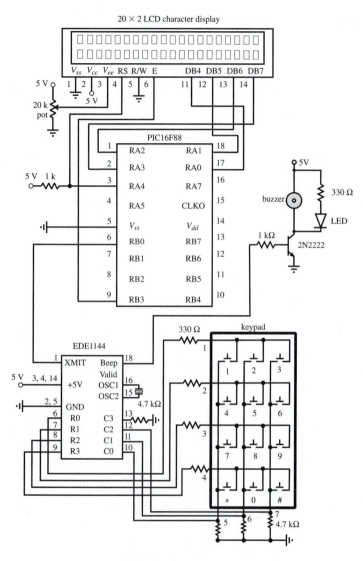

20 × 2 LCD character display

```
     Else
     If (key_value = key_3) Then
          Gosub adjust_gains
     Endif : Endif : Endif
Goto main      ' continue polling keypad buttons
End ' end of main program

' Subroutine to display the main menu on the LCD
main_menu:
     Lcdout $FE, 1, "Main Menu:"
     Lcdout $FE, $C0, "1:pos. 2:speed 3:gain"
Return
```

7.8 INTERFACING TO THE PIC

In this section, we discuss interfacing the PIC to a variety of input and output devices. As we saw in Section 7.5.1, each pin in the I/O ports may be configured in software as an input or an output. In addition, the port pins may be multiplexed with other functions to use additional features of the PIC. In this section, we examine the electronic schematics of the different input and output ports of the PIC16F84. The ports are different combinations of TTL and CMOS devices and have voltage and current limitations that must be considered when interfacing other devices to the PIC. You should first refer to Section 6.11 to review details of TTL and CMOS equivalent output circuits and open drain outputs.

We begin by looking at the architecture and function of each of the ports individually. PORTA is a 5-bit-wide latch with the pins denoted by RA0 through RA4. The block diagram for pins RA0 through RA3 is shown in Figure 7.14, and the block diagram for pin RA4 is shown in Figure 7.15. The five LSBs of the TRISA register configure the 5-bit-wide latch for input or output. Setting a TRISA bit high causes the corresponding PORTA pin to function as an input, and the CMOS output driver is in high impedance mode, essentially removing it from the circuit. Reading the PORTA register accesses the pin values. Clearing a TRISA bit low causes the corresponding PORTA pin to serve as an output, and the data on the data latch appears on the pin. RA4 is slightly different in that it has a Schmitt trigger input buffer that triggers with a distinct transition even for a slowly changing and/or noisy input (see Section 6.12.2). Also, the output configuration of RA4 is open drain, and external components (e.g., a pull-up resistor to power) are required to complete the output circuit.

PORTB is also bidirectional but is 8 bits wide. Its data direction register is denoted by TRISB. Figure 7.16 shows the schematic for pins RB4 through RB7, and Figure 7.17 shows pins RB0 through RB3. A high on any bit of the TRISB register sets the tristate gate to the high impedance mode, which disables the output driver. A low on any bit of the TRISB register places the contents of the data latch on the selected output pin. Furthermore, all of the PORTB pins have **weak pull-up** FETs. These FETs are controlled by a single control bit called $\overline{RBPU}$ (active low register B pull-up). When this bit is low the FET acts like a weak pull-up resistor. This pull-up is automatically disabled when the port pin is configured as an output. $\overline{RBPU}$ can be set in software through the OPTION_REG special purpose register (see Section 7.6).

7.8.1 Digital Input to the PIC

Figure 7.18 illustrates how to properly interface different types of components and digital families of devices as inputs to the PIC. All I/O pins of the PIC that are configured as inputs interface through a TTL input buffer (pins RA0 through RA3 and pins RB0 through RB7) or Schmitt trigger input buffer (RA4). The Schmitt trigger enhances noise immunity for a slowly changing input signal. Because an input pin is TTL buffered in the PIC, interfacing a TTL gate or device to the PIC can be done directly unless it is has an open-collector output. In this case, an external pull-up resistor is required. Because the output of a 5 V powered CMOS device swings

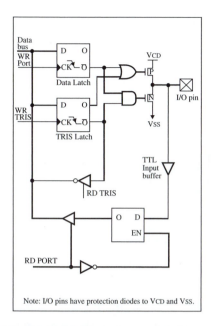

Note: I/O pins have protection diodes to VCD and VSS.

Figure 7.14 Block diagram for pins RA0 through RA3. *(Courtesy of Microchip Technology, Inc., Chandler, AZ)*

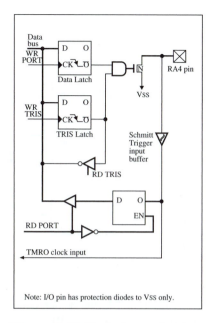

Note: I/O pin has protection diodes to VSS only.

Figure 7.15 Block diagram for pin RA4. *(Courtesy of Microchip Technology, Inc., Chandler, AZ)*

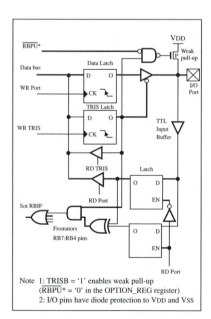

Note 1: TRISB = '1' enables weak pull-up
($\overline{\text{RBPU}}$* = '0' in the OPTION_REG register)
2: I/O pins have diode protection to VDD and VSS

Figure 7.16 Block diagram for pins RB4 through RB7. *(Courtesy of Microchip Technology, Inc., Chandler, AZ)*

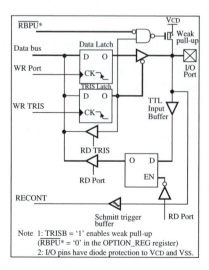

Note 1: TRISB = '1' enables weak pull-up
($\overline{\text{RBPU}}$* = '0' in the OPTION_REG register)
2: I/O pins have diode protection to VCD and VSS.

Figure 7.17 Block diagram for pins RB0 through RB3. *(Courtesy of Microchip Technology, Inc., Chandler, AZ)*

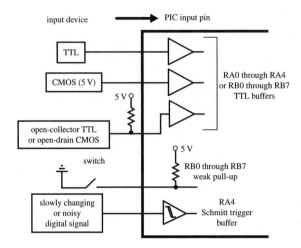

Figure 7.18 Interface circuits for input devices

nearly from 0 to 5 V, the device will drive a PIC input directly. The weak pull-up option on pins RB0 through RB7 is useful when using mechanical switches or keypads for input (see Section 7.7). The pull-up FET maintains a 5 V input until the switch is closed, bringing the input low. Although a TTL input usually floats high if it is open, the FET pull-up option is useful, because it simplifies the interface to external devices (e.g., keypad input). Finally, one must be aware of the current specifications of the PIC input and output pins. For the PIC16F84, there is a 25 mA sink maximum per pin with a 80 mA maximum for the entire PORTA and a 150 mA maximum for PORTB.

7.8.2 Digital Output from the PIC

Figure 7.19 illustrates how to properly interface different types of components and digital families of devices to outputs from the PIC. Pins RA0 through RA3 have full CMOS output drivers, and RA4 has an open-drain output. RB0 through RB7 are TTL buffered output drivers. A 20 mA maximum current is sourced per pin with a 50 mA maximum current sourced by the entire PORTA and a 100 mA maximum for PORTB. CMOS outputs can drive single CMOS or TTL devices directly. TTL outputs can drive single TTL devices directly but require a pull-up resistor to provide an adequate high-level voltage to a CMOS device. To drive multiple TTL or CMOS devices, a buffer can be used to provide adequate current for the fan-out. Because pin RA4 is an open-drain output, external power is required. Note that when RA4 is high, the output pin is grounded to V_{ss}, switching the small current load on, and when RA4 is low, the output is an open circuit, switching the load off. When interfacing transistors, power transistors, and relays, current requirements must be considered for a proper interface. If the PIC contains a D/A converter, it can be used with an amplifier to drive an analog load directly. Otherwise, as shown in the figure, an external D/A IC can be used with the digital I/O ports.

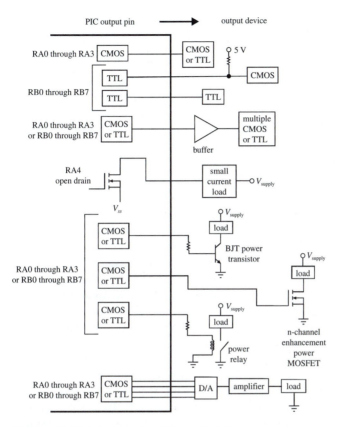

Figure 7.19 Interface circuits for output devices.

As an example of using a PIC digital output to control a high-current load, a power MOSFET is used to switch power to a DC motor in Lab Exercise 11. The voltage is actually switched on and off very quickly. To vary the speed of the motor, the percentage of on and off times (i.e., the **duty cycle**) is changed. This is called **pulse-width modulation** (see Section 10.5.3 for more information).

Lab Exercise

Lab 11 Pulse-width-modulation motor speed control with a PIC

7.9 METHOD TO DESIGN A MICROCONTROLLER-BASED SYSTEM

In all the examples presented in this chapter, the problems were simple, yielding short solutions to illustrate fundamental coding structures. Also, many design decisions were included as part of the problem statements. In conceiving an altogether new design, it is advisable to follow a methodical design process that will take you from the initial problem statement to a programmed microcontroller that can be embedded in application hardware. A design procedure we recommend follows. To illustrate the application of the procedure, we apply it to the problem presented in Design Example 7.2. Lab Exercise 11 also shows the procedure applied to an example.

1. *Define the problem.*   State the problem in words to explain the desired functionality of the device (i.e., what is the device supposed to do?).

2. *Draw a functional diagram*   Draw a block diagram that illustrates all of the major components of the design and shows how they are interconnected. Each component can be shown as a square with a descriptive label inside or, preferably, as a pictorial representation (e.g., a clipart image or photograph). Use single lines to connect the components (regardless of the number of wires involved), and include arrowheads to indicate the direction of signal flow.

3. *Identify I/O requirements.*   List the types of inputs and outputs required and what functions need to be performed by the microcontroller. You need to identify the number of each type of I/O line you require, including digital inputs, digital outputs, A/D converters, D/A converters, and serial ports.

4. *Select appropriate microcontroller models.*   Based on the types and number of inputs and outputs identified in the previous step, choose one or more microcontrollers that have sufficient on-chip resources. Another factor that influences this choice is the anticipated amount of program and data memory required. If the program is very complex and the application requires significant data storage, then choose a microcontroller with ample memory capacity. If multiple PICs are required (due to I/O and/or memory constraints), the PICs can communicate with each other through I/O lines by using simple handshaking (e.g., wait for a signal from another PIC to go high before doing something and then send another signal back when done) or PicBasic Pro's *Serout* and *Serin* statements for serial communication (e.g., to share data between the PICs). Refer to manufacturer literature for a list of available models and capacities. Information for Microchip's entire line of PIC microcontroller products can be found online (see Internet link 7.10, which points to the line of reprogrammable flash-memory microcontrollers listed at *www.microchip.com*).

Internet Link

7.10 Microchip PIC flash product line

5. *Identify necessary interface circuits.*   Refer to the microcontroller input and output circuit specifications and use the information in Section 7.8 to design appropriate interface circuitry utilizing pull-up resistors, buffers, transistors, relays, and amplifiers where required. Also, in cases that require many digital I/O lines, where the PIC(s) selected do not provide enough I/O pins, there are ways to interface to a large number of lines with a smaller set of pins. One approach is to use shift registers (e.g., the 74164, 74594, or 74595 for output, and the 74165 or 74597 for input), where a small set of PIC I/O pins (two for the nonlatched type and three for the latched type) can be used to transmit bits serially to or from an 8-bit register, providing eight lines of I/O. Another alternative when expanding your I/O capability is to use a device providing multiplexed programmable I/O ports (e.g., the Intel 82C55A programmable peripheral interface, or PPI). This type of device allows one I/O port to switch access among several I/O ports. With Intel's 82C55A, 5 control lines and 8 data lines provide access to 24 lines of general purpose, user-configurable I/O.

6. *Decide on a programming language.*   You can write the code in assembly language or in a high-level language such as C or PicBasic Pro. We recommend PicBasic Pro for most applications. Assembly language is a better option only when extremely fast execution speed is required or if memory capacity is a limiting factor. C might be a better choice if the solution requires complicated calculations, algorithms, or data structures.

7. *Draw the schematic.*   Draw a detailed schematic showing required components, input and output interface circuitry, and wire connections. If using the PIC16F84, Figure 7.4 serves as a good starting point.

8. *Draw a program flowchart.*   A **flowchart** is a graphical representation of the required functionality of your software. Figure 7.20 illustrates a set of building blocks that can be used to construct a flowchart. The flow control block is used as a destination label for a *goto* branch or a loop (e.g., *For . . . Next* or *While . . . Wend*). The functional block represents one or more instructions that perform some task. The decision block is used to represent logic decisions. Design Example 7.2 illustrates how a typical flowchart is constructed.

9. *Write the code.*   Implement the flowchart in software by writing code to create the desired functionality.

10. *Build and test the system.*   Compile your code into machine code and download the resulting hex file to the microcontroller. This can be done using a programming device available from the manufacturer (e.g., the PicStart Plus serial programmer available from Microchip). Internet link 7.11 points to the detailed list of steps required to create, compile, and download code using PicBasic Pro, MPLAB software, and the PicStart Plus programmer. The procedure is also presented and used in Lab Exercise 9. After downloading the code, assemble the system hardware, including the microcontroller and interface circuitry. Then, fully test the system for the desired functionality. We recommend you do steps 9 and 10 incrementally as you build functionality, carefully testing each addition before continuing. For example, make sure you can read and process an input first. Then add and test additional inputs and incrementally add and test output functionality. In other words, do not try to write the complete program on the first attempt!

Internet Link

7.11 How to
program a PIC

Lab Exercise

Lab 9 Pro-
gramming a PIC
microcontroller—
part I

For more advice on designing, building, and testing microcontroller-based systems, including advice on how to select a power supply, see Section 7.10.

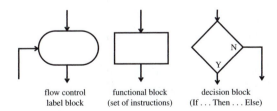

flow control   functional block   decision block
label block   (set of instructions)   (If . . . Then . . . Else)

Figure 7.20 Software flowchart building blocks.

PIC Solution to an Actuated Security Device

A few years ago, we assigned an interesting class project in our Introduction to Mechatronics and Measurement Systems course at Colorado State University. Internet Link 7.12 points to a generic project description that we currently use in the course. Internet Link 7.13 points to the specific guidelines for a combination security device project described here. Lab Exercise 15 also introduces the course project and provides useful reference information for powering a PIC project, using batteries, using relays vs. transistors, soldering, and dealing with other practical considerations.

We now illustrate the procedure just presented to generate a basic design. Then we describe some of the student solutions, because an interesting part of the project was the creativity that the class groups exhibited in solving this problem.

1. *Define the problem.*   The goal of the project is to use the PIC16F84 microcontroller in the design of a combination security lock device. The device requirements include switches to enter a combination, a pushbutton switch to process the combination, LEDs and a buzzer to indicate the success of a combination attempt, a digital display to indicate the number of failed combination attempts, and an actuator to perform a useful output function. In the simple design presented here, we use three toggle switches to enter the combination allowing for eight possible combinations.

2. *Draw a functional diagram.*   The figure below illustrates the major components in our basic design example.

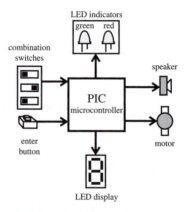

3. *Identify I/O requirements.*   The inputs and outputs are all digital for this problem.
   Inputs:

   - three switches for the combination
   - one pushbutton switch to serve as an enter key

   Outputs:

   - two LEDs to indicate the combination status
   - one seven-segment LED digit display
   - one small speaker
   - one small DC motor

4. *Select appropriate microcontroller models.* If we drive the seven-segment display with four digital outputs (see Design Example 7.1), the required number of digital I/O lines is 12. This is the only I/O we require (i.e., we do not need A/D, D/A, or serial ports). The PIC16F84 is adequate for this problem because it provides 13 digital I/O pins.

5. *Identify necessary interface circuits.* A 7447 interface is required to decode the four digital outputs to control the seven-segment display (see Design Example 7.1). A small audio speaker can be driven directly through a series capacitor, as recommended in the description of the *Sound* statement in the PicBasic Pro compiler manual. The PIC cannot source enough current to drive the motor directly, so a digital output is used to bias a power transistor connected to the motor. The only I/O pin we will not use is RA4, the special purpose Schmitt trigger input and open drain output pin.

6. *Decide on a programming language.* We use PicBasic Pro. This or some other high-level language should always be your first choice, unless you have memory or speed constraints.

7. *Draw the schematic.* The following figure shows the necessary components. It does not matter which I/O pins are used for the different functions, but we kept them organized according to function.

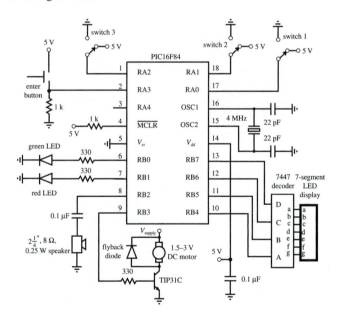

8. *Draw a program flowchart.* The next figure illustrates the logic and flow necessary to perform the desired functions. The program checks the state of the switches when the pushbutton switch is pressed and compares the switch states to a pre-stored combination. If the combination is valid, a green LED and a DC motor turn on and stay on while the pushbutton switch is held down. If the combination is

(continued)

(continued) invalid, a red LED turns on, a buzzer sounds for 3 sec, and the digital display of failures is incremented by 1. When a valid combination is entered, the counter display resets to 0.

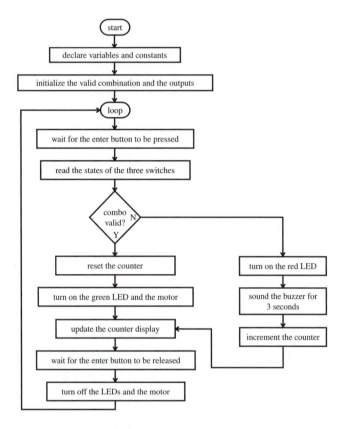

9. *Write the code.* The PicBasic Pro code to implement the logic and flow illustrated by the flowchart follows. Note that multiple PicBasic Pro statements can be included on a single line if they are separated by colons (e.g., the multiple *Low* statements). Also, a long statement can be continued on a second line by ending the first line with an underscore (e.g., the long *If* statement). Because the byte variable *number_invalid* is constrained to vary between 0 and 9, its four least significant bits represent the number in the binary coded decimal form required by the display decoder.

```
' project.bas
' This program checks the state of three switches when a pushbutton switch is
'  pressed and compares the switch states to a prestored combination. If the
'  combination is valid, a green LED and a DC motor turn on and stay on while the
'  pushbutton switch is held down. If the combination is invalid, a red LED turns
'  on, a buzzer sounds for 3 seconds, and the displayed digit (representing the
'  number of failed attempts) increments by one. When a valid combination is
'  entered, the counter display resets to zero.
```

```
' Declare all variables
switch_1 Var PORTA.0        ' first combination switch
switch_2 Var PORTA.1        ' second combination switch
switch_3 Var PORTA.2        ' third combination switch
enter_button Var PORTA.3    ' combination enter key
green_led Var PORTB.0       ' green LED indicating a valid combination
red_led Var PORTB.1         ' red LED indicating an invalid combination
speaker Var PORTB.2         ' speaker signal for sounding an alarm
motor Var PORTB.3           ' signal to bias the motor power transistor
a Var PORTB.4               ' bit 0 for the 7447 BCD input
b Var PORTB.5               ' bit 1 for the 7447 BCD input
c Var PORTB.6               ' bit 2 for the 7447 BCD input
d Var PORTB.7               ' bit 3 for the 7447 BCD input
combination Var BYTE        ' stores the valid combination in the 3 LSBs
number_invalid Var BYTE     ' counter used to keep track of the number of bad
                            ' combinations

' Initialize the valid combination and turn off all output functions
combination = %101                    ' valid combination (switch 3:on, switch
                                      ' 2:off, switch 1:on)
Low green_led : Low red_led           ' make sure the LEDs are off
Low motor                             ' make sure the motor is off
Low a : Low b : Low c : Low d         ' display zero on the digit display
number_invalid = 0                    ' reset the number of invalid combinations
                                      ' to zero

' Beginning of the main polling loop
loop:
        ' Wait for the pushbutton switch to be pressed
        If (enter_button == 0) Then loop

        ' Read switches and compare their states to the valid combination
        If ((switch_1 == combination.0) AND (switch_2 == combination.1)_
            AND (switch_3 == combination.2)) Then
                ' Turn on the green LED
                High green_led

                ' Turn on the motor
                High motor

                ' Reset the combination attempt counter
                number_invalid = 0
        Else
                ' Turn on the red LED
                High red_led

                ' Sound the alarm
                Sound speaker, [80,100]

                ' Increment the combination attempt counter and check for overflow
                number_invalid = number_invalid + 1
```

(continued)

(continued)

```
        If (number_invalid > 9) Then
                number_invalid = 0
        Endif
    Endif

    ' Update the invalid combination attempt counter digit display
    a = number_invalid.0 : b = number_invalid.1 : c = number_invalid.2
    d = number_invalid.3

    ' Wait for the pushbutton switch to be released
loop2: If (enter_button == 1) Then loop2

    ' Turn off the LEDs and the motor
    Low green_led : Low red_led
    Low motor

    ' Loop back to the beginning of the polling loop to continue the process
    Goto loop

    End    ' end of program
```

10. *Build and test the system.* Now that we have a detailed schematic and a complete program, all that remains is to build and test the system. When first testing the system, comment out secondary parts of the code (by placing comment apostrophes in front of selected lines to temporarily disable them), in order to test the remaining parts. In the example, we could test the combination input and green LED but comment out the motor driver, the alarm, and the count and digital display. This way, we could ensure that the basic I/O and logic of the program function properly when the programmed PIC is inserted in the circuit. Then, additional functionality can be added a piece at a time to achieve the complete solution. We recommend you create a first prototype on a solderless breadboard until all of the bugs have been worked out. Then, a more permanent version can be created on a protoboard or printed circuit board.

When we assigned this design problem, as a class project, we had 30 groups of three or four students creating unique designs. Some of the more interesting designs included a wall safe, where the students fabricated a section of drywall with a face plate containing three light switches. Externally it appeared to be a set of switches to control lights in a room, but when the switches were set in the correct combination and a small pushbutton switch on the side was pressed, a solenoid released a spring-loaded door exposing a hidden wall safe. Another design was a rocket launcher. When the correct switch combination was entered, interesting sound effects were created (by using various *For . . . Next* loops and the PicBasic Pro *Sound* statement), and then the digital display performed a countdown. When the count reached 0, the device used a relay to fire a model rocket, which rose several hundred feet and landed softly with parachute assist. This was demonstrated to the whole class and a curious crowd on the campus grounds outside our building. The most popular design was affectionately called the *Beer-Bot* shown in the following image. This device dispensed a glass of liquid to the user if he or she knew the correct combination.

When the correct combination was entered, the platform (lower right) translated out of the device with the aid of a DC motor driving a rack and pinion mechanism. The end of travel was detected by a limit switch. The platter was spring loaded so a simple switch could detect when a glass of adequate weight had been placed on it. Then the platter retracted and a pump was turned on to draw fluid from a concealed reservoir. When the liquid reached a certain level, a circuit was completed between two metal leads (top right) that pivot into the glass when the platter is retracted. At this point, the pump was turned off and the platter extended to present the full glass to the user, accompanied by delightful sound effects. Video Demo 7.6 shows a demonstration of the Beer-Bot in action, and Internet Link 7.14 points to numerous video demonstrations of other student design project solutions. These clips represent some of the best student projects at Colorado State University since 2001.

Video Demo

7.6 Beer-Bot—secure liquid dispensing system

Internet Link

7.14 PIC microcontroller student design projects

Below, we present the complete hardware and software solutions to the three threaded design examples (A, B, and C). Details for various parts of the solution are presented throughout the book. Refer to the list of Threaded Design Examples on page xiii for page number references to the various solution portions presented. All electrical components and devices used to build all of the Threaded Design Example projects are listed with ordering information at Internet Link 1.4.

Internet Link

1.4 Threaded Design Example components

THREADED DESIGN EXAMPLE

DC motor power-op-amp speed controller—Full solution **A.4**

The figure below shows the functional diagram for Threaded Design Example A (see Section 1.3 and Video Demo 1.6). Here, we include the entire solution to this problem. Some of the details can be found in Threaded Design Example A.2 (the potentiometer interface), A.3 (the power amp motor driver), and A.5 (the D/A converter).

(continued)

(continued)

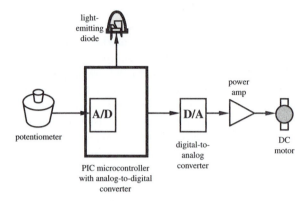

light-emitting diode

potentiometer

A/D

PIC microcontroller with analog-to-digital converter

D/A

digital-to-analog converter

power amp

DC motor

Video Demo

1.6 DC motor power-op-amp speed controller

Internet Link

7.15 TLC7524C D/A converter

The entire circuit schematic and software listing for a PIC16F88 are shown below. The code is commented, so you should be able to follow the logic as it relates to the functional diagram and circuit schematic. Specific information for the TLC7524C D/A converter can be found at Internet Link 7.15.

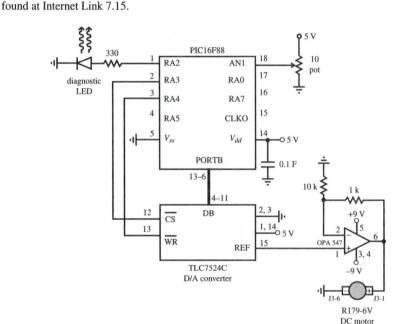

diagnostic LED

PIC16F88

TLC7524C D/A converter

OPA 547

R179-6V DC motor

```
' poweramp.bas (PIC16F88 microcontroller)

' Design Example
' Power amp motor driver controlled by a potentiometer

' A potentiometer is attached to an A/D input in the PIC. The PIC
' outputs the corresponding voltage as a digital word to a TI TLC7524
' external D/A converter, which is attached to a TI OPA547 power-op-amp
' circuit. The amplifier circuit can provide up to 500 mA of current
' to a DC motor (e.g., R179-6V-ENC-MOTOR)
```

```
' Configure the internal 8MHz internal oscillator
DEFINE OSC 8
OSCCON.4 = 1 : OSCCON.5 = 1 : OSCCON.6 = 1

' Turn on and configure AN1 (the A/D converter on pin 18)
ANSEL.1 = 1 : TRISA.1 = 1
ADCON1.7 = 1             ' have the 10 bits be right-justified
DEFINE ADC_BITS 10       ' AN1 is a 10-bit A/D

' Define I/O pin names
led Var PORTA.2          ' diagnostic LED
da_cs Var PORTA.3        ' external D/A converter chip select (low: activate)
da_wr Var PORTA.4        ' external D/A converter write (low: write)

' Declare Variables
key_value Var BYTE       ' code byte from the keypad
ad_word Var WORD         ' word from the A/D converter (10 bits padded with 6 0's)
ad_byte Var BYTE         ' byte representing the pot position

' Define constants
blink_pause Con 200      ' 1/5 second (200 ms) pause between LED blinks

' Initialize I/O
TRISB = 0                ' initialize PORTB pins as outputs
High da_wr               ' initialize the A/D converter write line
Low da_cs                ' activate the external D/A converter

' Main program (loop)
main:
    ' Read the potentiometer voltage with the A/D converter
    ADCIN 1, ad_word
    ' Scale the A/D word value down to a byte
    ad_byte = ad_word/4

    ' Send the potentiometer byte to the external D/A
    PORTB = ad_byte
    Low da_wr
    Pauseus 1    ' wait 1 microsec for D/A to settle
    High da_wr

    ' Blink the LED to indicate voltage output
    Gosub blink

Goto main    ' continue polling keypad buttons

End    ' end of main program

' Subroutine to blink the speed control indicator LED
blink:
        Low led
        Pause blink_pause
        High led
        Pause blink_pause
Return
```

THREADED DESIGN EXAMPLE

Video Demo

1.7 Stepper
motor position
and speed
controller

Internet Link

7.16 EDE1200
unipolar stepper
motor driver

B.2   *Stepper motor position and speed controller—Full solution*

The figure below shows the functional diagram for Threaded Design Example B (see Section
1.3 and Video Demo 1.7). Here, we show the complete solution to this problem.

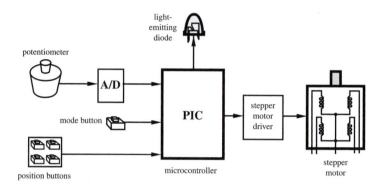

The circuit schematic and software listing are shown below. The code is commented, so
you should be able to follow the logic as it relates to the functional diagram and circuit schematic. A special integrated circuit available from E-Lab (see Internet Link 7.16) called the
EDE1200 is used to generate the proper coil sequences for the stepper motor (see Threaded
Design Example B.3).

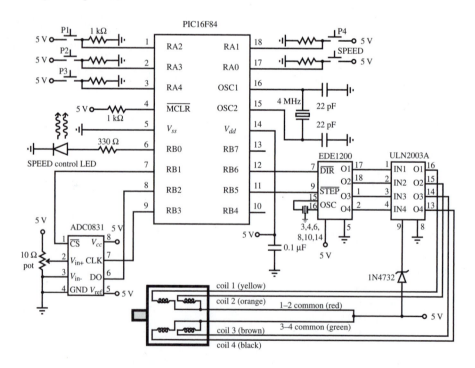

```
' stepper.bas (PIC16F84 microcontroller)

' Design Example
' Position and Speed Control of a Stepper Motor

' Four pushbutton switches are used to index to four different
' positions (0, 45, ' 90, and 180 degrees). Another pushbutton switch
' is used to toggle in and out of speed control mode (indicated by
' an LED). When in speed control mode, a potentiometer is used to
' control the speed. When to the right of the center position,
' the motor is turned CW at a speed proportional to the pot position.
' The motor turns CCW for pot positions to the left of center. The pot
' position is read from an external A/D converter (National
' Semiconductor ADC0831). The PIC retrieves the bits from the A/D
' converter via a clock signal generated by the PIC. The stepper
' motor is controlled via an E-lab EDE1200 unipolar driver IC and
' a ULN2003A Darlington driver.

' Define I/O pin names
led Var PORTB.0           ' speed control indicator LED
AD_start Var PORTB.1      ' A/D converter conversion start bit
                          ' (must be held low during A/D conversion)
AD_data Var PORTB.2       ' A/D converter data line
                          ' (for serial transmission of data bits)
AD_clock Var PORTB.3      ' A/D converter clock signal (400 kHz maximum)
P1 Var PORTA.2            ' position 1 NO button (0 degrees)
P2 Var PORTA.3            ' position 2 NO button (45 degrees)
P3 Var PORTA.4            ' position 3 NO button (90 degrees)
P4 Var PORTA.1            ' position 4 NO button (180 degrees)
SPD Var PORTA.0           ' speed control NO button to toggle speed control mode
motor_dir Var PORTB.6     ' stepper motor direction bit (0:CW 1:CCW)
motor_step Var PORTB.5    ' stepper motor step driver (1 pulse = 1 step)

' Declare Variables
motor_pos Var BYTE        ' current angle position of the motor (0, 45, 90, or 180)
new_motor_pos Var Byte    ' desired angle position of the motor
delta Var BYTE            ' required magnitude of angular motion required
num_steps Var BYTE        ' number of steps required for the given angular motion
step_period Var BYTE      ' millisecond width of step pulse (1/2 of period)
i Var Byte                ' counter used for For loops
AD_value Var BYTE         ' byte used to store the 8-bit value from the A/D converter
AD_pause Var BYTE         ' clock pulse width for the A/D converter
blink_pause Var BYTE      ' millisecond pause between LED blinks
bit_value Var BYTE        ' power of 2 value for each bit used in the A/D conversion

' Define Constants
CW Con 0                  ' clockwise motor direction
CCW Con 1                 ' counterclockwise motor direction
```

(continued)

(continued)

```
' Initialize I/O and variables
TRISA = $FF                    ' configure all PORTA pins as inputs
TRISB = %00000100              ' configure all PORTB pins as outputs except RB2
High AD_start                  ' disable A/D converter
Low motor_step                 ' start motor step signal in low state
motor_pos = 0                  ' assume the current position is the 0 degree position
step_period = 10               ' initial step speed (1/100 second between steps)
AD_pause = 10                  ' 10 microsecond pulsewidth for the A/D clock
blink_pause = 200              ' 1/5 second pause between LED blinks

' Blink the speed control LED to indicate start-up
Gosub blink : Gosub blink

' Wait for a button to be pressed (i.e., polling loop)
main:
        If (P1 == 1) Then
                ' Move motor to the 0 degree position
                new_motor_pos = 0
                Gosub move
        Else
        If (P2 == 1) Then
                ' Move motor to the 45 degree position
                new_motor_pos = 45
                Gosub move
        Else
        If (P3 == 1) Then
                ' Move motor to the 90 degree position
                new_motor_pos = 90
                Gosub move
        Else
        If (P4 == 1) Then
                ' Move motor to the 180 degree position
                new_motor_pos = 180
                Gosub move
        Else
        If (SPD == 1) Then
                ' Enter speed control mode
                Gosub speed
        EndIf : EndIf : EndIf : EndIf : EndIf
Goto main       ' continue polling buttons

End ' end of main program

' Subroutine to blink the speed control indicator LED
blink:
        High led
        Pause blink_pause
        Low led
        Pause blink_pause
Return
```

```
' Subroutine to move the stepper motor to the position indicated by motor_pos
' (the motor step size is 7.5 degrees)
move:
    ' Set the correct motor direction and determine the required displacement
    If (new_motor_pos > motor_pos) Then
        motor_dir = CW
        delta = new_motor_pos - motor_pos
    Else
        motor_dir = CCW
        delta = motor_pos - new_motor_pos
    EndIf

    ' Determine the required number of steps (given 7.5 degrees per step)
    num_steps = 10*delta / 75

    ' Step the motor the appropriate number of steps
    Gosub move_steps

    ' Update the current motor position
    motor_pos = new_motor_pos
Return

' Subroutine to move the motor a given number of steps (indicated by num_steps)
move_steps:
    For i = 1 to num_steps
        Gosub step_motor
    Next
Return

' Subroutine to step the motor a single step (7.5 degrees) in the motor_dir
'   direction
step_motor:
    Pulsout motor_step, 100*step_period      ' (100 * 10microsec = 1 millisec)
    Pause step_period
    ' Equivalent code:
      ' High motor_step
      ' Pause step_period
      ' Low motor_step
      ' Pause step_period
Return

' Subrouting to poll the POT for speed control of the stepper motor
speed:
    ' Turn on the speed control LED indicator
    High LED

    ' Wait for the SPEED button to be released
    Gosub button_release

    ' Polling loop for POT speed control
    pot_speed:
        ' Check if the SPEED button is down
```

(continued)

(continued)

```
        If (SPD == 1) Then
            ' Wait for the SPEED button to be released
            Gosub button_release

            ' Turn off the speed control LED indicator
            Low led

            ' Assume the new position is the new 0 position
            motor_pos = 0

            ' Exit the subroutine
            Return
        EndIf

        ' Sample the POT voltage via the A/D converter
        Gosub get_AD_value

        ' Adjust the motor speed and direction based on the POT value and
        '    step the motor a single step.
        '    Enforce a deadband at the center of the range
        '    Have the step period range from 100 (slow) to 1 (fast)
        If (AD_value > 150) Then
            motor_dir = CW
            step_period = 100 - (AD_value - 150)*99/(255 - 150)
            Gosub step_motor
        Else
            If (AD_value < 100) Then
                motor_dir = CCW
                step_period = 100 - (100 - AD_value)*99/100
                Gosub step_motor
            EndIf
        EndIf

  ' Continue polling
  goto pot_speed

Return ' end of subroutine, but not reached (see the SPD If statement above)

' Subroutine to wait for the speed control button to be released
button_release:
    Pause 50     ' wait for switch bounce to settle
    While (SPD == 1) : WEND
    Pause 50     ' wait for switch bounce to settle
Return

  ' Subroutine to sample the POT voltage from the A/D converter
  '    The value (0 to 255) is returned in the variable AD_value and corresponds
  '    to the original 0 to 5V analog voltage range.
  get_AD_value:
     ' Initialize the A/D converter
     Low AD_clock          ' initialize the clock state
     Low AD_start          ' enable the A/D converter
     Gosub pulse_clock     ' send initialization pulse to A/D clock
```

```
' Get each converted bit from the A/D converter (at 50 kHz)
bit_value = 128          ' value of the MSB
AD_value = 0
FOR I = 7 To 0 Step -1 ' for each bit from the MSB to the LSB
    ' Output clock pulse
    Gosub pulse_clock
    AD_value = AD_value + AD_data*bit_value
    bit_value = bit_value / 2
Next i

' Disable the A/D converter
High AD_start
Return

' Subroutine to send a pulse to the A/D clock line
pulse_clock:
    Pulsout AD_clock, 1 : PauseUS 10    ' 20 microsecond pulse
Return
```

THREADED DESIGN EXAMPLE

DC motor position and speed controller—Full solution with serial interface   **C.3**

The figure below shows the functional diagram for Threaded Design Example C (see Section 1.3 and Video Demo 1.8). Presented here is the entire solution to this problem. This solution utilizes two PIC microcontrollers. The main PIC is referred to as the "master" PIC, because it controls most of the system functions; the secondary PIC is referred to as the "slave" PIC, because it simply provides information to the master PIC upon command.

Video Demo

1.8 DC motor position and speed controller

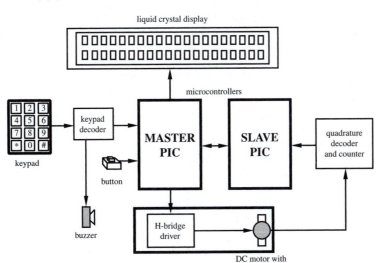

The circuit schematic and software listings are shown below. The code is commented, so you should be able to follow the logic as it relates to the functional diagram and circuit

(continued)

(continued) schematic. There are two software listings. One is for the master PIC (a PIC16F88) that monitors the keypad (see Threaded Design Example C.2), provides a menu-driven user interface on the LCD (see Threaded Design Example C.2), and drives the motor (see Threaded Design Examples C.4 and C.5). The other listing is for the slave PIC (a PIC16F84) that monitors the digital encoder sensor on the motor shaft and transmits the position information to the master PIC.

As with the other threaded design examples, details covering the different components of the design can be found throughout the book. This solution is a good example of how to communicate among multiple PICs using a serial interface. The specific code designed to implement the communication can be found in the *get_encoder* subroutine in the master PIC code and the main loop in the slave PIC code. One I/O line is simply set high or low by the master PIC to command the slave PIC when to send data. A second I/O line then receives the data through a standard serial communication protocol.

master PIC code:

```
' dc_motor.bas (PIC16F88 microcontroller)

' Design Example
' Position and Speed Control of a DC Servo Motor.

'    The user interface includes a keypad for data entry and an LCD for text
'    messages. The main menu offers three options: 1 - position control,
'    2 - speed control, and 3 - position control gain and motor PWM control.
'    When in position control mode, pressing a button moves to indexed positions
'    (1 - 0 degrees, 2 - 45 degrees, 3 - 90 degrees, and 4 - 180 degrees). When
'    in speed control mode, pressing 1 decreases the speed, pressing 2 reverses
'    the motor direction, pressing 3 increases the speed, and pressing 0 starts
'    the motor at the indicated speed and direction. The motor is stopped with
'    a separate pushbutton switch. When in gain and PWM control mode,
'    pressing 1/4 increases/decreases the proportional gain factor (kP)
'    and pressing 3/6 increases/decreases the number of PWM cycles sent
'    to the motor during each control loop update.

'    Pressing the "#" key from the position, speed, or gain menus returns control
'    back to the main menu. E-Lab's EDE1144 keypad encoder is used to detect
'    when a key is pressed on the keypad and transmit data (a single byte per
'    keypress) to the PIC16F88. Acroname's R179-6V-ENC-MOTOR servo motor is
'    used with their S17-3A-LV H-bridge for PWM control. A second PIC (16F84),
'    running dc_enc.bas, is used to communicate to an Agilent HCTL-2016
'    quadrature decoder/counter to track the position of the motor encoder.
'    The 16F88 communicates to the 16F84 via handshake (start) and serial
'    communication lines.

' Configure the internal 8MHz internal oscillator
DEFINE OSC 8
OSCCON.4 = 1 : OSCCON.5 = 1 : OSCCON.6 = 1
```

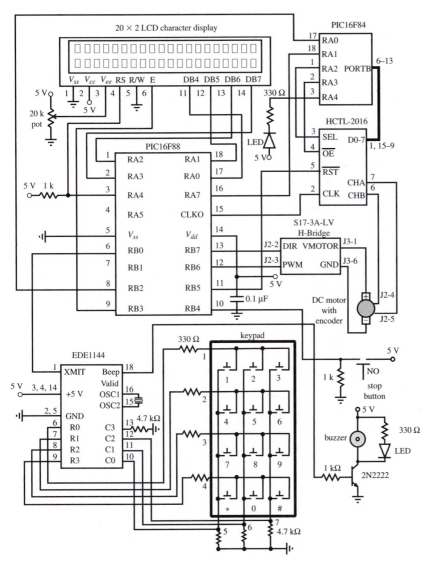

```
' Turn off A/D converters (thereby allowing use of pins for I/O)
ANSEL = 0

' Define I/O pin names
key_serial Var PORTB.0      ' keypad serial interface input
motor_dir Var PORTB.7       ' motor H-bridge direction line
motor_pwm Var PORTB.6       ' motor H-bridge pulse-width-modulation line
stop_button Var PORTB.4     ' motor stop button
enc_start Var PORTB.2       ' signal line used to start encoder data transmission
enc_serial Var PORTA.7      ' serial line used to get encoder data from the 16F84
enc_rst Var PORTB.5         ' encoder counter reset signal (active low)
```

(continued)

(continued)

```
' Declare Variables
key_value Var BYTE         ' code byte from the keypad
motor_pos Var Word         ' current motor position in degrees
new_motor_pos Var Word     ' desired motor position (set point) in degrees
error Var Word             ' error magnitude between current and desired positions
motor_speed Var BYTE       ' motor speed as percentage of maximum (0 to 100)
motion_dir Var BIT         ' motor direction (1:CW/Forward 0:CCW/Reverse)
on_time Var WORD           ' PWM ON pulse width
off_time Var WORD          ' PWM OFF pulse width
enc_pos Var WORD           ' motor encoder position (high byte and low byte)
i Var Byte                 ' counter variable for FOR loops
kp Var BYTE                ' proportional gain factor position control
pwm_cycles Var BYTE        ' # of PWM pulses sent during the position control loop

' Define constants
key_mode Con 0             ' 2400 baud mode for serial connection to keypad
key_1 Con $30              ' hex code for the 1-key on the keypad
key_2 Con $31              ' hex code for the 2-key on the keypad
key_3 Con $32              ' hex code for the 3-key on the keypad
key_4 Con $34              ' hex code for the 4-key on the keypad
key_5 Con $35              ' hex code for the 5-key on the keypad
key_6 Con $36              ' hex code for the 6-key on the keypad
key_7 Con $38              ' hex code for the 7-key on the keypad
key_8 Con $39              ' hex code for the 8-key on the keypad
key_9 Con $41              ' hex code for the 9-key on the keypad
key_star Con $43           ' hex code for the *-key on the keypad
key_0 Con $44              ' hex code for the 0-key on the keypad
key_pound Con $45          ' hex code for the #-key on the keypad
CW Con 1                   ' motor clockwise (forward) direction
CCW Con 0                  ' motor counterclockwise (reverse) direction
pwm_period Con 50          ' period of each motor PWM signal cycle (in microsec)
                           '   (50 microsec corresponds to 20kHz)
enc_mode Con 2             ' 9600 baud mode for serial connection to the encoder IC

' Initialize I/O and variables
TRISB.6 = 0                ' configure H-bridge DIR pin as an output
TRISB.7 = 0                ' configure H-bridge PWM pin as an output
motion_dir = CW            ' starting motor direction: CW (forward)
motor_pos = 0              ' define the starting motor position as 0 degrees
motor_speed = 50           ' starting motor speed = 50% duty cycle
kp = 50                    ' starting proportional gain for position control
pwm_cycles = 20            ' starting # of PWM pulses sent during the
                           '   position control loop
Low motor_pwm              ' make sure the motor is off to begin with
Low enc_start              ' disable encoder reading to begin with
Gosub reset_encoder        ' reset the encoder counter
```

```
' Wait 1/2 second for everything (e.g., LCD) to power up
Pause 500

' Receive dummy byte from the PIC16F84 to ensure proper communication
SERIN enc_serial, enc_mode, key_value

' Wait for a keypad button to be pressed (i.e., polling loop)
Gosub main_menu      ' display the main menu on the LCD
main:
     Serin key_serial, key_mode, key_value
     If (key_value = key_1) Then
        Gosub reset_encoder
        Gosub position
     Else
     If (key_value = key_2) Then
        motor_speed = 50    ' initialize to 50% duty cycle
        Gosub speed
     Else
     If (key_value = key_3) Then
        Gosub adjust_gains
     Endif : Endif : Endif
Goto main      ' continue polling keypad buttons

End ' end of main program

' Subroutine to display the main menu on the LCD
main_menu:
     Lcdout $FE, 1, "Main Menu:"
     Lcdout $FE, $C0, "1:pos. 2:speed 3:gain"
Return

' Subroutine to reset the motor encoder counter to 0
reset_encoder:
     Low enc_rst      ' reset the encoder counter
     High enc_rst     ' activiate the encoder counter
Return

' Suroutine for position control of the motor
position:
' Display the position control menu on the LCD
     Lcdout $FE, 1, "Position Menu:"
     Lcdout $FE, $C0, "1:0 2:45 3:90 4:180 #:<"

     ' Wait for a keypad button to be pressed
     Serin key_serial, key_mode, key_value

     ' Take the appropriate action based on the key pressed
     If (key_value == key_1) Then
        new_motor_pos = 0
     Else
```

(continued)

(continued)

```
If (key_value == key_2) Then
    new_motor_pos = 45
Else
If (key_value == key_3) Then
    new_motor_pos = 90
Else
If (key_value == key_4) Then
    new_motor_pos = 180
Else
If (key_value == key_pound) Then
    Gosub main_menu
    Return
Else
    Goto position
Endif : Endif : Endif : Endif : Endif

' Position control loop
While (stop_button == 0)      ' until the stop button is pressed
    ' Get the encoder position (enc_pos)
    Gosub get_encoder

    ' Calculate the error signal magnitude and sign and set the motor direction
    ' Convert encoder pulses to degrees.  The encoder outputs 1230 pulses
    '    per 360 degrees of rotation
    motor_pos = enc_pos * 36 / 123
    If (new_motor_pos >= motor_pos) Then
      error = new_motor_pos - motor_pos
      motor_dir = CW
    Else
      error = motor_pos - new_motor_pos
      motor_dir = CCW
    Endif

    ' Set the PWM duty cycle based on the current error
    If (error > 20) Then    ' use maximum speed for large errors
        motor_speed = kp
    Else
        ' Perform proportional position control for smaller errors
        motor_speed = kp * error / 20
    Endif

    ' Output a series of PWM pulses with the speed-determined duty cycle
    Gosub pwm_periods        ' calculate the on and off pulse widths
    For I = 1 to pwm_cycles
        Gosub pwm_pulse       ' output a full PWM pulse
    Next I
```

```
' Display current position and error on the LCD
Lcdout $FE, 1, "pos:", DEC motor_pos, " error:", DEC error
Lcdout $FE, $C0, "exit: stop button"

Wend

Goto position      ' continue the polling loop
Return             ' end of subroutine, not reached (see the #-key If above)

' Subroutine to get the encoder position (enc_pos) from the counter
get_encoder:

     ' Command the PIC16F84 to transmit the low byte
     High enc_start

     ' Receive the high byte
     SERIN enc_serial, enc_mode, enc_pos.HighBYTE

     ' Command the PIC16F84 to transmit the high byte
     Low enc_start

     ' Receive the low byte
     SERIN enc_serial, enc_mode, enc_pos.LowBYTE
Return

' Subroutine to calculate the PWM on and off pulse widths based on the desired
'     motor speed (motor_speed)
pwm_periods:
     ' Be careful to avoid integer arithmetic and
     '   WORD overflow [max=65535] problems
     If (pwm_period >= 655) Then
        on_time = pwm_period/100 * motor_speed
        off_time = pwm_period/100 * (100-motor_speed)
     Else
        on_time = pwm_period*motor_speed / 100
        off_time = pwm_period*(100-motor_speed) / 100
     Endif
Return

' Subroutine to output a full PWM pulse based on the data from pwm_periods
pwm_pulse:
     ' Send the ON pulse
     High motor_pwm
     Pauseus on_time

     ' Send the OFF pulse
     Low motor_pwm
     Pauseus off_time
Return
```

(continued)

(continued)

```
' Subroutine for speed control of the motor
speed:
      ' Display the speed control menu on the LCD
      Gosub speed_menu

      ' Wait for a keypad button to be pressed
      Serin key_serial, key_mode, key_value

      ' Take the appropriate action based on the key pressed
      If (key_value == key_1) Then
          ' Slow the speed by 10%
         If (motor_speed > 0) Then        ' don't let speed go negative
             motor_speed = motor_speed - 10
         Endif
      Else
      If (key_value == key_2) Then
           ' Reverse the motor direction
           motion_dir = ~motion_dir
      Else
      If (key_value == key_3) Then
          ' Increase the speed by 10%
         If (motor_speed < 100) Then ' don't let speed exceed 100
            motor_speed = motor_speed + 10
         EndIf
      Else
      If (key_value == key_pound) Then
          Gosub main_menu
          Return
      Else
      If (key_value == key_0) Then
           ' Run the motor until the stop button is pressed
           Gosub run_motor
      Else
           ' Wrong key pressed
          Goto speed
      Endif : Endif : Endif : Endif : Endif

Goto speed      ' continue the polling loop
Return          ' end of subroutine, not reached (see the #-key If above)

' Subroutine to display the speed control menu on the LCD
speed_menu:
      Lcdout $FE, 1, "speed:", DEC motor_speed, " dir:", DEC motion_dir
      Lcdout $FE, $C0, "1:- 2:dir 3:+ 0:start #:<"
Return
```

```
' Subroutine to run the motor at the desired speed and direction until the
'    stop button is pressed. The duty cycle of the PWM signal is the
'    motor_speed percentage
run_motor:
     ' Display the current speed and direction
     Lcdout $FE, 1, "speed:", DEC motor_speed, " dir:", DEC motion_dir
     Lcdout $FE, $C0, "exit: stop button"

     ' Set the motor direction
     motor_dir = motion_dir

     ' Output the PWM signal
     Gosub pwm_periods      ' calculate the on and off pulse widths
     While (stop_button == 0)      ' until the stop button is pressed
       Gosub pwm_pulse      ' send out a full PWM pulse
     Wend

     ' Return to the speed menu
     Gosub speed_menu
Return

' Subroutine to wait for the stop button to be pressed and released
' (used during program debugging)
button_press:
     While (stop_button == 0) : Wend   ' wait for button press
     Pause 50                          ' wait for switch bounce to settle
     While (stop_button == 1) : Wend    wait for button release
     Pause 50                          ' wait for switch bounce to settle
Return

' Subroutine to allow the user to adjust the proportional and derivative
'    gains used in position control mode
adjust_gains:
     ' Display the gain values and menu on the LCD
     Gosub gains_menu

     ' Wait for a keypad button to be pressed
     Serin key_serial, key_mode, key_value

     ' Take the appropriate action based on the key pressed
     If (key_value == key_1) Then
        ' Increase the proportional gain by 10%
        If (kp < 100) Then
           kp = kp + 10
        Endif
     Else
```

(continued)

(continued)

```
    If (key_value == key_4) Then
        ' Decrease the proportional gain by 10%
        If (kp > 0) Then    ' don't allow negative gain
            kp = kp - 10
        Endif
    Else
    If (key_value == key_3) Then
        ' Increase the number of PWM cycles sent each position control loop
        pwm_cycles = pwm_cycles + 5
    Else
    If (key_value == key_6) Then
        ' Decrease the number of PWM cycles sent each position control loop
        If (kp > 5) Then    ' maintain positive number of pulses
            pwm_cycles = pwm_cycles - 5
        Endif
    Else
    If (key_value == key_pound) Then
        Gosub main_menu
        Return
    Else
        Goto adjust_gains
    Endif : Endif : Endif : Endif : Endif

Goto adjust_gains       ' continue the polling loop
Return                  ' end of subroutine, not reached (see the #-key If above)

' Subroutine to display the position control gain, the number of PWM
'    cycles/loop, and the adjustment menu on the LCD
gains_menu:
    Lcdout $FE, 1, "kp:", DEC kp, " PWM:", DEC pwm_cycles
    Lcdout $FE, $C0, "1:+P 4:-P 3:+C 6:-C #:<"
Return
```

slave PIC code:

```
' dc_enc (PIC16F84 microcontroller)

' Design Example
' Position and Speed Control of a dc Servo Motor.

' Slave program to send encoder data, upon request, to the a PIC16F88
' microcontroller running dc_motor.bas

' Define I/O pin names and constants
enc_start Var PORTA.0     ' signal line used to start encoder data transmission
enc_serial Var PORTA.1    ' serial line used to get encoder data from the 16F84
enc_sel Var PORTA.2       ' encoder data byte select (0:high 1:low)
enc_oe Var PORTA.3        ' encoder output enable latch signal (active low)
led Var PORTA.4           ' diagnostic LED (open drain output: 1:OC, 0:ground)
enc_mode Con 2            ' 9600 baud mode for serial connection to encoder IC
blink_pause Con 200       ' 1/5 second (200 ms) pause between LED blinks
```

```
' Turn off the diagnostic LED
High led

' Wait to ensure the PIC16F88 is initialized
PAUSE 500

' Initialize I/O signals
High enc_oe              ' disable encoder output
Low enc_sel             ' select the encoder counter high byte initially
                        ' (to prevent transparent latch on low byte)

' Blink the LED to indicate proper operation
Gosub blink : Gosub blink : Gosub blink
' Send dummy byte (66) to ensure proper communication
SEROUT enc_serial, enc_mode, [66]
' Main loop
start:
        ' Wait for the start signal from the PIC16F88 to go high
        While (enc_start == 0) : Wend

        ' Enable the encoder output (latch the counter values)
        Low enc_oe

        ' Send out the high byte of the counter
        SEROUT enc_serial, enc_mode, [PORTB]

        ' Wait for the start signal from the PIC16F88 to go low
        While (enc_start == 1) : Wend

        ' Send out the low byte of the counter
        High enc_sel
        SEROUT enc_serial, enc_mode, [PORTB]

        ' Disable the encoder output
        High enc_oe
        Low enc_sel
goto start  ' wait for next request

End     ' end of main program (not reached)

' Subroutine to blink the diagnostic LED on and back off again
blink:
        Low led
        Pause blink_pause
        High led
        Pause blink_pause
Return
```

Lab Exercise

Lab 9 Programming a PIC microcontroller—Part I

Lab 10 Programming a PIC microcontroller—Part II

Lab 11 Pulse-width-modulation motor speed control with a PIC

■ **CLASS DISCUSSION ITEM 7.10**
Negative logic LED

Why does the *blink* subroutine in the slave PIC code in Threaded Design Example
C.3 use negative logic (i.e., *Low* turns the LED on, and *High* turns it off)?

7.10 PRACTICAL CONSIDERATIONS

Throughout this chapter we've seen how to design and program microcontroller-based systems. As with most things in the "real world," when you actually construct circuits and implement software, things don't always work perfectly. Lab Exercises 9 through 11 provide experiences that will help you develop some of the skills necessary to be successful. The best advice anyone could give concerning designing a complicated microcontroller-based system is to follow a methodical design procedure as documented and demonstrated in Section 7.9. It also helps to have and follow a detailed procedure for the specific development system you are using. Internet Link 7.11 provides the detailed procedure necessary for working with Microchip PIC microcontrollers programmed in PicBasic Pro, using the PicStart Plus programming hardware.

Internet Link

7.11 How to program a PIC

7.10.1 PIC Project Debugging Procedure

The following check list and items of advice can be helpful when **debugging** software and trying to get a PIC circuit to function properly:

1. Before writing and testing the entire code for your project, always start with a very simple program (e.g., the *flash.bas* program in Section 7.5.1) on your PIC(s) to first make sure all necessary components and wiring are in place to allow the PIC(s) to run.

2. Once the PIC is known to be running properly, incrementally add and test portions of your code, one functional component at a time. In other words, modularize your software and independently develop and test each module (i.e., don't write the entire program at once, expecting it to work).

3. Use LEDs to indicate status and location within your program when it is running and to indicate input and output states.

4. Be aware of the different characteristics of the I/O pins on the PIC. Refer to Section 7.8 to see how to properly interface to the different pins for different purposes.

5. Be aware that PicBasic Pro commands totally occupy the processor while they are running (e.g., the line after a SOUND command is not reached or processed until the SOUND command has completely finished its action).

6. If you are using a PIC with no internal oscillator (e.g., the PIC16F84), make sure your circuit includes the proper clock crystal and capacitor components. If you are using a PIC with an internal oscillator (e.g., the PIC16F88), make sure you have included the necessary initialization code (e.g., see the code template for the PIC16F88 available in Internet Link 7.17).

7. Make sure you carefully select all of the configuration bit settings when downloading code to a PIC, so the oscillator, timers, and selected pin functions are defined properly. This can be done manually each time a PIC is reprogrammed (as documented in Internet Link 7.11), or you can define all of the settings in your code as described in Internet Link 7.18.

Internet Link

7.17 PIC16F88 code template

7.18 Setting configuration bits in code

8. Follow all of the recommendations in Section 2.10.2 for prototyping circuits.
9. Always follow the PIC programming procedure in Internet Link 7.11 to ensure that you don't miss any important steps or details.

Much more advice for developing and debugging PIC-based systems can be found at Internet Link 7.19, which provides a summary of "lessons learned" by many past students in our mechatronics course at Colorado State University.

Internet Link

7.19 Lessons learned by past students

7.10.2 Power Supply Options for PIC Projects

There are a number of ways to provide the DC power required by a PIC and any ancillary digital integrated circuits. Actuators may also be powered by the same DC supply if their drive voltages match that of the digital circuitry, and if the current demands don't exceed the supply's capacity. We begin by assuming that TTL digital ICs are being used, which require a regulated 5 V DC source. If CMOS is used exclusively, there are fewer restrictions on the regulation of the DC voltage.

Figure 7.21 shows several low-cost options for powering systems requiring a 5 V supply. These and other options include:

1. A 6 V, 9 V, or 12 V wall transformer with a 5 V regulator
2. A potted power supply with AC input and 5 V regulated output
3. Four AA batteries (6 V) in series with a 5 V regulator
4. A 9 V battery with a 5 V regulator
5. A rechargeable battery (or batteries in series) with a 5 V regulator
6. A full-featured instrumentation power supply

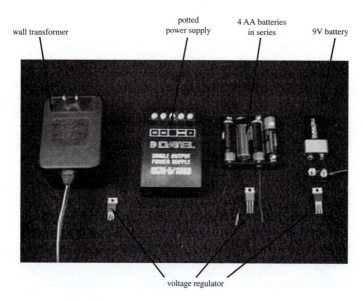

Figure 7.21 Low-cost power supply options.

Other alternatives for powering projects include a computer power supply, or large batteries (e.g., car or motorcycle lead-acid batteries), especially if you have high current demands.

A wall transformer (6 V, 9 V, or 12 V) will provide current up to its rating and must be used with a 5 V regulator to control the level of the output voltage. Be sure that the current rating of the wall transformer exceeds (with a healthy margin) the maximum current your circuit and actuators will draw. A potted power supply also has AC inputs and may provide one or more regulated DC outputs at its rated current. No voltage regulator is required if a 5 V output is provided. Four AA batteries can be connected in series with the 6 V output regulated down to 5 V with a voltage regulator. A 9 V battery must also be connected to a 5 V regulator. The battery options provide portability for your design but may not be able to supply enough current. Section 7.10.3 presents more information on different types of batteries and their characteristics. Generally, actuators such as motors and solenoids, as well as high-current LEDs, can draw substantial current. You should test your batteries to ensure that they will provide sufficient supply (do not simply assume that they will). Digital circuitry, on the other hand, usually draws very little current.

Figure 7.22 shows an example of a full-featured instrumentation power supply. This particular model (HP 6235A) is a triple-output power supply, with three adjustable voltage outputs, each independently current rated. A full-featured instrumentation power supply provides the easiest solution, but these are expensive, heavy, and generally not portable.

Except for the 5 V potted supply and the adjustable instrumentation power supply, voltage regulators are required to convert the output voltage down to the 5 V level. If your system is entirely CMOS, the regulation of the DC voltage is not required. Figure 7.23 illustrates a standard 7805 5 V voltage regulator and shows how it is properly connected to your unregulated power-supply output and your system. There must be a common ground from the power supply to your system. The mounting hole on the heat sink allows you to easily connect to the common ground.

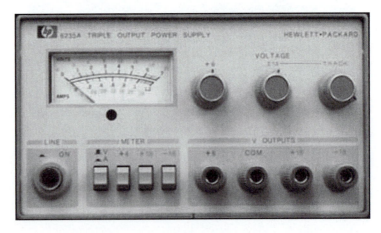

Figure 7.22 Example of a full-featured instrumentation power supply.

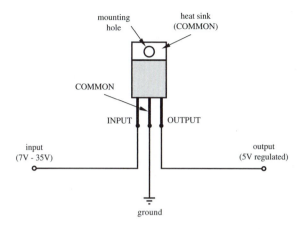

Figure 7.23 7805 voltage regulator connections.

Table 7.7 5 V power supply options summary

Device	Typical current	Relative size	Relative cost
instrumentation power supply	1 A – 5 A	large	very expensive (~ $1,000) but many features
small potted, open frame, or enclosed power supply	1 A – 10 A	medium	moderately expensive (~ $20–$100)
wall transformer	1 A	small	cheap
9 V battery	100 mA	small	cheap
4 AA batteries	100 mA	small	cheap
rechargeable battery	See Section 7.10.3	small	moderate

Table 7.7 provides a summary of how the various power supply options compare in terms of current ratings, size, and cost. Figure 7.24 shows an example specification sheet for an enclosed power supply. Before selecting or purchasing a supply for your design, it is important to first review the specifications, especially the current rating (2.5 A in this case).

7.10.3 Battery Characteristics

This section presents some of the important terms, considerations, and specifications in the proper selection of a battery as a power source. The most important specification for a battery (besides its rated voltage) is the **amp-hour capacity.** This is defined as the current a battery can provide for one hour before it reaches its end-of-life point. The current that a battery can deliver is limited by its **equivalent series resistance,** which is the internal resistance that can be considered in series with an "ideal voltage source." The load current times the internal resistance results in a voltage drop reducing the effective voltage of the battery. Furthermore, there will

PWR
SPLY,SW,16W,5VDC/2.5A,

Jameco #208952
Mfg Ref # PRK15U-0512W

16-Watt Switching Power Supply
Dual Outputs

- Output voltage: +5VDC @ 2.5A, +12VDC @ 0.7A
- Power: 16.0 Watts
- Input voltage: 120VAC @ 50-60Hz
- Size: 4.6"L x 3.2"W x 1.1"H
- Mounting holes: 4.0"L x 2.5"W x 0.08"Dia.
- Power density: 0.99W/in3
- Load reg.: ±1.0%
- Line reg.: ±0.4%
- Inrush current: 12A@120VAC
- Leakage current: 1.0mA@240V
- Rise time 100ms
- Hold-up time: 20ms
- Vibration: 10-55Hz, 20G
- Voltage ajustment: 5%
- Eficiency: 64%
- Weight: 0.5 lbs.
- UL/CSA approved

Figure 7.24 Specifications for an example closed-frame power supply.

be power dissipated by the internal resistance, which at high currents may result in considerable heat production.

Batteries are composed of **cells**, the electrochemical construct that supplies voltage and current. Cells can be combined in series or parallel within a battery for larger current and voltage capacities. The voltage of a cell will differ among the types of batteries, due to their chemistry. **Primary cell batteries** are not rechargeable and are meant for one-time use. Devices that are used infrequently or that require very low drain currents are good candidates for primary cells. **Secondary cell batteries** are rechargeable, and their effectiveness may be replenished many times. Devices that require daily use with higher drain currents are good candidates for secondary cells.

The plot of a **battery discharge curve** is important in determining the stability of the voltage output. Figure 7.25 shows a typical shape for a discharge curve. One desires a broad plateau representing a long effective life.

The salient factors a designer must consider in the selection of a power source for a mechatronic design are:

- Voltage required by the load
- Current required by the load
- Duty cycle of the system
- Cost
- Size and weight (specific energy)
- Need for rechargeability

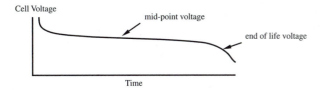

Figure 7.25 Example battery discharge curve.

As shown in Table 7.8, the chemistry of the cell will determine its open circuit voltage. High peak-current-demand devices are good candidates for lead-acid and nickel-cadmium (NiCd) batteries. If a device is in storage most of the time, alkaline batteries are appropriate. Because batteries may be the heaviest component of a mechatronic system design, the very light lithium-ion (Li-ion) and lithium-polymer chemistries may be good candidates. Lithium chemistries provide the highest energy per unit weight (specific energy) and per volume (energy density) of all types of batteries.

Most rechargeable batteries will function well even after hundreds of charge-discharge cycles. They are significantly more expensive than primary cell batteries. Nickel-metal-hydride (NiMH) batteries should be deep-discharged several times

Table 7.8 Characteristics for various types of batteries

Type	Voltage (open circuit)	Type	Typical Ah Capacity	R internal (W)
9 V (heavy duty)	9 V	primary	0.30 @ 1 mA 0.15 @ 10 mA	35
9 V alkaline	9 V	primary	0.60 @ 25 mA	2
9 V lithium	9 V	primary	1.0 @ 25 mA	18
alkaline D	1.5 V	primary	17.1 @ 25 mA 0.95 @ 80 mA	0.1
alkaline C	1.5 V	primary	7.9 @ 25 mA	0.2
alkaline AA	1.5 V	primary	2.7 @ 25 mA	0.4
alkaline AAA	1.5 V	primary	1.2 @ 25 mA	0.6
BR-C PCMF-Li	3 V	primary	5.0 @ 5 mA	
CR-V3 MnLi	3 V	primary	3.0 @ 100 mA	
NiCd D	1.3 V	secondary	4.0 @ 800 mA 3.5 @ 4 A	0.009
NiCd 9 V	8.1 V	secondary	0.1 @ 10 mA	0.84
Lead-acid D	2.0 V	secondary	2.5 @ 25 mA 2.0 @ 1 A	0.006
NiMH AAA	1.2 V	secondary	0.55 @ 200 mA	
NiMH AA	1.2 V	secondary	1.3 @ 200 mA	
NiMH C	1.2 V	secondary	3.5 @ 200 mA	
NiMH D	1.2 V	secondary	7.0 @ 200 mA	
NiMH 9 V	8.4 V	secondary	0.13 @ 200 mA	
ML2430 MnLi	3 V	secondary	0.12 @ 300 mA	
Lithium Ion	3.7 V	secondary	0.76 @ 200 mA	

Internet Link

7.20 Battery University

7.21 Battery information resource page

7.22 All about circuits—batteries and power systems

when put into service, for best performance. NiCd batteries can suffer from an effect called "memory" where the battery capacity can diminish over time. This is caused by shallow charge cycles where the battery is only partially discharged and then fully charged repeatedly. Sometimes, you need to give the battery a deep discharge periodically for best performance.

Excellent resources for battery information can be found online at Internet Links 7.20 and 7.21. Also, Internet Link 7.22 offers good practical advice for using batteries effectively.

7.10.4 Other Considerations for Project Prototyping and Design

For basic prototype circuit assembly and troubleshooting advice, see Section 2.10.2. For advice concerning debugging circuits with PIC microcontrollers, see Section 7.10.1. Here are some other practical suggestions when prototyping and designing projects:

- When ordering ICs, make sure you specify DIP (dual in-line packages) and not surface mount packages (e.g., SOP). DIP chips are well suited to use in breadboards and protoboards. Surface mount ICs require printed circuit boards (PCBs) and special soldering equipment.

- Make sure your power supply can provide adequate current for the entire design. If necessary, use separate power supplies for your signal and power circuits.

- Use breadboards with caution and care because connections can be unreliable, and the base plate adds capacitance to your circuits. Hardwired and soldered protoboards or printed circuit boards (PCBs) can be much more reliable (see Section 2.10.4 for more information). For soldered prototypes, be sure to use sockets for all ICs, to prevent damage during soldering and to allow easy replacement of the ICs. Also, if you have a working breadboard circuit, it is advisable to use duplicate components (where possible) for the soldered board (i.e., don't cannibalize components from a working prototype circuit, in case something goes wrong or gets damaged when soldering your board).

- Use a storage capacitor (e.g., 100 µf) across the main power and ground lines of a power supply that does not have built-in output capacitance (e.g., batteries, wall transformers, and regulated voltages) to minimize voltage swings during output current spikes. Also, use "bypass" or "decoupling" capacitors (e.g., 0.1 µf) across the power and ground lines of all individual ICs to suppress any current and voltage spikes.

- Be careful with grounding and electromagnetic interference (EMI). Section 2.10.6 presents various methods to reduce EMI, specifically using opto-isolators, single point grounding, ground planes, coaxial or twisted pair cables, and decoupling capacitors.

- Don't leave IC pins floating (especially with CMOS devices). In other words, connect all functional pins to signals or power or ground. As an example, do not assume that leaving a microcontroller's reset pin disconnected will keep a microcontroller from resetting itself. You should connect the reset pin to 5 V for an active-low reset or ground for an active-high reset, and not leave the pin floating where its state can be uncertain.

■ Be aware of possible switch bounce in your digital circuits and add debounce circuits or software to eliminate the bounce (see Section 6.10.1).

■ Use flyback diodes on motors, solenoids, and other high-inductance devices that are being switched (see Section 3.3).

■ Use buffers, line drivers, and inverters where current demand is large for a digital output.

■ Use Schmitt triggers (see Section 6.12.2) on all noisy digital inputs (e.g., a Hall-effect proximity sensor or photo-interrupter).

■ Use a common-emitter configuration with transistors (i.e., put the load on the high side) to avoid voltage biasing difficulties and emitter degeneration (see Section 3.4.2).

■ Be careful to identify and properly interface any open-collector or open-drain outputs (see Section 6.11.2) on digital ICs (e.g., pin RA4 on the PIC16F84).

■ For reversible DC motors, use "off-the-shelf" commercially available H-bridge drivers (e.g., National Semiconductor's LMD 18200) instead of building your own. For more information, see Internet Link 10.3.

Internet Link

10.3 H-bridge resource

QUESTIONS AND EXERCISES

Section 7.4   Programming a PIC

7.1. Document a complete and thorough answer to Class Discussion Item 7.2.

Section 7.5   PicBasic Pro

7.2. Write an assembly language program to turn an LED on and off at 0.5 Hz. Draw the schematic required for your solution.

7.3. Write a PicBasic Pro program to turn an LED on and off at 1 Hz while a pushbutton switch is held down. Draw the schematic required for your solution.

7.4. Write a PicBasic Pro program to perform the functionality of the *Pot* statement (see Class Discussion Item 7.7).

7.5. Write a PicBasic Pro subroutine to provide a software debounce on pin RB0 (see Class Discussion Item 7.8). The code should wait for the pushbutton switch to be pressed and released while ignoring the switch bounce.

7.6. Microcontrollers usually do not include D/A converters, but you can easily create a crude version of one using a single digital I/O pin. This can be done by outputting a variable-width pulse train to an RC circuit. See the documentation for the PWM statement in the PicBasic Pro compiler manual for more information. Write a PicBasic Pro subroutine that uses pin RA0 to output a constant voltage (between 0 and 5 V) that is proportional to the value of a byte variable called *digital_value,* whose value can range from 0 to 255. The subroutine should hold this voltage for approximately 1 sec.

7.7. For the security system in Example 7.5, explain what would happen if a burglar closes an opened door or window after setting off the alarm? Consider all pertinent operating states. How could the design be improved to overcome any perceived limitations?

Section 7.6   Using Interrupts

7.8. Write a PicBasic Pro program to implement an interrupt-driven solution to the security system presented in Example 7.9.

Section 7.7   Interfacing Common PIC Peripherals

7.9. Write a PicBasic Pro program to display the value of a potentiometer as a percentage on an LCD display. The message should have the form *pot value = X%,* where *X* is the percentage value ranging from 0 to 100. Draw the schematic required for your solution.

7.10. Write a PicBasic Pro program that allows the user to enter a set of a multidigit numbers (up to five digits) on a numeric keypad. Have the # key serve as an enter key; and when a number is entered, it should appear on a two-line LCD display. The first number should appear on the first line, the second number should appear on the second line, and subsequent numbers should appear on the second line with the previous number moving up to the first line. Draw the schematic required for your solution.

7.11. In Figure 7.13, explain why there is a 1 k resistor attached from 5 V to RA4. How does this interface differ from connecting RA4 directly to the V_{ee} pin of the LCD (with no resistor or 5 V connection)?

Section 7.9   Method to Design a Microcontroller-Based System

7.12. Apply the procedure presented in Section 7.9 to solving the problem stated in Question 6.44. Perform every step in detail. Do as much as possible in software (e.g., instead of using flip-flops).

7.13. Solve Question 7.12 based on the revised requirements in Question 6.58.

7.14. Using a PIC16F84 and two 7447 display decoders (see Design Example 7.1), write a PicBasic Pro program to implement a counter with a two-digit display controlled by three pushbutton switches: one to reset the count to 0, one to increment the count by 1, and the third to decrement the count by 1. If the count is less than 10, the first digit should be blank. Incrementing past 99 should reset the count to 0, and decrementing below 0 should not be allowed. Make sure you debounce the switch input where necessary and prevent repeats while the pushbutton switches are held down. Use the design procedure presented in Section 7.9 and show the results for each step.

7.15. Visit Microchip's website and select a flash-memory PIC microcontroller available as a DIP package that has at least 16 lines of digital I/O and 2 A/D converters. Select the lowest cost model that meets these requirements, and list the approximate cost, the number of I/O pins, the number of A/D converters (along with their resolutions), and the total number of pins on the package.

7.16. Draw a detailed flowchart illustrating all of the functionality of Threaded Design Example A.

7.17. Draw a flowchart illustrating the functionality of the main program (not including details for the subroutines) for Threaded Design Example B.

7.18. Draw a detailed flowchart illustrating all of the functionality of the *move* subroutine in Threaded Design Example B. Include details for the *move_steps* and *step_motor* subroutines.

7.19. Draw a detailed flowchart illustrating all of the functionality of the *speed* subroutine in Threaded Design Example B. Include details for the *get_AD_value* subroutine.

7.20. Draw a detailed flowchart illustrating all of the functionality of the *position* subroutine in Threaded Design Example C. Include details for the *get_encoder* subroutine.

7.21. Draw a detailed flowchart illustrating all of the functionality of the *slave* PIC code in Threaded Design Example C.

BIBLIOGRAPHY

Gibson, G. and Liu, Y., *Microcomputers for Engineers and Scientists,* Prentice-Hall, Englewood Cliffs, NJ, 1980.

Herschede, R., "Microcontroller Foundations for Mechatronics Students," master's thesis, Colorado State University, summer 1999.

Horowitz, P. and Hill, W., *The Art of Electronics,* 2nd Edition, Cambridge University Press, New York, 1989.

Microchip Technology, Inc., www.microchip.com, 2010.

Microchip Technology, Inc., *PIC16F8X Data Sheet,* Chandler, AZ, 1998.

Microchip Technology, Inc., *MPASM User's Guide,* Chandler, AZ, 1999.

Microchip Technology, Inc., *MPLAB User's Guide,* Chandler, AZ, 2000.

microEngineering Labs, Inc., www.melabs.com, 2010.

microEngineering Labs, Inc., *PicBasic Pro Compiler,* Colorado Springs, CO, 2008.

Motorola Technical Summary, "MC68HC11EA9/MC68HC711EA9 8-bit Microcontrollers," document number MC68HC11EA9TS/D, Motorola Advanced Microcontroller Division, Austin, TX, 1994.

Peatman, J., *Design with Microcontrollers,* McGraw-Hill, New York, 1988.

Predko, M., *Programming and Customizing PICmicro Microcontrollers,* McGraw-Hill, New York, 2001

Stiffler, A., *Design with Microprocessors for Mechanical Engineers,* McGraw-Hill, New York, 1992.

Texas Instruments, *TTL Linear Data Book,* Dallas, TX, 1992.

Texas Instruments, *TTL Logic Data Book,* Dallas, TX, 1988.

8
CHAPTER

Data Acquisition

his chapter presents the concepts involved with converting between analog and digital signals. This is important when interfacing digital circuits to analog components in mechatronic systems. The concept of virtual instrumentation is also introduced along with LabVIEW software examples from National Instruments. ∎

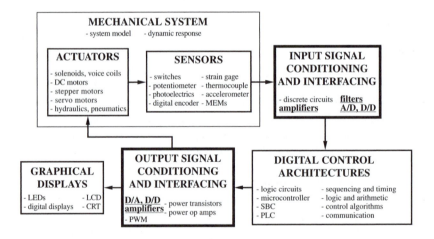

CHAPTER OBJECTIVES

After you read, discuss, study, and apply ideas in this chapter, you will:

1. Understand how to properly sample a signal for digital processing

2. Understand how digitized data are coded

The authors express their thanks to Jim Cahow of National Instruments (Austin, TX) for providing resources and support for the editing of this chapter.

3. Know the components of an A/D converter

4. Understand how A/D and D/A converters function and recognize their limitations

5. Be aware of commercially available hardware and software tools for data acquisition and control

6. Understand the basics of LabVIEW programming and data acquisition

7. Understand the effects of sampling rate and resolution on music sampling

8.1 INTRODUCTION

Microprocessors, microcontrollers, single-board computers, and personal computers are in widespread use in mechatronic and measurement systems. It is increasingly important for engineers to understand how to directly access information and analog data from the surrounding environment with these devices. As an example, consider a signal from a sensor as illustrated by the analog signal in Figure 8.1. One could record the signal with an analog device such as a chart recorder, which physically plots the signal on paper, or display it with an oscilloscope. Another option is to store the data using a microprocessor or computer. This process is called computer **data acquisition,** and it provides more compact storage of the data (magnetic, optical, or flash media vs. long rolls of paper), can result in greater data accuracy, allows use of the data in a real-time control system, and enables data processing long after the events have occurred.

To be able to input analog data to a digital circuit or microprocessor, the analog data must be transformed into coded digital values. The first step is to numerically evaluate the signal at discrete instants in time. This process is called **sampling,** and the result is a **digitized signal** composed of discrete values corresponding to each sample, as illustrated in Figure 8.1. Therefore, a digitized signal is a sequence of numbers that is an approximation to an analog signal. Note that the time relation between the numbers is an inherent property of the sampling process and need not be recorded separately. The collection of sampled data points forms a data array, and although this representation is no longer continuous, it can accurately describe the original analog signal.

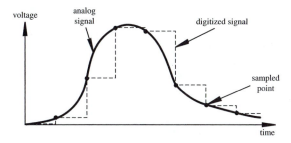

Figure 8.1 Analog signal and sampled equivalent.

An important question is how fast or often the signal should be sampled to obtain an accurate representation. The naive answer might be "as fast as you possibly can." The problems with this conclusion are that specialized, high-speed hardware is required and a large amount of computer memory is needed to store the data. A better answer is to select the minimal sampling rate required for a given application that retains all important signal information.

The **sampling theorem,** also called **Shannon's sampling theorem,** states that we need to sample a signal at a rate more than two times the maximum frequency component in the signal to retain all frequency components. In other words, to faithfully represent the analog signal, the digital samples must be taken at a frequency f_s such that

$$f_s > 2f_{max} \qquad (8.1)$$

where f_{max} is the highest frequency component in the input analog signal. The term f_s is referred to as the **sampling rate,** and the limit on the minimum required rate ($2f_{max}$) is called the **Nyquist frequency.** If we approximate a signal by a truncated Fourier series, the maximum frequency component is the highest harmonic frequency. The time interval between the digital samples is

$$\Delta t = 1/f_s \qquad (8.2)$$

As an example, if the sampling rate is 5000 Hz, the time interval between samples would be 0.2 ms.

If a signal is sampled at less than two times its maximum frequency component, **aliasing** can result. Figure 8.2 illustrates an example of this with an analog sine wave sampled regularly at the points shown. Twelve equally spaced samples are taken over 10 cycles of the original signal. Therefore, the sampling frequency is $1.2f_0$, where f_0 is the frequency of the original sine wave. Because the sampling frequency is not greater than $2f_0$, we do not capture the frequency in the original signal. Furthermore, the apparent frequency in the sampled signal is $0.2f_0$ (2 aliased signal cycles for ten original signal cycles). You can think of this as a "phantom" frequency, which is an alias of the true frequency. Therefore, undersampling not only results in errors but also creates information that is not really there! Video Demo 8.1 demonstrates the concepts of sampling rate, Nyquist frequency and aliasing, applied to a sine wave signal, and Video Demo 8.2 provides an interesting visual example of aliasing.

Video Demo

8.1 Sampling rate, Nyquist frequency, and aliasing

8.2 Tuning fork super-slow-motion video aliasing

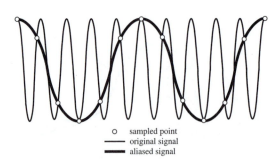

o   sampled point
—— original signal
━━ aliased signal

Figure 8.2 Aliasing.

The method of Fourier decomposition presented in Chapter 4 provides a means to determine the frequency components of an arbitrary analog signal. Representing a signal in terms of its frequency components allows us to identify the bandwidth and correctly apply the sampling theorem.

■ **CLASS DISCUSSION ITEM 8.1**
Wagon Wheels and the Sampling Theorem

Relate the sampling theorem and its implications about correct and incorrect data representation to the use of a movie camera to film a rotating wagon wheel in a Western movie. The camera shutter can be considered a sampling device, where the shutter speed is the sampling rate. Typically, the sampling rate is 30 Hz to provide a flicker-free image to the human eye. What effects would you expect in the movie, when viewed, if the wagon wheel is turning rapidly?

Sampling Theorem and Aliasing — **EXAMPLE 8.1**

Consider the function

$$F(t) = \sin(at) + \sin(bt)$$

Using a trigonometric identity for the sum of two sinusoidal functions, we can rewrite $F(t)$ as the following product:

$$F(t) = \left[2 \cos\left(\frac{a-b}{2}t\right)\right] \cdot \sin\left(\frac{a+b}{2}t\right)$$

If frequencies a and b are close in value, the bracketed term has a very low frequency in comparison to the sinusoidaal term on the right. Therefore, the bracketed term modulates the amplitude of the higher frequency sinusoidal term. The resulting waveform exhibits what is called a **beat frequency** that is common in optics, mechanics, and acoustics when two waves close in frequency add. For more information and an audio example of beat frequency, see Internet Link 8.1 and Video Demo 8.3.

To illustrate aliasing associated with improper sampling, the waveform is plotted in the following figures using two different sampling frequencies. If a and b are chosen as

$$a = 1 \text{ Hz} = 2\pi \frac{\text{rad}}{\text{sec}} \qquad b = 0.9a$$

then to sample the signal $F(t)$ properly, the sampling rate must be more than twice the highest frequency in the signal:

$$f_s > 2a = 2 \text{ Hz}$$

Therefore, the time interval between samples (plotted points) must be

$$\Delta t = \frac{1}{f_s} < 0.5 \text{ sec}$$

(continued)

Internet Link

8.1 Beat frequency

Video Demo

8.3 Beat frequency from mixing signals of similar frequency

(continued)

The first data set is plotted with a time interval of 0.01 sec (100 Hz sampling rate), providing an adequate representation of the waveform. The second data set is plotted with a time interval of 0.75 sec (1.33 Hz sampling rate), which is less than twice the maximum frequency of the waveform (2 Hz). Therefore, the signal is undersampled, and aliasing results. The sampled waveform is an incorrect representation, and its observed maximum frequency appears to be approximately 0.4 Hz because there are approximately four cycles over 10 sec.

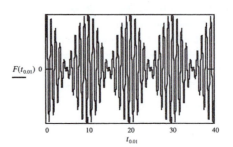

MathCAD Example

8.1 Sampling theorem, aliasing, beat frequency

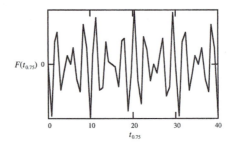

You might argue that neither plot is perfect. However, unlike the second plot, the first plot retains all the frequency information in the analog signal. MathCAD Example 8.1 contains the analysis used to generate the plots above. You can easily edit the file to see the effects of different sampling rates.

Lab Exercise

Lab 12 Data acquisition

■ CLASS DISCUSSION ITEM 8.2
Sampling a Beat Signal

What is the minimum sampling rate required to adequately represent the signal in Example 8.1?

Lab Exercise 12 explores the effects of using different sampling rates when digitizing a sine wave signal and audio from a music player. Video Demos 8.4 and 8.5 demonstrate the results. Video Demo 8.6, which contains only audio, and Video Demo 8.7 illustrate the effects of sampling rate on various tones and music recordings. For many more video demonstrations of various music, sound, and vibration concepts, see Internet Link 8.2.

Video Demo

8.4 Data acquisition experiment

8.5 LabVIEW data acquisition and music sampling

8.6 Sampling rate and its effect on sound and music (audio only)

8.7 Music sampling rate and resolution effects

8.2 QUANTIZING THEORY

Now we look at the process required to change a sampled analog voltage into digital form. The process, called analog-to-digital conversion, conceptually involves two steps: quantizing and coding. **Quantizing** is defined as the transformation of a continuous analog input into a set of discrete output states. **Coding** is the assignment of a digital code word or number to each output state. Figure 8.3 illustrates how a continuous voltage range is divided into discrete output states, each of which is assigned a unique code. Each output state covers a subrange of the overall voltage range. The stair-step signal represents the states of a digital signal generated by sampling a linear ramp of an analog signal occurring over the voltage range shown.

An **analog-to-digital converter** is an electronic device that converts an analog voltage to a digital code. The output of the A/D converter can be directly interfaced to digital devices such as microcontrollers and computers. The **resolution** of an A/D converter is the number of bits used to digitally approximate the analog value of the input. The number of possible states N is equal to the number of bit combinations that can be output from the converter:

Internet Link

8.2 Music, sound, and vibration demonstrations

$$N = 2^n \tag{8.3}$$

where n is the number of bits. For the example illustrated in Figure 8.3, the 3-bit device has 2^3 or 8 output states as listed in the first column. The output states are

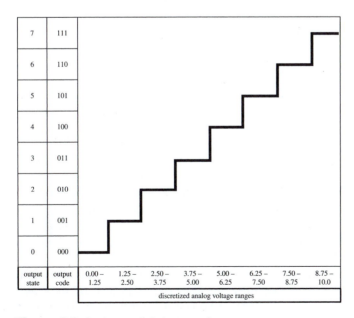

Figure 8.3 Analog-to-digital conversion.

usually numbered consecutively from 0 to ($N - 1$). The corresponding code word for each output state is listed in the second column. Most commercial A/D converters are 8-, 10-, or 12-bit devices that resolve 256, 1024, and 4096 output states, respectively.

The number of analog **decision points** that occur in the process of quantizing is ($N - 1$). In Figure 8.3, the decision points occur at 1.25 V, 2.50 V, . . . , and 8.75 V. The **analog quantization size** Q, sometimes called the **code width**, is defined as the full-scale range of the A/D converter divided by the number of output states:

$$Q = (V_{max} - V_{min})/N \tag{8.4}$$

It is a measure of the analog change that can be resolved by the converter. Although the term resolution is defined as the number of output bits from an A/D converter, sometimes it is used to refer to the analog quantization size. For our example, the analog quantization size is 10/8 V = 1.25 V. This means that the amplitude of the digitized signal has an error of at most 1.25 V. Therefore, the A/D converter can only resolve a voltage to within 1.25 V of the exact analog voltage.

Video Demo 8.7 demonstrates how both sampling rate and A/D resolution affect the fidelity of digitized music.

Video Demo

8.7 Music sampling rate and resolution effects

■ **CLASS DISCUSSION ITEM 8.3**
Laboratory A/D Conversion

Why is a 12-bit A/D converter satisfactory for most laboratory measurements?

8.3 ANALOG-TO-DIGITAL CONVERSION

8.3.1 Introduction

To properly acquire an analog voltage signal for digital processing, the following components must be properly selected and applied in this sequence:

1. buffer amplifier
2. low-pass filter
3. sample and hold amplifier
4. analog-to-digital converter
5. computer

The components required for A/D conversion along with an illustration of their respective outputs are shown in Figure 8.4. The buffer amplifier isolates the output from the input (i.e., it draws negligible current and power from the input) and provides a signal in a range close to but not exceeding the full input voltage range of the A/D converter. The low-pass filter is necessary to remove any undesirable high-frequency components in the signal that could produce aliasing. The cutoff frequency of the low-pass filter should be no greater than 1/2 the sampling rate. The sample and hold amplifier (see Section 5.12) maintains a fixed input value (from an instantaneous sample) during the short conversion time of the A/D converter.

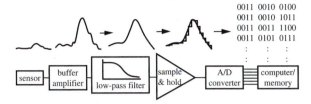

Figure 8.4 Components used in A/D conversion.

Internet Link

8.3 Data acquisition online resources and vendors

Internet Link

8.4 National Instruments' LabVIEW software

Figure 8.5 Various form factors of data acquisition products. (*Courtesy of National Instruments, Austin, TX*)

The converter should have a resolution and analog quantization size appropriate to the system and signal. The computer must be properly interfaced to the A/D converter system to store and process the data. The computer must also have sufficient memory and permanent storage media to hold all of the data.

The A/D system components described above can be found packaged in a variety of commercial products called data acquisition (DAC or DAQ) cards or modules. As shown in Figure 8.5, DAC products are available in a variety of form factors including PC and instrument panel plug-in cards, laptop computer PCMCIA cards, and stand-alone external units with standard interfaces (e.g., USB). Internet Link 8.3 provides links to vendors and sources of information for various products available on the market.

Data acquisition cards and modules usually support various high-level language interfaces (e.g., C++, Visual Basic, FORTRAN) that give easy access to the product's features. Easy-to-use software applications are also available to provide graphical interfaces to program a DAC module to acquire, process, and output signals. National Instruments' LabVIEW software is an example (see Internet Link 8.4 and Section 8.5 for more information). With a commercial DAC system, interfacing signals between a mechatronic system or laboratory experiment and a desktop computer is a simple matter of calling a function from a custom-written program or clicking and dragging icons in a graphical interface. Lab Exercise 12 explores the basics of how to use the LabVIEW software to sample and store an external voltage signal, including sampled music. Video Demo 8.5 demonstrates all of the steps in the process, and Section 8.6 explores the details.

In addition to A/D converters, commercial DAC systems usually provide other input and output functionality including binary (TTL-compatible) I/O, D/A

Lab Exercise

Lab 12 Data
acquisition

Video Demo

8.5 LabVIEW data
acquisition and
music sampling

converters, counters and timers, and signal conditioning circuitry. Important features when selecting a DAC system include the A/D and D/A resolutions (the number of bits used) and the maximum sampling rate supported. These features are key to accuracy and reliability in computer monitoring and control applications. An example data acquisition and control card that can be inserted into a PC motherboard slot is shown in Figure 8.6. Its internal architecture is illustrated in Figure 8.7. The card's functions include two 12-bit D/A converters, one 12-bit A/D converter, 24 lines of digital I/O, and three 16-bit counter-timers.

The process of analog-to-digital conversion requires a small but finite interval of time that must be taken into consideration when assessing the accuracy of the results. The **conversion time,** also called **settling time,** depends on the design of the converter, the method used for conversion, and the speed of the components used in the electronic design. Because analog signals change continuously, the uncertainty about when in the sample time window the conversion occurs causes corresponding uncertainty in the digital value. This is of particular concern if there is no sample and hold amplifier on the A/D input. The term **aperture time** refers to the duration of the time window and is associated with any error in the digital output due to changes in the input during this time. The relationship between the aperture time and the uncertainty in the input amplitude is shown in Figure 8.8. During the aperture time ΔT_a, the input signal changes by ΔV, where

$$\Delta V \approx \frac{dV(t)}{dt} \Delta T_a \qquad (8.5)$$

Sampling at or above the Nyquist frequency will yield the correct frequency components in a signal. However, to also obtain accurate amplitude resolution, we must have an A/D converter with a sufficiently small aperture time. It is often in the nanosecond range for 10- and 12-bit resolution.

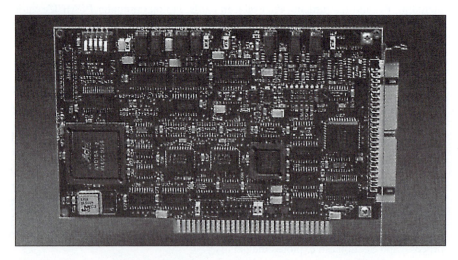

Figure 8.6 Example data acquisition and control card. (*Courtesy of National Instruments, Austin, TX*)

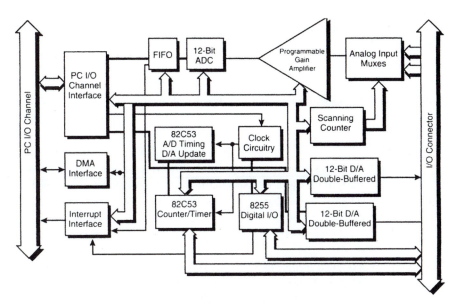

Figure 8.7 Example data acquisition and control card architecture. (*Courtesy of National Instruments, Austin, TX*)

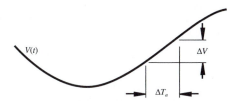

Figure 8.8 A/D conversion aperture time.

Aperture Time   **EXAMPLE 8.2**

Consider a sinusoidal signal $A\sin(\omega t)$ as an input to an A/D converter. The time rate of change of the signal $A\omega\cos(\omega t)$ has a maximum value of $A\omega$. Using Equation 8.5, the maximum change in the input voltage during an aperture time ΔT_a is

$$\Delta V = A\omega\Delta T_a$$

To eliminate uncertainty in the value of the digital output, we need to ensure that ΔV is less than the analog quantization size:

$$\Delta V < \frac{2A}{N}$$

(*continued*)

(continued) where $2A$ is the total voltage range and N is the number of output states. At the limit of this constraint,

$$\Delta V = \frac{2A}{N}$$

Therefore, the required aperture time, if a sample and hold amplifier is not being used, is

$$\Delta T_a = \frac{\Delta V}{A\omega} = \frac{2}{N\omega}$$

Consider, for example, converting a signal using 10-bit resolution, which provides 2^{10} (1024) output states. If the signal were speech (from a microphone) with a bandwidth of 10 kHz, ΔT_a would have to be less than

$$\Delta T_a = \frac{2}{N\omega} = \frac{2}{1024\,(2\pi \cdot 10{,}000)} = 3.2 \times 10^{-8} = 32 \text{ nsec}$$

This is a very short required aperture time compared to the required minimum sample period given by $1/[2(10 \text{ kHz})] = 50$ μsec. Even for this low-resolution converter, the required aperture time (32 nsec) is much smaller than the required sample period (50,000 nsec).

8.3.2 Analog-to-Digital Converters

A/D converters are designed based on a number of different principles: successive approximation, flash or parallel encoding, single-slope and dual-slope integration, switched capacitor, and delta sigma. We consider the first two because they occur most often in commercial designs. The **successive approximation** A/D converter is very widely used because it is relatively fast and cheap. As shown in Figure 8.9, it uses a D/A converter in a feedback loop. D/A converters are described in the next section. When the start signal is applied, the sample and hold (S&H) amplifier latches the analog input. Then the control unit begins an iterative process, where the digital value is approximated, converted to an analog value with the D/A converter, and compared to the analog input with the comparator. When the D/A output equals the analog input, the end signal is set by the control unit, and the correct digital output is available at the output.

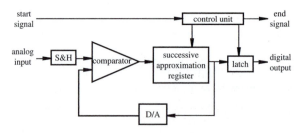

Figure 8.9 Successive approximation A/D converter.

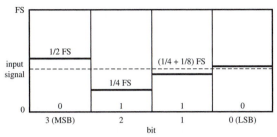

Figure 8.10 4-bit successive approximation A/D conversion.

If n is the resolution of the A/D converter, it takes n steps to complete the conversion. More specifically, the input is compared to combinations of binary fractions $(1/2, 1/4, 1/8, \ldots, 1/2^n)$ of the full-scale (FS) value of the A/D converter. The control unit first turns on the most significant bit (MSB) of the register, leaving all lesser bits at 0, and the comparator tests the DAC output against the analog input. If the analog input exceeds the DAC output, the MSB is left on (high); otherwise, it is reset to 0. This procedure is then applied to the next lesser significant bit and the comparison is made again. After n comparisons have occurred, the converter is down to the least significant bit (LSB). The output of the DAC then represents the best digital approximation to the analog input. When the process terminates, the control unit sets the end signal signifying the end of the conversion.

As an example, a 4-bit successive approximation procedure is illustrated graphically in Figure 8.10. The MSB is 1/2 FS, which in this case is greater than the signal; therefore, the bit is turned off. The second bit is 1/4 FS and is less than the signal, so it is left on. The third bit gives $1/4 + 1/8$ of FS, which is still less than the analog signal, so the third bit is left on. The fourth provides $1/4 + 1/8 + 1/16$ of FS and is greater than the signal, so the fourth bit is turned off and the conversion is complete. The digital result is 0110. Higher resolution would produce a more accurate value.

An n-bit successive approximation A/D converter has a conversion time of $n\Delta T$, where ΔT is the cycle time for the D/A converter and control unit. Typical conversion times for 8-, 10-, and 12-bit successive approximation A/D converters range from 1 to 100 μs.

The fastest type of A/D converter is known as a **flash converter.** As Figure 8.11 illustrates, it consists of a bank of input comparators acting in parallel to identify the signal level. The output of the latches is in a coded form easily converted to the required binary output with combinational logic. The flash converter illustrated in Figure 8.11 is a 2-bit converter having a resolution of four output states. Table 8.1

■ **CLASS DISCUSSION ITEM 8.4**
Selecting an A/D Converter

What are some reasons why a designer might select a 10-bit A/D converter instead of a 12-bit or higher resolution converter?

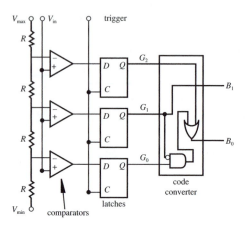

Figure 8.11 A/D flash converter.

Table 8.1 2-bit flash converter output

State	Code ($G_2 G_1 G_0$)	Binary ($B_1 B_0$)	Voltage Range
0	000	00	0–1
1	001	01	1–2
2	011	10	2–3
3	111	11	3–4

lists the comparator output codes and corresponding binary outputs for each of the states, assuming an input voltage range of 0 to 4 V. The voltage range is set by the V_{min} and V_{max} supply voltages shown in Figure 8.11 (0 V and 4 V in this example). The code converter is a simple combinational logic circuit. For the 2-bit converter, the relationships between the code bits G_i and the binary bits B_i (Question 8.11) are

$$B_0 = G_0 \cdot \overline{G_1} + G_2 \tag{8.6}$$

$$B_1 = G_1 \tag{8.7}$$

Adding more resolution is a simple matter of adding more resistors, comparators, and latches. The combinational logic code converter would also be different. Unlike with the successive approximation converter, adding resolution does not increase the time required for a conversion.

Several analog signals can be digitized by a single A/D converter if the analog signals are multiplexed at the input to the A/D converter. An analog **multiplexer** simply switches among several analog inputs using transistors or relays and control signals. This can significantly reduce the cost of a system's design. In addition to cost, other parameters important in selecting an A/D converter are the input voltage range, output resolution, and conversion time.

8.4 DIGITAL-TO-ANALOG CONVERSION

Often we need to reverse the process of A/D conversion by changing a digital value to an analog voltage. This is called **digital-to-analog (D/A) conversion.** A D/A converter allows a computer or other digital device to interface with external analog circuits and devices.

The simplest type of D/A converter is a resistor ladder network connected to an inverting summer op amp circuit as shown in Figure 8.12. This particular converter is a 4-bit R-$2R$ resistor ladder network, which requires only two precision resistance values (R and $2R$). The digital input to the DAC is a 4-bit binary number represented by bits b_0, b_1, b_2, and b_3, where b_0 is the least significant bit and b_3 is the most significant bit. Each bit in the circuit controls a switch between ground and the inverting input of the op amp. To understand how the analog output voltage V_{out} is related to the input binary number, we can analyze the four different input combinations 0001, 0010, 0100, and 1000 and apply the principle of superposition for an arbitrary 4-bit binary number.

If the binary number is 0001, the b_0 switch is connected to the op amp, and the other bit switches are grounded. The resulting circuit is as shown in Figure 8.13. Because the noninverting input of the op amp is grounded, the inverting input is also at ground. The equivalent resistance between node V_0 and ground is R, which is the parallel combination of two $2R$ values. Therefore, V_0 is the result of voltage division of V_1 across two series resistors of equal value R:

$$V_0 = \frac{1}{2}V_1 \tag{8.8}$$

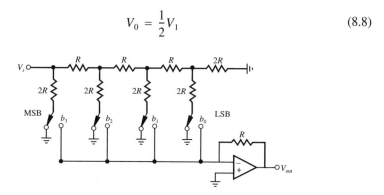

Figure 8.12 4-bit resistor ladder D/A converter.

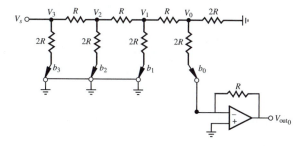

Figure 8.13 4-bit resistor ladder D/A with digital input 0001.

Similarly, we can show that

$$V_1 = \frac{1}{2}V_2 \quad \text{and} \quad V_2 = \frac{1}{2}V_3 \tag{8.9}$$

Therefore,

$$V_0 = \frac{1}{8}V_3 = \frac{1}{8}V_s \tag{8.10}$$

V_0 is the input to the inverting amplifier circuit, which has a gain of

$$-\frac{R}{2R} = -\frac{1}{2} \tag{8.11}$$

Therefore, the analog output voltage corresponding to the binary input 0001 is

$$V_{out_0} = -\frac{1}{16}V_s \tag{8.12}$$

Similarly, we can show (Questions 8.16 through 8.18) that, for the input 0010,

$$V_{out_1} = -\frac{1}{8}V_s \tag{8.13}$$

and for the input 0100,

$$V_{out_2} = -\frac{1}{4}V_s \tag{8.14}$$

and for the input 1000,

$$V_{out_3} = -\frac{1}{2}V_s \tag{8.15}$$

The output for any combination of bits comprising the input binary number can now be found using the principle of superposition:

$$V_{out} = b_3 V_{out_3} + b_2 V_{out_2} + b_1 V_{out_1} + b_0 V_{out_0} \tag{8.16}$$

If V_s is 10 V, the output ranges from 0 V to $(-15/16)10$ V for the 4-bit binary input, which has 16 values ranging from 0000 (0) to 1111 (15). A negative reference voltage V_s can be used to produce a positive output voltage range. Either case yields a **unipolar** output, which is either positive or negative but not both. A **bipolar** output, which ranges over negative and positive values, can be produced by replacing all ground references in the circuit with a reference voltage whose sign is opposite to V_s.

■ **CLASS DISCUSSION ITEM 8.5**
Bipolar 4-Bit D/A Converter

If $V_s = 10$ V and the ground reference is replaced by a -10 V reference, what output voltage would correspond to each possible binary input applied to a 4-bit D/A R-$2R$ ladder network?

THREADED DESIGN EXAMPLE

DC motor power-op-amp speed controller — D/A converter interface   **A.5**

The figure below shows the functional diagram for Threaded Design Example A (see Section 1.3 and Video Demo 1.6), with the portion described here highlighted.

Video Demo

1.6 DC motor power-op-amp speed controller

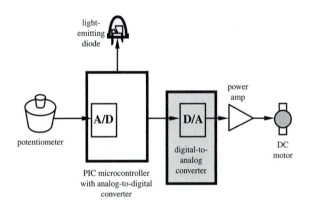

light-emitting diode

power amp

potentiometer

A/D

D/A

DC motor

digital-to-analog converter

PIC microcontroller with analog-to-digital converter

The D/A used in this design is an external 8-bit TLC7524 converter. Detailed information about the device can be found in the data sheet at Internet Link 7.15. As shown in the figure on the next page, the PIC interfaces to the D/A converter via two control lines and 8 data lines (the entire PORTB register). One control line is the chip select pin that is used to activate the D/A converter. The other is a write line that signals the D/A when to perform the conversion and update the output voltage. Both lines are "active low."

The code required to initialize the PIC outputs and to activate the D/A converter follows:

Internet Link

7.15 TLC7524 D/A converter

```
da_cs Var PORTA.3    ' external D/A converter chip select
                       (low:activate)
da_wr Var PORTA.4    ' external D/A converter write (low:
                       write)
TRISB = 0            ' initialize PORTB pins as outputs
High da_wr           ' initialize the A/D converter write line
Low da_cs            ' activate the external D/A converter
```

(continued)

(continued)

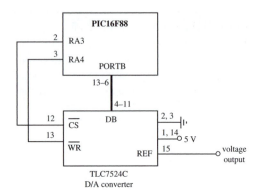

The code required to update the voltage output from the D/A converter follows. The voltage is stored as a scaled value in the PIC in a byte variable (*ad_byte*). The value ranges from 0 (corresponding to 0 V) to 255 (corresponding to 5 V). The entire byte (all 8 bits) is transmitted in parallel to the D/A converter through PORTB. Note the small pause after the write command to ensure that the conversion has completed before the write is disabled.

```
' Send the potentiometer byte to the external D/A
PORTB = ad_byte
Low da_wr
Pauseus 1     ' wait 1 microsec for D/A to settle
High da_wr
```

Figure 8.14 illustrates the roles that A/D and D/A converters play in a mechatronic control system. An analog voltage signal from a sensor (e.g., a thermocouple) is converted to a digital value, the computer uses this value in a control algorithm, and the computer outputs an analog signal to an actuator (e.g., an electric motor) to cause some change in the system being controlled. This topic is explored more in the next section and in Chapter 11. Sensors and actuators are the topics of the next two chapters.

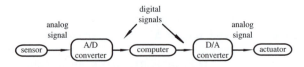

Figure 8.14 Computer control hardware.

> ■ **CLASS DISCUSSION ITEM 8.6**
> **Audio CD Technology**
>
> A compact disk (CD) stores music (an analog signal) in a digital format. How is this done? How is the digital data reconverted to the music you hear? Given that audible frequencies range from 20 Hz to 20 kHz, what is an appropriate sampling frequency? For this sampling rate, how many bits must be stored on a CD to produce 45 minutes of listening pleasure, assuming the music is stored uncompressed (e.g., as a WAV file and not as an MP3 file)? Assume just the raw data is stored, without the sophisticated error correction and compensation schemes described in Internet Link 8.5.

Internet Link

8.5 Audio CD format information

> ■ **CLASS DISCUSSION ITEM 8.7**
> **Digital Guitar**
>
> A digital guitar is a standard electric guitar equipped with additional components that send MIDI signals to a digital synthesizer. MIDI stands for musical instrument digital interface, and a MIDI signal consists of digital bytes containing amplitude and frequency codes for musical notes. What system components are required to do this?

8.5 VIRTUAL INSTRUMENTATION, DATA ACQUISITION, AND CONTROL

A **virtual instrument** consists of a personal computer equipped with data acquisition hardware and software to perform the functions of traditional instruments. Stand-alone traditional instruments such as oscilloscopes and waveform generators are very powerful, but they can be very expensive and sometimes limiting. The user generally cannot extend or customize the instrument's functionality. The knobs and buttons on the instrument, the built-in circuitry, and the functions available to the user are all specific to the instrument. Virtual instrumentation provides an alternative.

Software is the most important component of a virtual instrument. It can be used to create custom applications by designing and integrating the routines that a particular process requires. The software also allows one to create an appropriate user interface that best suits the purpose of the application and the users who will interact with it. The LabVIEW software from National Instruments (see Internet Link 8.4) is an example of an easy-to-use application development environment designed specifically for creating virtual instruments. LabVIEW offers powerful features that make connecting to a wide variety of hardware and other software easy.

One of the most powerful features of LabVIEW is its graphical programming environment. Custom graphical user interfaces can be designed to allow a user to

Internet Link

8.4 National Instruments' LabVIEW software

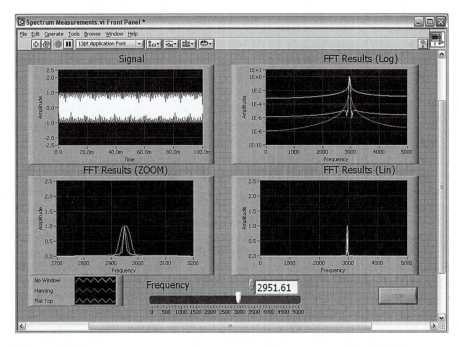

Figure 8.15 Example custom graphical user interface created with LabVIEW.

operate the instrumentation program, control selected hardware, analyze acquired data, and display results graphically, all on the PC screen. One can customize virtual front panels with knobs, buttons, dials, and graphs to emulate control panels of traditional instruments, create custom test panels, or visually represent the control and operation of processes. Figure 8.15 shows an example graphical user interface that can easily be created. Figure 8.16 shows the block diagram used to create the user interface and to perform the acquisition, analysis, and display functions. LabVIEW is known as a visual programming environment, because to create a custom application and user interface, you simply drag, drop, and connect icons on the screen. Computer programming skills are not required.

LabVIEW, in addition to its virtual instrumentation and data acquisition capabilities, provides many tools for model-based control system design. Tools are available for characterizing (modeling) a system through data acquisition of test data (a process called **system identification**), designing a controller for a system, simulating how the system will respond to various control inputs, and embedding the controller into hardware to perform real-time control. Video Demo 8.8 shows a demonstration of how these tools can be used to develop a speed controller for a simple DC motor system. LabVIEW is used to send a command signal to the DC motor and to monitor the resulting motion by acquiring a feedback signal from a tachometer. Video Demo 8.9 provides much more background and detailed demonstrations of the individual steps in the process. Also, the basics of control theory are presented in Section 11.3.

Video Demo

8.8 National Instruments DC motor demo

8.9 National Instruments LabVIEW DC motor controller design

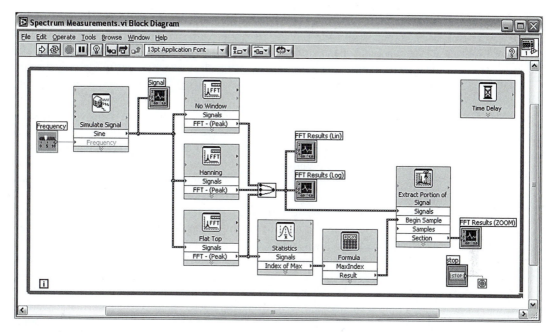

Figure 8.16 Example LabVIEW block diagram.

8.6 PRACTICAL CONSIDERATIONS

This section introduces detailed procedures required to use the **LabVIEW** software and a USB data acquisition card to sample and display analog signals. Specific tools for sampling, displaying, and replaying audio signals are also presented. Detailed instructions are provided for the National Instruments USB 6009, a specific USB data acquisition module, to show how the software is used in practice. Lab Exercise 12 provides even more detail and explores the effects of sampling rate.

Lab Exercise

Lab 12 Data Acquisition

8.6.1 Introduction to LabVIEW Programming

LabVIEW is a user-friendly graphical programming environment. It has many built-in features for data acquisition and works well with many commercial DAC cards. A brief description of the basics of LabVIEW is presented below.

There are two primary windows in LabVIEW, the **Block Diagram** and **Front Panel** window (see Figures 8.17 and 8.18). The Block Diagram window contains the graphical program that you create, and the Front Panel window contains the user interface. The user interface is used to input control parameters, run the program, and visualize the results (e.g. the plot of a waveform).

Additional windows called **palettes** contain the libraries of built-in LabVIEW functions. The **Functions** palette is a library of blocks that can be used in the Block Diagram window (it is only available when the Block Diagram window is active). The **Controls** palette is a library of the functions available for the front panel (it is available only when the Front Panel window is active). The **Tools** palette is used to modify the function of the cursor. Different tools are used to perform different functions. For example, *connect wire* 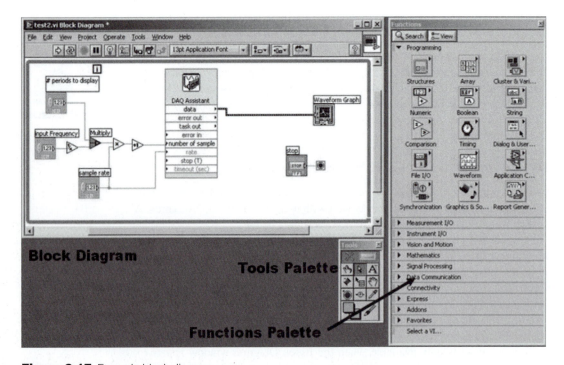 is used to connect blocks, and *operate value* is used to used to change the value of a control (described below). Alternatively, the *automatic tool selection* will automatically change which tool you are using depending on the location of the cursor.

A LabVIEW VI (pronounced "vee eye") file is made up of **objects** (or blocks) with connections between the objects. There are two types of objects: **nodes** and **terminals**. The nodes perform functions such as acquiring a digital signal from a data acquisition card, multiplication, and signal processing. Terminals are the connections between the block diagram and the front panel. Everything in the front panel appears as a terminal on the block diagram. Every object has inputs, outputs, and parameters that determine its function. For example, an analog-to-digital conversion block will have an analog signal as its input (hardware input), a digital-value signal as its output, and parameters such as sampling rate. The output of an A/D conversion block can be an input to a block that graphically displays the waveform (terminal). The parameters of a block can be set in different ways. One way is to open the properties window for the block (by right-clicking and selecting properties) and enter the values for the parameters. Some of the parameters cannot be changed independently of other parameters (e.g., parameters that define the configuration or mode of the block) and can be set only within the properties window. The parameters that can be

Figure 8.17 Example block diagram.

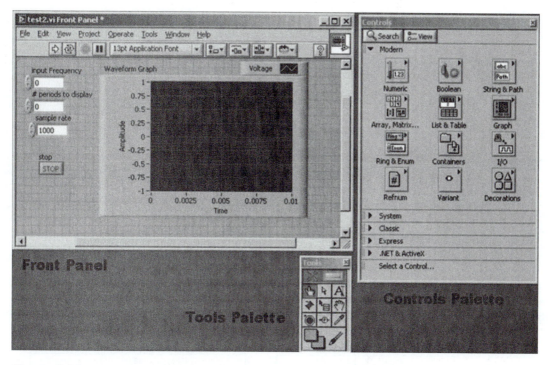

Figure 8.18 Example front panel.

set independently of the others (such as sample rate and number of samples for an A/D converter block) may be set using inputs. This can be done using **constant** or **control** terminals. A constant is set in the block diagram, and a control is set in the front panel. Figure 8.17 contains control blocks on the left labeled *# periods to display, input frequency,* and *sample rate.* Figure 8.18 shows the corresponding controls in the front panel. These figures define the VI file to perform A/D conversion using the National Instruments USB 6009 DAC card.

8.6.2 The USB 6009 Data Acquisition Card

The National Instruments USB 6009 is a typical small external data acquisition card that is connected to a computer through a USB port. The device is pictured in Figure 8.19. It has A/D conversion capabilities as well as D/A conversion, digital I/O, and counters/timers. The I/O lines are connected with wire (16-28 AWG) to the detachable screw terminals.

As shown in Tables 8.2 and 8.3, screw terminals 1–16 are used for analog I/O, and terminals 17–32 are used for digital I/O and counter/timer functions. Notice that the analog terminals are different depending on which mode the device is in, single-ended mode (also known as RSE) or differential mode. In single-ended mode the positive voltage signal is connected to an AI terminal and the negative voltage or ground signal is connected to a GND terminal. This mode uses two terminals,

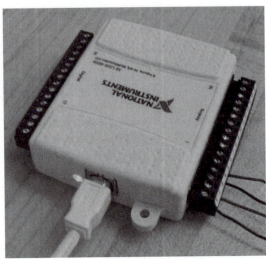

Figure 8.19 The USB 6009 connected to the computer with the screw terminals attached.

Table 8.2 Analog (1–16) and Digital (17–32) pin assignments of the USB 2009

Module	Terminal	Signal, Single-Ended Mode	Signal, Differential Mode
	1	GND	GND
	2	AI 0	AI 0+
	3	AI 4	AI 0−
	4	GND	GND
	5	AI 1	AI 1+
	6	AI 5	AI 1−
	7	GND	GND
	8	AI 2	AI 2+
	9	AI 6	AI 2−
	10	GND	GND
	11	AI 3	AI 3+
	12	AI 7	AI 3−
	13	GND	GND
	14	AO 0	AO 0
	15	AO 1	AO 1
	16	GND	GND

Module	Terminal	Signal
	17	P0.0
	18	P0.1
	19	P0.2
	20	P0.3
	21	P0.4
	22	P0.5
	23	P0.6
	24	P0.7
	25	P1.0
	26	P1.1
	27	P1.2
	28	P1.3
	29	PFI 0
	30	+2.5 V
	31	+5 V
	32	GND

meaning there are eight analog inputs available (AI0–AI7). The maximum voltage range in this mode is −10V to 10V. Differential mode can be used to get a larger voltage range. This mode measures the difference between two signals, AI+ and AI−, each referenced to GND. A voltage range of −20V to 20V can be achieved, but the maximum voltage on one pin (AI+ or AI−) referenced to ground is ±10V. Differential mode uses one more wire than single-ended mode, so only four analog inputs are available in this mode.

Table 8.3 Signal descriptions for the USB 6009

Signal Name	Reference	Direction	Description
GND	—	—	**Ground**—The reference point for the single-ended AI measurements, bias current return point for differential mode measurements, AO voltages, digital signals at the I/O connector, +5 VDC supply, and the +2.5 VDC reference.
AI <0..7 >	Varies	Input	**Analog Input Channels 0 to 7**—For single-ended measurements, each signal is an analog input voltage channel. For differential measurements, AI 0 and AI 4 are the positive and negative inputs of differential analog input channel 0. The following signal pairs also form differential input channels: <AI 1, AI 5 > , <AI 2, AI 6> , and <AI 3, AI 7> .
AO 0	GND	Output	**Analog Channel 0 Output**—Supplies the voltage output of AO channel 0.
AO 1	GND	Output	**Analog Channel 1 Output**—Supplies the voltage output of AO channel 1.
P1.<0..3> P0.<0..7>	GND	Input or Output	**Digital I/O Signals**—You can individually configure each signal as an input or output.
+2.5 V	GND	Output	**+2.5 V External Reference**—Provides a reference for wrap-back testing.
+5 V	GND	Output	**+5 V Power Source**—Provides +5 V power up to 200 mA.
PFI 0	GND	Input	**PFI 0**—This pin is configurable as either a digital trigger or an event counter input.

Another difference between differential and single-ended mode is the resolution of the analog inputs. Differential mode has a resolution of 14 bits whereas single-ended mode has a resolution of 13 bits.

The analog input converter type is successive approximation, and the maximum sampling rate is 48 thousand samples per second (kS/s). The device contains one analog-to-digital converter that is multiplexed to each input.

8.6.3 Creating a VI and Sampling Music

This example is based on LabVIEW version 8.0, but should be fairly compatible with previous or later versions. The procedure also assumes that the USB 6009 is already set up to interface with the computer according to the instructions that come with the device.

Opening a blank VI file

1. Start LabVIEW with *Start > Programs > National Instruments > LabVIEW 8.0 > LabVIEW.*

2. Click *Blank VI* to start a new project. The *Block Diagram* window and the *Front Panel* window should appear. If only one is open, then under the Windows menu click *Show Block Diagram* or *Show Front Panel.* Some other small windows may also be open.

3. Open the *Functions* palette if it is not open. From the *Block Diagram* window, under the View menu, click *Functions palette* to open the *Functions* palette.

Creating node blocks

1. From the *Functions* palette, select *Measurement I/O > NI - DAQmx*. Drag the *DAQ Assist* icon onto the *Block Diagram.* A DAQ Assistant window should appear.

2. Connect the USB 6009 device to the computer. The green light should start blinking. From the DAQ Assistant window, select *Analog Input > Voltage > ai0 > Finish*. If the ai0 does not appear, then press the plus next to "Dev1 (USB-6009)" to display the available analog input channels.

3. A new window will open displaying the properties of the DAQ Assistant block (see below).

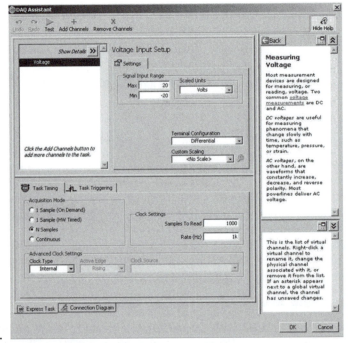

a.

4. Under *Settings,* set the maximum and minimum values for the *Signal Input Range* based on the amplitude of the input and the desired quantization size.

5. Under *Settings,* set the *Terminal Configuration* to *RSE* (single-ended mode).

6. Under the *Task Timing* tab, set the acquisition mode to *N samples*. The wiring diagram can be view by selecting the *Connection Diagram* tab towards the bottom of the window, assuming an appropriate range was selected (max ≤ 10 and min≥ −10).

7. Select *Ok* to close the DAQ Assistant properties window. This may be opened later by right-clicking on the *DAQ Assistant* block.

Creating terminal blocks

1. Select the wire spool icon 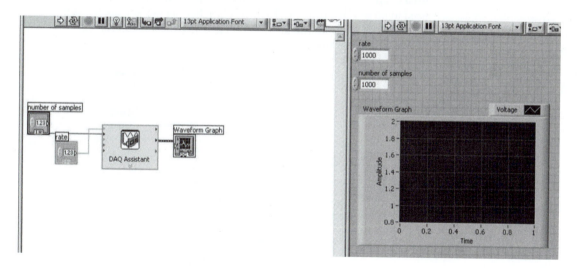 from the *Tools* palette (or the automatic icon). Right-click on the *rate* input (arrow on the side of the block) on the *DAQ Assistant* block and select *create > control*. A block labeled *rate* should appear with a wire connected to the *DAQ Assistant* block.

2. Repeat this to create a control for the *number of samples* input. These two controls will appear in the *Front Panel* window.

3. Activate the *Front Panel* window and open the *controls* palette if it is not open. From the *Front Panel* window, under the *View* menu, select *Controls Palette* to open the *Controls* palette.

4. From the *Controls* palette, select *modern > graph* and drag the *Waveform Graph* icon onto the *Front Panel* window. A block labeled "Waveform graph" will appear in the *Block Diagram* window.

5. Right-click on the graph and select *properties.* Under the *Scales* tab select *Amplitude (Y-axis)* in the top pull-down menu. Deselect *Autoscale* and set the maximum and minimum to the values used for the signal input range on the DAQ Assistant block. Click *Ok* to close the properties window.

6. Select the *Block Diagram* window and select the wire spool icon (or automatic icon) on the *Tools* palette. Click on the *data* output of the *DAQ Assistant* block and then on the *Waveform Graph* block. A wire will now connect the two blocks and it should look like the window below.

Sampling an analog signal

1. Connect an analog signal to the USB 6009. The positive side is connected to screw terminal 2 (AI0), and the negative side is connected to screw terminal 1 (GND).

2. Select the *Front Panel* window and select the *operate value* 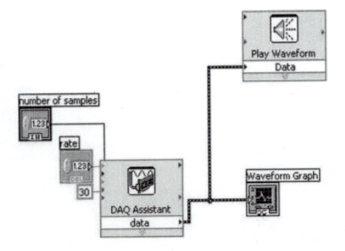 icon from the *Tools* palette (or automatic icon).

3. Set the *rate* and *number of samples* controls to appropriate values.

4. Under the *Operate* menu, select *run* to run the program. A waveform should appear on the Waveform Graph. A picture of the waveform can be saved to a file by right-clicking on the waveform and selecting *Data Operations > Export Simplified Image. . . .*

Sampling music

1. Connect the output of an audio device (e.g., speaker wires from an amplifier, or the headphone wires of an MP3 player) to the inputs of the USB 6009.

2. Select a sampling rate of 44,000 kHz (the standard high-fidelity music sampling rate) to capture all audio frequencies without aliasing. Because human hearing is generally limited to the 20 Hz to 20 kHz range, any rate less than or equal to 40,000 (from Shannon's Sampling Theorem) can result in aliasing (poor fidelity).

3. Add the *Play Waveform* block to the block diagram, which can be found in the *Functions* palette: *Programming, Graphics and Sound, Sound, Output*. Press *OK* when the configuration dialog window appears. Wire the *data* input of the *Play Waveform* block to the *data* output of the *DAQ Assistant* block. Create a constant for the *timeout* input of the *DAQ Assistant* (using the same method used to create a control for the rate input) and set it to 30 (the letter icon will need to be selected from the Tools palette). Setting the timeout to 30 allows up to 30 seconds of music to be recorded. The block diagram should now look like the following figure.

4. Sample the music and listen to the recorded waveform.

Video Demo 8.5 demonstrates all of the steps in the process, including the playback of sampled music at different sampling rates. Video Demo 8.7 illustrates the effects of sampling in more detail and also demonstrates the effects of reduced resolution. Excellent online tutorials dealing with various aspects of LabVIEW can be found at Internet Link 8.6.

Video Demo

8.5 LabVIEW data acquisition and music sampling

8.7 Music sampling rate and resolution effects

QUESTIONS AND EXERCISES

Section 8.1   Introduction

8.1. Why is it not possible to connect sensors such as thermocouples, strain gages, and accelerometers directly to a digital computer or microprocessor?

8.2. Assuming that the bandwidth of an audio music track is 15 kHz, what is the minimum sampling rate necessary to digitize the track with high fidelity for storage on an audio CD?

8.3. Suppose you want to digitize the raw signals coming from the following sources:

a. thermocouple sensing room temperature

b. stereo amplifier output

What is the minimum sampling frequency that you would choose in each of these cases? Be sure to state and justify any assumptions you make.

8.4. Plot two periods of the function $V(t) = \sin(t) + \sin(2t)$ at time intervals corresponding to 1/3, 2/3, 2, and 10 times the Nyquist frequency. Comment on the results.

8.5. Generate plots for the function in Example 8.1 with sample intervals of 0.5 sec, 1 sec, and 10 sec. Explain the results in light of the sampling theorem.

8.6. Document a complete and thorough answer to Class Discussion Item 8.2.

Internet Link

8.6 LabVIEW Graphical Programming online course

Section 8.2   Quantizing Theory

8.7. Given a 12-bit A/D converter operating over a voltage range from -5 V to 5 V, how much does the input voltage have to change, in general, in order to be detectable?

8.8. A signal's values have a range of ±5 V, and you wish to make measurements with an analog quantization size no more than 5 mV. What minimum resolution A/D converter is required to perform this task?

8.9. If an 8-bit A/D converter with a 0 to 10 V voltage range is used to sample an analog sensor's voltage, what digital output code would correspond to each of the following sensor values:

a. 0.0 V

b. 1.0 V

c. 5.0 V

d. 7.5 V

Section 8.3   Analog-to-Digital Conversion

8.10. How much computer memory (in bytes) would be required to store 10 seconds of a sensor signal sampled by a 12-bit A/D converter operating at a sampling rate of 5kHz?

8.11. Derive Equations 8.6 and 8.7 from the truth table in Table 8.1.

8.12. Generate a plot similar to Figure 8.10 for a 5-bit converter if the input signal is 2.25 V and the input range is -5 V to 5 V. What is the correct digital output?

8.13. As a mechatronic designer moves progressively from an 8-bit to 10-bit to 12-bit A/D converter, what problems might be incurred in the design of the data acquisition system?

8.14. Visit Microchip's website (*www.microchip.com*) and find specifications for the MCP32X series of A/D converters. What is their resolution? What type of architecture is used in obtaining the binary representation of the analog value?

8.15. Using the Internet, look up the specifications for the National Semiconductor (*www.national.com*) ADC0800 8-bit A/D converter. Determine the maximum sampling rate and the method for performing the conversion. Also define each of the inputs and outputs.

Section 8.4   Digital-to-Analog Conversion

8.16. Prove Equation 8.13.

8.17. Prove Equation 8.14.

8.18. Prove Equation 8.15.

8.19. Document a complete and thorough answer to Class Discussion Item 8.5.

8.20. Document a complete and thorough answer to Class Discussion Item 8.6.

BIBLIOGRAPHY

Datel Intersil, *Data Acquisition and Conversion Handbook*, Mansfield, MA, 1980.

Gibson, G. and Liu, Y., *Microcomputers for Engineers and Scientists*, Prentice-Hall, Englewood Cliffs, NJ, 1980.

Horowitz, P. and Hill, W., *The Art of Electronics*, 2nd Edition, Cambridge University Press, New York, 1989.

O'Connor, P., *Digital and Microprocessor Technology*, 2nd Edition, Prentice-Hall, Englewood Cliffs, NJ, 1989.

CHAPTER 9

Sensors

T his chapter describes various sensors important in mechatronic system design. ∎

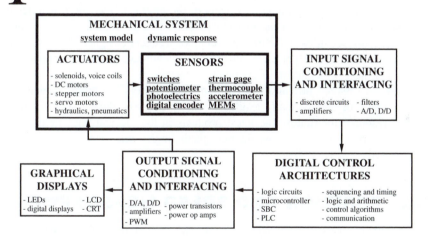

CHAPTER OBJECTIVES

After you read, discuss, study, and apply ideas in this chapter, you will:

1. Understand the fundamentals of simple electromechanical sensors, including proximity sensors and switches, potentiometers, linear variable differential transformers, optical encoders, strain gages, load cells, thermocouples, and accelerometers

2. Be able to describe how natural and binary codes are used to encode linear and rotational position in digital encoders

3. Be able to apply engineering mechanics principles to interpret data from a single strain gage or strain gage rosette

4. Be able to make accurate temperature measurements using thermocouples

5. Know how to measure acceleration and understand the frequency response of accelerometers

6. Understand what a microelectromechanical (MEM) system is

9.1 INTRODUCTION

A **sensor** is an element in a mechatronic or measurement system that detects the magnitude of a physical parameter and changes it into a signal that can be processed by the system. Often the active element of a sensor is referred to as a **transducer.** Monitoring and control systems require sensors to measure physical quantities such as position, distance, force, strain, temperature, vibration, and acceleration. The following sections present devices and techniques for measuring these and other physical quantities.

Sensor and transducer design always involves the application of some law or principle of physics or chemistry that relates the quantity of interest to some measurable event. Appendix B summarizes many of the physical laws and principles that have potential application in sensor and transducer design. Some examples of applications are also provided. This list is useful to a transducer designer who is searching for a method to measure a physical quantity. Practically every transducer applies one or more of these principles in its operation.

Internet Link

9.1 Sensor online resources and vendors

Internet Link 9.1 provides links to numerous vendors and online resources for a wide range of commercially available sensors and transducers. The Internet is a good resource for finding the latest products in the mechatronics field. This is especially true for sensors, where new technologies and improvements evolve continuously.

9.2 POSITION AND SPEED MEASUREMENT

Other than electrical measurements (e.g., voltage, current, resistance), the most commonly measured quantity in mechatronic systems is position. We often need to know where various parts of a system are in order to control the system. Section 9.2.1 presents proximity sensors and limit switches that are a subset of position sensors that detect whether or not something is close or has reached a limit of travel. Section 9.2.2 presents the potentiometer, which is an inexpensive analog device for measuring rotary or linear position. Section 9.2.3 presents the linear variable differential transformer, which is an analog device capable of accurately measuring linear displacement. Finally, Section 9.2.4 presents the digital encoder, which is useful for measuring a position with an output in digital form suitable for direct interface to a computer or other digital system.

Because most applications involve measuring and controlling shaft rotation (e.g., in robot joints, numerically controlled lathe and mill axes, motors, and generators), rotary position sensors are more common than linear sensors. Also, linear motion can often be easily converted to rotary motion (e.g., with a belt, gear, or wheel mechanism), allowing the use of rotary position sensors in linear motion applications.

Speed measurements can be obtained by taking consecutive position measurements at known time intervals and computing the time rate of change of the position values. A tachometer is an example of a speed sensor that does this for a rotating shaft.

9.2.1 Proximity Sensors and Switches

A proximity sensor consists of an element that changes either its state or an analog signal when it is close to, but often not actually touching, an object. Magnetic, electrical capacitance, inductance, and eddy current methods are particularly suited to the design of a proximity sensor. Video Demo 9.1 shows an example application for a magnetic proximity sensor. A **photoemitter-detector pair** represents another approach, where interruption or reflection of a beam of light is used to detect an object in a noncontact manner. The emitter can be a laser or focused LED, and the detector is usually a phototransistor or photodiode. Various configurations for photoemitter-detector pairs are illustrated in Figure 9.1. In the opposed and retroreflective configurations, the object interrupts the beam; and in the proximity configuration, the object reflects the beam. Figure 9.2 shows a commercial sensor that can be used in the retroreflective or proximity configurations. Video Demo 9.2 shows an interesting student project example of the proximity configuration. Common applications for proximity sensors and limit switches are detecting the presence of an object (e.g., a man in front of a public urinal), counting moving objects (e.g., passing by on a conveyor belt), and in limiting the traverse of a mechanism (e.g., by detecting the end of travel of a slider or joint).

There are many designs for limit **switches,** including pushbutton and levered microswitches. All switches are used to open or close connections within circuits. As illustrated in Figure 9.3, switches are characterized by the number of poles (P)

Video Demo

9.1 Magnetic pickup tachometer used in a PID speed controller test-stand

9.2 Automated laboratory rat exercise machine with infrared sensor and stepper motor

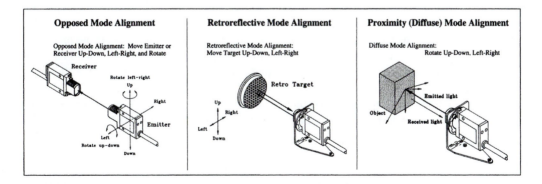

Figure 9.1 Various configurations for photoemitter-detector pairs. *(Courtesy of Banner Engineering, Minneapolis, MN)*

Figure 9.2 Example of a photoemitter-detector pair in a single housing. *(Courtesy of Banner Engineering, Minneapolis, MN)*

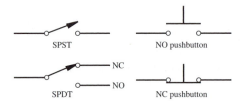

Figure 9.3 Switches.

Video Demo

9.3 Switches
9.4 Thermostat with bimetallic strip and mercury switch

and throws (T) and whether connections are **normally open (NO)** or **normally closed (NC).** A **pole** is a moving element in the switch that makes or breaks connections, and a **throw** is a contact point for a pole. The SPST switch is a single-pole (SP), single-throw (ST) device that opens or closes a single connection. The SPDT switch changes the pole between two different throw positions. There are many variations on the pole and throw configurations of switches, but their function is easily understood from the basic terminology. Figure 9.4 and Video Demo 9.3 show various types of switches with the appropriate designations. Video Demo 9.4 shows an interesting example of a normally open mercury switch that is used to turn on or off an air conditioning or heating unit when a bimetallic strip coil (see Section 9.4.2) rotates a certain amount.

When mechanical switches are opened or closed, they exhibit switch bounce, where many break-reconnect transitions occur before a new state is established. If a switch is connected to a digital circuit that requires a single transition, the switch output must be debounced using a circuit or software as described in Section 6.10.1.

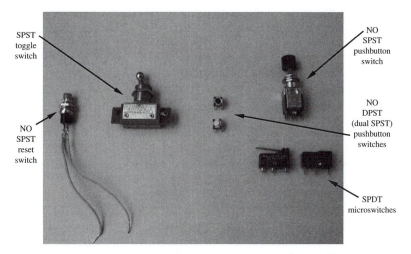

Figure 9.4 Photograph of various types of switches.

■ **CLASS DISCUSSION ITEM 9.1**
Household Three-Way Switch

A household three-way switch is used to allow lights to be controlled from two locations (e.g., at the top and bottom of a stairwell). Note that the term three-way refers to the number of terminals (3) on each switch and not the number of switches (2). A three-way switch is the SPDT variety. Draw a schematic of how AC power and the two switches can be wired to a light fixture to achieve the desired functionality, where either switch can be used to turn the light on or off.

9.2.2 Potentiometer

The rotary **potentiometer** (aka **pot**) is a variable resistance device that can be used to measure angular position. It consists of a wiper that makes contact with a resistive element, and as this point of contact moves, the resistance between the wiper and end leads of the device changes in proportion to the angular displacement. Figure 9.5 illustrates the form and internal schematic for a typical rotary potentiometer. Figure 9.6 shows two common types of potentiometers. The one on the left is called a **trim pot.** It has a small screw on the left side that can be turned with a

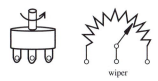

wiper

Figure 9.5 Potentiometer.

Figure 9.6 Photograph of a trim pot and a rotary pot.

screwdriver to accurately make small changes in resistance (i.e. "trim" or adjust the resistance). On the right is a standard rotary pot with a knob allowing a user to easily make adjustments by hand. Through voltage division, the change in resistance of a pot can be used to create an output voltage that is directly proportional to the input displacement. This relationship was derived in Section 4.8.

9.2.3   Linear Variable Differential Transformer

The **linear variable differential transformer (LVDT)** is a transducer for measuring linear displacement. As illustrated in Figure 9.7, it consists of primary and secondary windings and a movable iron core. It functions much like a transformer, where voltages are induced in the secondary coil in response to excitation in the primary coil. The LVDT must be excited by an AC signal to induce an AC response in the secondary. The core position can be determined by measuring the secondary response.

   With two secondary coils connected in the series-opposing configuration as shown, the output signal describes both the magnitude and direction of the core motion. The primary AC excitation V_{in} and the output signal V_{out} for two different core positions are shown in Figure 9.7. There is a midpoint in the core's position where the voltage induced in each coil is of the same amplitude and 180° out of phase, producing a "null" output. As the core moves from the null position, the output amplitude increases a proportional amount over a linear range around the null as shown in Figure 9.8. Therefore, by measuring the output voltage amplitude, we can easily and accurately determine the magnitude of the core displacement. Internet Link 9.2 points to an interesting animation that illustrates how the output voltage of the LVDT changes with core displacement.

   To determine the direction of the core displacement, the secondary coils can be connected to a demodulation circuit as shown in Figure 9.9. The diode bridges in this circuit produce a positive or negative rectified sine wave, depending on which side of the null position the core is located (see Class Discussion Item 9.2).

Internet Link

9.2 Animation of LVDT function

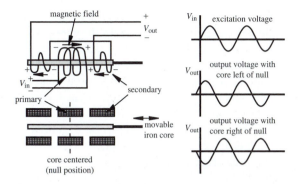

Figure 9.7 Linear variable differential transformer.

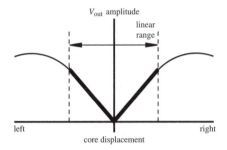

Figure 9.8 LVDT linear range.

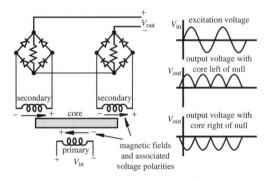

Figure 9.9 LVDT demodulation.

■ **CLASS DISCUSSION ITEM 9.2**
LVDT Demodulation

Trace the currents through the diodes in the demodulation circuit shown in Figure 9.9 for different core positions (null, left of null, and right of null) and explain why the output voltage behaves as shown. Assume ideal diodes. Also, explain why the output is 0 when the core is in the null or center position.

As illustrated in Figure 9.10, a low-pass filter may also be used to convert the rectified output into a smoothed signal that tracks the core position. The cutoff frequency of this low-pass filter must be chosen carefully to filter out the high frequencies in the rectified wave but not the frequency components associated with the core motion. The excitation frequency is usually chosen to be at least 10 times the maximum expected frequency of the core motion to yield a good representation of the time-varying displacement.

Commercial LVDTs, such as the one shown in Figure 9.11, are available in cylindrical forms with different diameters, lengths, and strokes. Often, they include internal circuitry that provides a DC voltage proportional to displacement.

The advantages of the LVDT are accuracy over the linear range and an analog output that may not require amplification. Also, it is less sensitive to wide ranges in temperature than other position transducers (e.g., potentiometers, encoders, and semiconductor devices). The LVDT's disadvantages include limited range of motion and limited frequency response. The overall frequency response is limited by inertial effects associated with the core's mass and the choice of the primary excitation frequency and the filter cutoff frequency.

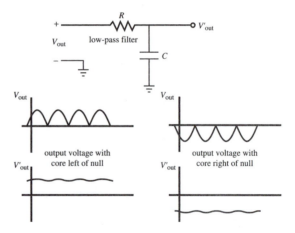

Figure 9.10 LVDT output filter.

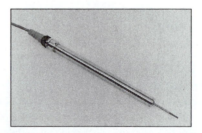

Figure 9.11 Commercial LVDT. *(Courtesy of Sensotec, Columbus, OH)*

> ■ **CLASS DISCUSSION ITEM 9.3**
> *LVDT Signal Filtering*
>
> Given the spectrum of a time-varying core displacement, what effect does the choice of the primary excitation frequency have, and how should the low-pass filter be designed to produce an output most representative of the displacement?

A **resolver** is an analog rotary position sensor that operates very much like the LVDT. It consists of a rotating shaft (rotor) with a primary winding and a stationary housing (stator) with two secondary windings offset by 90°. When the primary is excited with an AC signal, AC voltages are induced in the secondary coils, which are proportional to the sine and cosine of the shaft angle. Because of this, the resolver is useful in applications where trigonometric functions of position are required.

Two other types of linear position sensors that measure linear displacement directly, based on magnetic principles, are the **voice coil** and **magnetostrictive** position transducers. Video Demos 9.5 and 9.6 show two example devices and describe how they work.

Video Demo

9.5 Voice coil
9.6 Magneto-strictive position sensor

9.2.4 Digital Optical Encoder

A **digital optical encoder** is a device that converts motion into a sequence of digital pulses. By counting a single bit or decoding a set of bits, the pulses can be converted to relative or absolute position measurements. Encoders have both linear and rotary configurations, but the most common type is rotary. Rotary encoders are manufactured in two basic forms: the absolute encoder where a unique digital word corresponds to each rotational position of the shaft, and the incremental encoder, which produces digital pulses as the shaft rotates, allowing measurement of relative displacement of the shaft. As illustrated in Figure 9.12, most rotary encoders are composed of a glass or plastic code disk with a photographically deposited radial pattern organized in tracks. As radial lines in each track interrupt the beam between a photoemitter-detector pair, digital pulses are produced.

Video Demo 9.7 shows and describes all of the internal components of a small digital encoder. In this case the code disk is made of stamped sheet metal. Video Demos 9.8 and 9.9 describe two interesting applications of encoders: a computer mouse and an industrial robot. View Video Demos 1.1 and 1.2 to see a demonstration of how the robot works and how the encoders are incorporated into the internal design. Video Demo 1.5 shows another application of encoders where cost is a major concern and a custom design is necessary.

The optical disk of the **absolute encoder** is designed to produce a digital word that distinguishes N distinct positions of the shaft. For example, if there are eight tracks, the encoder is capable of measuring 256 (2^8) distinct positions corresponding to an angular resolution of 1.406° (360°/256). The most common types of numerical encoding used in the absolute encoder are gray and natural binary codes. To illustrate the action of an absolute encoder, the gray code and natural binary code disk

Video Demo

9.7 Encoder components
9.8 Computer mouse relative encoder
9.9 Adept robot digital encoder components
1.1 Adept One robot demonstration (8.0 MB)
1.2 Adept One robot internal design and construction (4.6 MB)
1.5 Inkjet printer components with DC motors and piezoelectric inkjet head

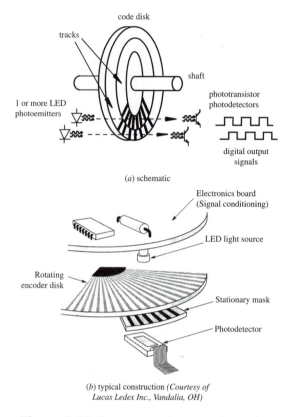

(a) schematic

(b) typical construction *(Courtesy of Lucas Ledex Inc., Vandalia, OH)*

Figure 9.12 Components of an optical encoder.

track patterns for a simple four-track (4-bit) encoder are illustrated in Figures 9.13 and 9.14. The linear patterns and associated timing diagrams are what the photodetectors sense as the code disk circular tracks rotate with the shaft. The output bit codes for both coding schemes are listed in Table 9.1.

The **gray code** is designed so that only one track (one bit) changes state for each count transition, unlike the binary code where multiple tracks (bits) can change during count transitions. This effect can be seen clearly in Figures 9.13 and 9.14 and in the last two columns of Table 9.1. For the gray code, the uncertainty during a transition is only one count, unlike with the binary code, where the uncertainty could be multiple counts.

■ CLASS DISCUSSION ITEM 9.4
Encoder Binary Code Problems

What is the maximum count uncertainty for a 4-bit gray code absolute encoder and a 4-bit natural binary absolute encoder? At what decimal code transitions does the maximum count uncertainty occur in a 4-bit natural binary absolute encoder?

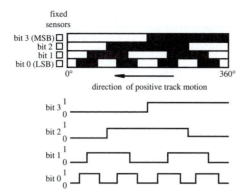

Figure 9.13 4-bit gray code absolute encoder disk track patterns.

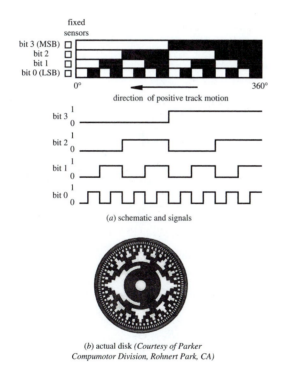

(a) schematic and signals

(b) actual disk *(Courtesy of Parker Compumotor Division, Rohnert Park, CA)*

Figure 9.14 4-bit natural binary absolute encoder disk track patterns.

Because the gray code provides data with the least uncertainty but the natural binary code is the preferred choice for direct interface to computers and other digital devices, a circuit to convert from gray to binary code is desirable. Figure 9.15 shows a simple circuit that utilizes Exclusive OR (XOR) gates to perform this

Table 9.1 4-bit gray and natural binary codes

Decimal Code	Rotation Range (°)	Natural binary code ($B_3B_2B_1B_0$)	Gray code ($G_3G_2G_1G_0$)
0	0–22.5	0000	0000
1	22.5–45	0001	0001
2	45–67.5	0010	0011
3	67.5–90	0011	0010
4	90–112.5	0100	0110
5	112.5–135	0101	0111
6	135–157.5	0110	0101
7	157.5–180	0111	0100
8	180–202.5	1000	1100
9	202.5–225	1001	1101
10	225–247.5	1010	1111
11	24 7.5–270	1011	1110
12	270–292.5	1100	1010
13	292.5–315	1101	1011
14	315–337.5	1110	1001
15	337.5–360	1111	1000

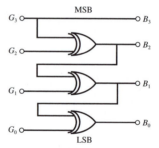

Figure 9.15 Gray-code-to-binary-code conversion.

function. The Boolean expressions that relate the binary bits (B_i) to the gray code bits (G_i) are

$$B_3 = G_3$$
$$B_2 = B_3 \oplus G_2$$
$$B_1 = B_2 \oplus G_1$$
$$B_0 = B_1 \oplus G_0$$

(9.1)

For a gray-code-to-binary-code conversion of any number of bits N (e.g., $N = 4$ as previously), the most significant bits of the binary and gray codes are always identical ($B_{N-1} = G_{N-1}$), and for each other bit, the binary bit is the XOR combination: $B_i = B_{i+1} \oplus G_i$ for $i = 0$ to $N - 2$. This pattern can be easily seen in the 4-bit example above (Equations 9.1).

■ **CLASS DISCUSSION ITEM 9.5**
Gray-to-Binary-Code Conversion

Examine the validity of Equations 9.1 by applying them to the last two columns in Table 9.1.

The **incremental encoder,** sometimes called a **relative encoder,** is simpler in design than the absolute encoder. It consists of two tracks and two sensors whose outputs are designated A and B. As the shaft rotates, pulse trains occur on A and B at a frequency proportional to the shaft speed, and the lead-lag phase relationship between the signals yields the direction of rotation as described in detail below. The code disk pattern and output signals A and B are illustrated in Figure 9.16. By counting the number of pulses and knowing the resolution of the disk, the angular motion can be measured. A and B are 1/4 cycle out of phase with each other and are known as **quadrature signals.** Often a third output, called INDEX, yields one pulse per revolution, which is useful in counting full revolutions. It is also useful to define a reference or zero position.

Figure 9.16a illustrates a configuration using two separate tracks for A and B, but a more common configuration uses a single track (see Figure 9.16b and Video Demo 9.7) with the A and B sensors offset a quarter cycle on the track to

Video Demo

9.7 Encoder components

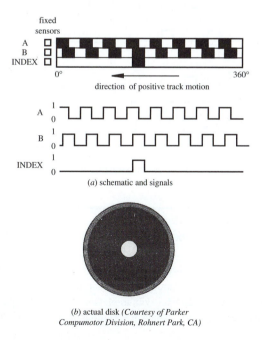

(*a*) schematic and signals

(*b*) actual disk *(Courtesy of Parker Compumotor Division, Rohnert Park, CA)*

Figure 9.16 Incremental encoder disk track patterns.

yield the same signal pattern. A single-track code disk is simpler and cheaper to manufacture.

The quadrature signals A and B can be decoded to yield angular displacement and the direction of rotation as shown in Figure 9.17. Pulses appear on one of two output lines (CW and CCW) corresponding either to clockwise (CW) rotation or counterclockwise (CCW) rotation. Decoding transitions of A and B using sequential logic circuits can provide three different resolutions: 1X, 2X, and 4X. The 1X resolution provides an output transition at each negative edge of signal A or B, resulting in a single pulse for each quadrature cycle. The 2X resolution provides an output transition at every negative or positive edge of signal A or B, resulting in two times the number of output pulses. The 4X resolution provides an output pulse at every positive and negative edge of signal A or B, resulting in four times the number of output pulses. The direction of rotation is determined by the level of one quadrature signal during an edge transition of the second quadrature signal. For example, in the 1X mode, A = ↓ with B = 1 implies clockwise rotation, and B = ↓ with A = 1 implies counterclockwise rotation. If we only had one signal instead of both A and B, it would be impossible to determine the direction of rotation. Furthermore, shaft jitter around an edge transition in the single signal would result in erroneous pulses (see Class Discussion Item 9.6).

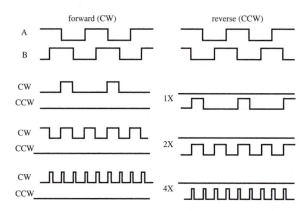

Figure 9.17 Quadrature direction sensing and resolution enhancement.

■ **CLASS DISCUSSION ITEM 9.6**
Encoder 1X Circuit with Jitter

An incremental encoder connected to a 1X quadrature decoder circuit is experiencing a small rotational vibration with an amplitude roughly equivalent to one quadrature pulse width. During this vibration, you observe many pulses on both the CW and CCW lines but no net change in the output of the up-down counter. Explain why this happens.

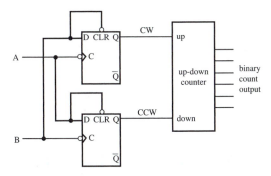

Figure 9.18 1X quadrature decoder circuit.

Figure 9.18 shows a circuit that will yield the 1X resolution by creating and counting pulses at specific negative edges of the quadrature signals. The D flip-flops decode whether the shaft is rotating clockwise or counterclockwise, and this information is used to drive an up-down counter to keep the current pulse count for the encoder rotation. In addition to the edges detected for the 1X resolution, circuits can be designed to detect other edges in the quadrature signals resulting in two times (2X) and four times (4X) the base resolution (1X). These quadrature decoder circuits can be constructed with discrete components, but they are also available on ICs (e.g., Hewlett Packard's HCTL-2016). Quadrature decoding can also be done with software running on a microcontroller (Question 9.11).

Incremental encoders provide more resolution at lower cost than absolute encoders, but they measure only relative motion and do not provide absolute position directly. However, an incremental encoder can be used in conjunction with a limit switch to define absolute position relative to a reference position defined by the switch. Absolute encoders are chosen in applications where establishing a reference position is impractical or undesirable.

■ **CLASS DISCUSSION ITEM 9.7**
Robotic Arm with Encoders

When a robotic arm with absolute encoders on its joints is powered up, the robot knows exactly where its links are relative to its base. If the absolute encoders were replaced with incremental encoders, would this remain true? If not, how would the robot establish a home or zero reference position for the arm?

THREADED DESIGN EXAMPLE

DC motor position and speed controller—Digital encoder interface   **C.4**

The figure below shows the functional diagram for Threaded Design Example C (see Section 1.3 and Video Demo 1.8), with the portion described here highlighted.

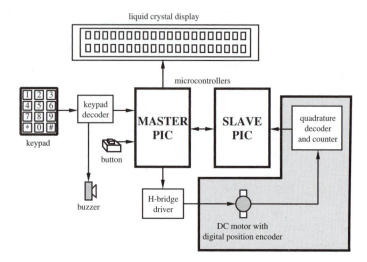

The following figure shows all components and interconnections required to read the digital encoder's position from the slave PIC. A commercially available quadrature decoder/counter interface IC, the HCTL-2016, is the main component in the design. Detailed information about this component can be found in the data sheet at Internet Link 9.3. The HCTL-2016 requires a clock signal to operate. In this design, it is provided by a clock output on the master PIC. The master PIC also resets the encoder counter at the appropriate time (via the $\overline{\text{RST}}$ line) to define a zero position. The digital encoder quadrature signals (channels A and B) are also connected to the HTCL-2016. Because the HTCL-2016 contains a 16-bit counter, and the interface to the slave PIC is via the 8-bit PORTB, the data must be retrieved one byte at a time. The thick line connecting PORTB on the PIC to D0-7 on the HCTL-2016 in the wiring diagram indicates multiple wires (in this case, an 8-wire cable). The SEL pin is used to indicate which byte is being read. Finally, the $\overline{\text{OE}}$ pin is used to latch the encoder values before they are read by the PIC.

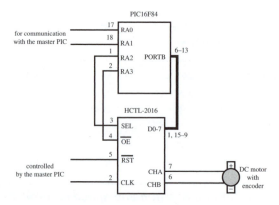

Shown below is the slave PIC code, which monitors the digital encoder position and transmits the data back to the master PIC upon request. The full code for both PICs is presented in Threaded Design Example C.3. The serial communication between the PICs is also described there.

```
' Define I/O pin names and constants

enc_start Var PORTA.0      ' signal line used to start encoder data transmission
enc_serial Var PORTA.1     ' serial line used to transmit encoder data
enc_sel Var PORTA.2        ' encoder data byte select (0:high 1:low)
enc_oe Var PORTA.3         ' encoder output enable latch signal (active low)
enc_mode Con 2             ' 9600 baud mode for serial communication

' Main loop
start:
      ' Wait for the start signal from the master PIC to go high
      While (enc_start == 0) : Wend

      ' Enable the encoder output (latch the counter values)
      Low enc_oe

      ' Send out the high byte of the counter
      SEROUT enc_serial, enc_mode, [PORTB]

      ' Wait for the start signal from the master PIC to go low
      While (enc_start == 1) : Wend

      ' Send out the low byte of the counter
      High enc_sel
      SEROUT enc_serial, enc_mode, [PORTB]

      ' Disable the encoder output
      High enc_oe
      Low enc_sel

goto start    ' wait for next request
```

9.3 STRESS AND STRAIN MEASUREMENT

Measurement of stress in a mechanical component is important when assessing whether or not the component is subjected to safe load levels. Stress and strain measurements can also be used to indirectly measure other physical quantities such as force (by measuring strain of a flexural element), pressure (by measuring strain in a flexible diaphragm), and temperature (by measuring thermal expansion of a material). The most common transducer used to measure strain is the electrical resistance strain gage. As we will see, stress values can be determined from strain measurements using principles of solid mechanics.

Basic stress and strain relations and planar stress analysis techniques are presented in Appendix C for your review if necessary.

9.3.1 Electrical Resistance Strain Gage

The most common transducer for experimentally measuring strain in a mechanical component is the bonded metal foil **strain gage** illustrated in Figure 9.19. It consists of a thin foil of metal, usually constantan, deposited as a grid pattern onto a thin plastic backing material, usually polyimide. The foil pattern is terminated at both ends with large metallic pads that allow leadwires to be easily attached with solder. The entire gage is usually very small, typically 5 to 15 mm long.

To measure strain on the surface of a machine component or structural member, the gage is adhesively bonded directly to the component, usually with epoxy or cyanoacrylate. The backing makes the foil gage easy to handle and provides a good bonding surface that also electrically insulates the metal foil from the component. Leadwires are then soldered to the solder tabs on the gage. When the component is loaded, the metal foil deforms, and the resistance changes in a predictable way (see below). If this resistance change is measured accurately, the strain on the surface of the component can be determined. Strain measurements allow us to determine the state of stress on the surface of the component, where stresses typically have their highest values. Knowing stresses at critical locations on a component under load can help a designer validate analytical or numerical results (e.g., from a finite element analysis) and verify that stress levels remain below safe limits for the material (e.g., below the yield strength). It is important to note that, because strain gages are finite in size, a measurement actually reflects an average of the strain over a small area. Hence, making measurements where stress gradients are large (e.g., where there is stress concentration) can yield poor results.

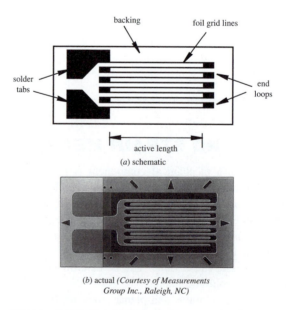

(a) schematic

(b) actual *(Courtesy of Measurements Group Inc., Raleigh, NC)*

Figure 9.19 Metal foil strain gage construction.

Experimental stress analysis (e.g., with strain gages) and analytical or numerical stress analysis (e.g., with finite element analysis) are both important to design reliable mechanical parts. The two approaches should be considered complements to each other and not replacements. Finite element analysis involves many assumptions about material properties, load application, and boundary conditions that may not accurately model the actual component when it is manufactured and loaded. Strain gage measurements may also have some inaccuracies due to imperfect bonding and alignment on the component surface and due to uncompensated temperature effects. Also, only specific locations can be checked with strain gages because space and access on the component can be limiting factors.

Effects that are easily measured with strain gages but difficult to model with finite element analysis include stresses resulting from mechanical assembly of components and complex loading and boundary conditions. These and other effects are often difficult to predict and model accurately with analytical and numerical methods.

Experimental stress applications usually involve mounting a large number of strain gages on a mechanical component, as illustrated in Figure 9.20, before it is loaded. Experimental strain values are usually acquired through an automated data acquisition system. The strain data can be converted to stresses in the object under different loading conditions, and the stresses can be compared to analytical and numerical finite element analysis results.

To understand how a strain gage is used to measure strain, we first look at how the resistance of the foil changes when deformed. The metal foil grid lines in the active portion of the gage (see Figure 9.19a) can be approximated by a single rectangular conductor as illustrated in Figure 9.21, whose total resistance is given by

$$R = \frac{\rho L}{A} \tag{9.2}$$

Figure 9.20 Strain gage application. *(Courtesy of Measurements Group Inc., Raleigh, NC)*

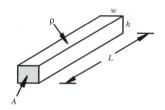

Figure 9.21 Rectangular conductor.

where ρ is the foil metal resistivity, L is the total length of the grid lines, and A is the grid line cross-sectional area. The gage end loops and solder tabs have negligible effects on the gage resistance because they typically have a much larger cross section than the foil lines.

To see how the resistance changes under deformation, we need to take the differential of Equation 9.2. If we first take the natural logarithm of both sides,

$$\ln R = \ln \rho + \ln L - \ln A \tag{9.3}$$

taking the differential yields the following expression for the change in resistance given material property and geometry changes in the conductor:

$$dR/R = d\rho/\rho + dL/L - dA/A \tag{9.4}$$

As we would expect, the signs in this equation imply that the resistance of the conductor increases ($dR > 0$) with increased resistivity and increased length and decreases with increased cross-sectional area. Because the cross-sectional area of the conductor is

$$A = wh \tag{9.5}$$

the area differential term is

$$\frac{dA}{A} = \frac{w \cdot dh + h \cdot dw}{w \cdot h} = \frac{dh}{h} + \frac{dw}{w} \tag{9.6}$$

From the definition of Poisson's ratio (see Appendix C),

$$\frac{dh}{h} = -v\frac{dL}{L} \tag{9.7}$$

and

$$\frac{dw}{w} = -v\frac{dL}{L} \tag{9.8}$$

so

$$\frac{dA}{A} = -2v\frac{dL}{L} = -2v\varepsilon_{axial} \tag{9.9}$$

where ε_{axial} is the axial strain in the conductor (see Appendix C). When the conductor is elongated ($\varepsilon_{axial} > 0$), the cross-sectional area decreases ($dA/A < 0$), causing the resistance to increase.

Using Equation 9.9, Equation 9.4 can be expressed as

$$dR/R = \varepsilon_{axial}(1 + 2\nu) + d\rho/\rho \tag{9.10}$$

Dividing through by ε_{axial} gives

$$\frac{dR/R}{\varepsilon_{axial}} = 1 + 2\nu + \frac{d\rho/\rho}{\varepsilon_{axial}} \tag{9.11}$$

The first two terms on the right-hand side, 1 and 2ν, represent the change in resistance due to increased length and decreased area. The last term $(d\rho/\rho)/(\varepsilon_{axial})$ represents the **piezoresistive effect** in the material, which explains how the resistivity of the material changes with strain. All three terms are approximately constant over the operating range of typical strain gage metal foils.

Commercially available strain gage specifications usually report a constant **gage factor F** to represent the right-hand side of Equation 9.11. This factor represents the material characteristics of the gage that relate the gage's change in resistance to strain:

$$F = \frac{\Delta R/R}{\varepsilon_{axial}} \tag{9.12}$$

Thus, when a gage of known resistance R and gage factor F is bonded to the surface of a component and the component is then loaded, we can determine the strain in the gage ε_{axial} simply by measuring the change in resistance of the gage ΔR:

$$\varepsilon_{axial} = \frac{\Delta R/R}{F} \tag{9.13}$$

This gage strain is the strain on the surface of the loaded component in the direction of the gage's long dimension.

For the bonded metal foil strain gage, the gage factor F is usually close to 2, and the gage resistance R is close to 120 Ω. Strain gage suppliers also report a **transverse sensitivity** for the gage, which is a measure of the resistance changes in the end loops and grid lines due to strain in the transverse direction. The transverse sensitivity for a bonded metal foil gage is usually close to 1%. This number predicts the gage's sensitivity to transverse strains: those perpendicular to the measuring axis of the gage. A gage experiencing 50 $\mu\varepsilon$ (50×10^{-6}, read as "50 microstrain") in the axial direction and 100 $\mu\varepsilon$ in the transverse direction with a transverse sensitivity of 1% will sense 51 $\mu\varepsilon$ (50 + 1% of 100), not 50 $\mu\varepsilon$.

Strain Gage Resistance Changes | EXAMPLE 9.1

If a 120 Ω strain gage with gage factor 2.0 is used to measure a strain of 100 $\mu\varepsilon$ (100×10^{-6}), how much does the resistance of the gage change from the unloaded state to the loaded state?

Equation 9.12 tells us that

$$\Delta R = R \cdot F \cdot \varepsilon$$

so the change in resistance would be

$$\Delta R = (120 \ \Omega)(2.0)(0.000100) = 0.024 \ \Omega$$

9.3.2 Measuring Resistance Changes with a Wheatstone Bridge

To use strain gages to accurately measure strains experimentally, we need to be able to accurately measure small changes in resistance. The most common circuit used to measure small changes in resistance is the Wheatstone bridge, which consists of a four-resistor network excited by a DC voltage. A Wheatstone bridge is better than a simple voltage divider because it can be easily balanced to establish an accurate zero position, it allows temperature compensation, and it can provide better sensitivity and accuracy. There are two different modes of operation of a Wheatstone bridge circuit: the static balanced mode and the dynamic unbalanced mode. For the **static balanced mode,** illustrated in Figure 9.22, R_2 and R_3 are precision resistors, R_4 is a precision potentiometer (variable resistor) with an accurate scale for displaying the resistance value, and R_1 is the strain gage resistance for which the change is to be measured. To balance the bridge, the variable resistor is adjusted until the voltage between nodes A and B is 0. In the balanced state, the voltages at A and B must be equal so

$$i_1 R_1 = i_2 R_2 \tag{9.14}$$

Also, because the high-input impedance voltmeter between A and B is assumed to draw no current,

$$i_1 = i_4 = \frac{V_{\text{ex}}}{R_1 + R_4} \tag{9.15}$$

and

$$i_2 = i_3 = \frac{V_{\text{ex}}}{R_2 + R_3} \tag{9.16}$$

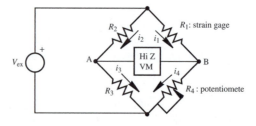

Figure 9.22 Static balanced bridge circuit.

where V_{ex} is the DC voltage applied to the bridge called the **excitation voltage.** Substituting these expressions into Equation 9.14 and rearranging gives

$$\frac{R_1}{R_4} = \frac{R_2}{R_3}$$ (9.17)

If we know R_2 and R_3 accurately and we note the value for R_4 on the precision potentiometer scale, we can accurately calculate the unknown resistance R_1 as

$$R_1 = \frac{R_4 R_2}{R_3}$$ (9.18)

Note that this result is independent of the excitation voltage, V_{ex} (see Class Discussion Item 9.9).

The static balanced mode of operation can be used to measure a gage's resistance under fixed load, but usually balancing is done only as a preliminary step to measuring changes in gage resistance. In **dynamic deflection operation** (see Figure 9.23), again with R_1 representing a strain gage and R_4 representing a potentiometer, the bridge is first balanced, before loads are applied, by adjusting R_4 until there is no output voltage. Then changes in the strain gage resistance R_1 that occur under time-varying load can be determined from changes in the output voltage.

The output voltage can be expressed in terms of the resistor currents as

$$V_{out} = i_1 R_1 - i_2 R_2 = -i_1 R_4 + i_2 R_3$$ (9.19)

and the excitation voltage can be related to the currents:

$$V_{ex} = i_1(R_1 + R_4) = i_2(R_2 + R_3)$$ (9.20)

Solving for i_1 and i_2 in terms of V_{ex} in Equation 9.20 and substituting these into the first expression in Equation 9.19 gives

$$V_{out} = V_{ex}\left(\frac{R_1}{R_1 + R_4} - \frac{R_2}{R_2 + R_3}\right)$$ (9.21)

When the bridge is balanced, V_{out} is 0 and R_1 has a known value. When R_1 changes value, as the strain gage is loaded, Equation 9.21 can be used to relate this voltage change ΔV_{out} to the change in resistance ΔR_1. To find this relation, we can

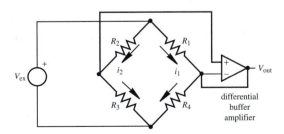

Figure 9.23 Dynamic unbalanced bridge circuit.

replace R_1 by its new resistance $R_1 + \Delta R_1$ and V_{out} by the output deflection voltage ΔV_{out}. Then Equation 9.21 gives

$$\frac{\Delta V_{out}}{V_{ex}} = \frac{R_1 + \Delta R_1}{R_1 + \Delta R_1 + R_4} - \frac{R_2}{R_2 + R_3} \qquad (9.22)$$

Rearranging this equation gives us the desired relation between the change in resistance and the measured output voltage:

$$\frac{\Delta R_1}{R_1} = \frac{\dfrac{R_4}{R_1}\left(\dfrac{\Delta V_{out}}{V_{ex}} + \dfrac{R_2}{R_2 + R_3}\right)}{\left(1 - \dfrac{\Delta V_{out}}{V_{ex}} - \dfrac{R_2}{R_2 + R_3}\right)} - 1 \qquad (9.23)$$

By measuring the change in the output voltage ΔV_{out}, we can determine the gage resistance change ΔR_1 from Equation 9.23 and can compute the gage strain from Equation 9.13. The differential buffer amplifier shown in Figure 9.23 provides high input impedance (i.e., it does not load the bridge) and high gain for the small change in voltage due to the small change in resistance.

■ **CLASS DISCUSSION ITEM 9.9**
Wheatstone Bridge Excitation Voltage

What undesirable effects can the magnitude of the excitation voltage have on the resistance change measurements made with a Wheatstone bridge?

Figure 9.24 illustrates the effects of leadwires when using a strain gage located far from the bridge circuit. Figure 9.24a illustrates a two-wire connection from a strain gage to a bridge circuit. With this configuration, each of the leadwire resistances R' adds to the resistance of the strain gage branch of the bridge. The problem with this is that, if the leadwire temperature changes, it causes changes in the resistance of the bridge branch. This effect can be substantial if the leadwires are long and extend through environments where the temperature changes. Figure 9.24b illustrates a three-wire connection that solves this problem. With this configuration, equal leadwire resistances are added to adjacent branches in the bridge so the effects of changes in the leadwire resistances offset each other. The third leadwire is connected to the high-input impedance voltage measuring circuit, and its resistance has a negligible effect because it carries negligible current. The three wires are usually in the form of a small ribbon cable to ensure they experience the same temperature changes and to minimize electromagnetic interference due to inductive coupling.

In addition to temperature effects in leadwires, temperature changes in the strain gage can cause significant changes in resistance, which would lead to erroneous measurements. A convenient method for eliminating this effect is to use

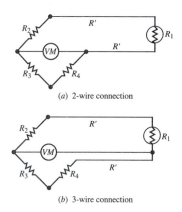

(a) 2-wire connection

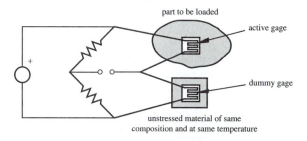

(b) 3-wire connection

Figure 9.24 Leadwire effects in 1/4 bridge circuits.

Figure 9.25 Temperature compensation with a dummy gage in a half bridge circuit.

■ CLASS DISCUSSION ITEM 9.10
Bridge Resistances in Three-Wire Bridges

What must be true about the bridge resistance R_4 for the three-wire configuration shown in Figure 9.24b to result in a balanced bridge ($V_o = 0$) in the no-strain condition? (Hint: Use Equation 9.21 and assume $R_2 = R_3$.)

---

the circuit (called a half bridge) illustrated in Figure 9.25, where two of the four bridge legs contain strain gages. The gage in the top branch is the active gage used to measure surface strains on a component to be loaded. The second "dummy" gage is mounted to an unloaded sample of material identical in composition to the component. If this sample is kept at the same temperature as the component by keeping it in close proximity, the resistance changes in the two gages due to temperature cancel because they are in adjacent branches of the bridge circuit. Therefore, the bridge generates an unbalanced voltage only in response to strain in the active gage.

9.3.3 Measuring Different States of Stress with Strain Gages

Mechanical components may have complex shapes and are often subjected to complex loading conditions. In these cases, it is difficult to predict the orientation of principal stresses at arbitrary points on the component. However, with some geometries and loading conditions, the principal axes are known, and measuring the state of stress is easier.

If a component is loaded uniaxially (i.e., loaded in only one direction in tension or compression), the state of stress in the component can be determined with a single gage mounted in the direction of the load. Figure 9.26 illustrates a bar in tension and the associated state of stress. By measuring the strain ε_x, the stress is obtained using Hooke's law (see Appendix C):

$$\sigma_x = E\varepsilon_x \tag{9.24}$$

where the axial stress in the σ_x is given by

$$\sigma_x = \frac{P}{A} \tag{9.25}$$

where A is the bar's cross-sectional area. Therefore, the force P applied to the bar can be determined from the strain gage measurement:

$$P = AE\varepsilon_x \tag{9.26}$$

If a component is known to be loaded biaxially (i.e., loaded in two orthogonal directions in tension or compression), the state of stress in the component can be determined with two gages aligned with the stress directions. Figure 9.27 illustrates a pressurized tank and the associated state of stress. By measuring the strains ε_x and ε_y, the stresses in the tank shell can be determined from Hooke's law generalized to two dimensions:

$$\varepsilon_x = \frac{\sigma_x}{E} - v\frac{\sigma_y}{E} \tag{9.27}$$

$$\varepsilon_y = \frac{\sigma_y}{E} - v\frac{\sigma_x}{E} \tag{9.28}$$

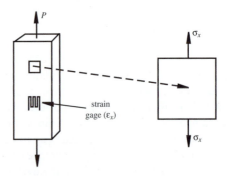

Figure 9.26 Bar under uniaxial stress.

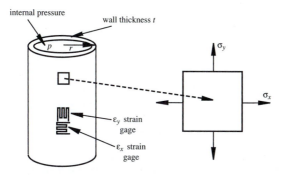

Figure 9.27 Biaxial stress in a long, thin-walled pressure vessel.

Solving for the stress components gives

$$\sigma_x = \frac{E}{1 - v^2}(\varepsilon_x + v\varepsilon_y) \tag{9.29}$$

$$\sigma_y = \frac{E}{1 - v^2}(\varepsilon_y + v\varepsilon_x) \tag{9.30}$$

For a thin-walled pressure vessel (i.e., $t/r < 1/10$), the stresses are approximated by

$$\sigma_x = \frac{pr}{t} \qquad \sigma_y = \frac{pr}{2t} \tag{9.31}$$

where p is the internal pressure, t is the wall thickness, and r is the radius of the vessel. The stress σ_x is the transverse or hoop stress, and σ_y is the axial or longitudinal stress. Either Equation 9.29 or 9.30 can be used to compute the pressure in the vessel based on the strain gage measurements, yielding

$$p = \frac{t\sigma_x}{r} = \frac{tE}{r(1 - v^2)}(\varepsilon_x + v\varepsilon_y) \tag{9.32}$$

or

$$p = \frac{2t\sigma_y}{r} = \frac{2tE}{r(1 - v^2)}(\varepsilon_y + v\varepsilon_x) \tag{9.33}$$

Either expression would yield the correct pressure value for an ideal thin-walled vessel and error-free measurements. In this example, the strain gages are serving as a pressure transducer.

For uniaxial and biaxial loading, we already know the directions of principal stresses in the component; hence, we need only one or two gages, respectively, to determine the stress magnitudes. However, when the loading is more complex or when the geometry is more complex, which is often the case in mechanical design, we have to use three gages mounted in three different directions as illustrated in Figure 9.28. This assembly of strain gages is referred to as a **strain gage rosette.**

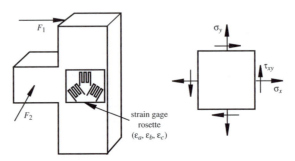

Figure 9.28 General state of planar stress on the surface of a component.

Figure 9.29 Assortment of different strain gage and rosette configurations. *(Courtesy of Measurements Group Inc., Raleigh, NC)*

There is a wide variety of commercially available rosettes with two or more grid patterns accurately oriented on a single backing in close proximity. An assortment of rosettes and single-element gages illustrating the variety of shapes and sizes is shown in Figure 9.29.

The most common rosette patterns for measuring a general state of planar stress are illustrated in Figure 9.30, where the grids are shown as single lines labeled by letters. Of these, the rectangular strain gage is the most common configuration, where the strain gages are positioned 45° apart (see Figure 9.31). Figure 9.32 shows several commercially available three-gage rosettes.

Using principles of solid mechanics, the magnitude and direction of the principal stresses can be determined directly from three simultaneous strain measurements using any of the rosette patterns shown in Figure 9.30. The magnitude and direction of the principal stresses for the rectangular rosette are

$$\sigma_{max,\,min} = \frac{E}{2}\left[\frac{\varepsilon_a + \varepsilon_c}{1 - v} \pm \frac{1}{1 + v}\sqrt{2(\varepsilon_a - \varepsilon_b)^2 + 2(\varepsilon_b - \varepsilon_c)^2}\right] \qquad (9.34)$$

$$\tau_{max} = \frac{E}{2(1 + v)}\sqrt{2(\varepsilon_a - \varepsilon_b)^2 + 2(\varepsilon_b - \varepsilon_c)^2} \qquad (9.35)$$

$$\tan 2\theta_p = \frac{2\varepsilon_b - \varepsilon_a - \varepsilon_c}{\varepsilon_a - \varepsilon_c} \qquad (9.36)$$

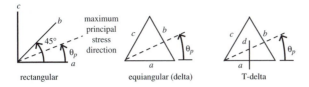

Figure 9.30 Most common strain gage rosette configurations.

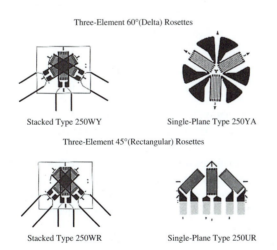

Figure 9.31 Rectangular strain gage rosette.

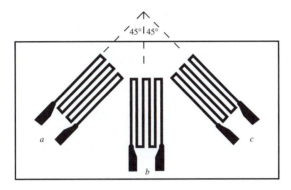

Three-Element 60°(Delta) Rosettes

Stacked Type 250WY   Single-Plane Type 250YA

Three-Element 45°(Rectangular) Rosettes

Stacked Type 250WR   Single-Plane Type 250UR

Figure 9.32 Various three-gage commercial rosettes. *(Courtesy of Measurements Group Inc., Raleigh, NC)*

where ε_a, ε_b, and ε_c are the strains in each of the rosette gages, and θ_p is the angle from gage "a" to the direction of maximum principal stress. When using Equation 9.36 to calculate θ_p, you must use a quadrant-sensitive inverse tangent. If the numerator is positive ($\varepsilon_b > (\varepsilon_a + \varepsilon_c)/2$), $2\theta_p$ lies in the first or second quadrant so $0 < \theta_p < 90°$. Otherwise, $2\theta_p$ lies in the third or fourth quadrant resulting in $-90° < \theta_p < 0$.

The relations for the equiangular (delta) rosette are

$$\sigma_{max,\ min} = \frac{E}{3}\left[\frac{\varepsilon_a + \varepsilon_b + \varepsilon_c}{1 - \nu} \pm \frac{1}{1 + \nu}\sqrt{2(\varepsilon_a - \varepsilon_b)^2 + 2(\varepsilon_b - \varepsilon_c)^2 + 2(\varepsilon_c - \varepsilon_a)^2}\right] \quad (9.37)$$

$$\tau_{max} = \frac{E}{3(1 + \nu)}\sqrt{2(\varepsilon_a - \varepsilon_b)^2 + 2(\varepsilon_b - \varepsilon_c)^2 + 2(\varepsilon_c - \varepsilon_a)^2} \quad (9.38)$$

$$\tan 2\theta_p = \frac{\sqrt{3}(\varepsilon_c - \varepsilon_b)}{2\varepsilon_a - \varepsilon_b - \varepsilon_c} \quad (9.39)$$

Lab Exercise

Lab 13 Strain gages

If the numerator in Equation 9.39 is positive ($\varepsilon_c > \varepsilon_b$), $2\theta_p$ lies in the first or second quadrant so $0 < \theta_p < 90°$. Otherwise, $2\theta_p$ lies in the third or fourth quadrant resulting in $-90° < \theta_p < 0$.

The relations for the T-delta rosette, which has four gages, are

$$\sigma_{max,\ min} = \frac{E}{2}\left[\frac{\varepsilon_a + \varepsilon_d}{1 - \nu} \pm \frac{1}{1 + \nu}\sqrt{(\varepsilon_a - \varepsilon_d)^2 + \frac{4}{3}(\varepsilon_b - \varepsilon_c)^2}\right] \quad (9.40)$$

Video Demo

9.10 Strain gage rosette experiment

9.11 Strain gage rosette experiment analysis discussion

$$\tau_{max} = \frac{E}{2(1 + \nu)}\sqrt{(\varepsilon_a - \varepsilon_d)^2 + \frac{4}{3}(\varepsilon_b - \varepsilon_c)^2} \quad (9.41)$$

$$\tan 2\theta_p = \frac{2(\varepsilon_c - \varepsilon_b)}{\sqrt{3}(\varepsilon_a - \varepsilon_d)} \quad (9.42)$$

If the numerator in Equation 9.42 is positive ($\varepsilon_c > \varepsilon_b$), $2\theta_p$ lies in the first or second quadrant so $0 < \theta_p < 90°$. Otherwise, $2\theta_p$ lies in the third or fourth quadrant resulting in $-90° < \theta_p < 0$.

Internet Link

9.4 Strain gage rosette experiment analysis

Lab Exercise 13 explores how to use a strain gage rosette and commercial strain gage instrumentation to make strain measurements. These measurements are used to compute corresponding stresses that can be compared to expected theoretical results. Video Demo 9.10 shows a demonstration of the experiment, Internet Link 9.4 points to a PDF file containing the underlying analysis with the theoretical results, and Video Demo 9.11 discusses the results of the analysis.

■ **CLASS DISCUSSION ITEM 9.11**
Strain Gage Bond Effects

Does a strain gage bonded to a component have any influence on the stresses that are being measured? If so, how? In what cases are these effects significant?

(a) Courtesy of MTS Systems Corp., Minneapolis, MN

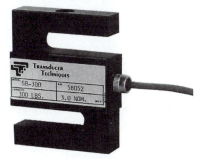

(b) Courtesy of Transducer Techniques, Temecula, CA

Figure 9.33 Typical axial load cells.

9.3.4 Force Measurement with Load Cells

A load cell is a sensor used to measure force. It contains an internal flexural element, usually with several strain gages mounted on its surface. The flexural element's shape is designed so that the strain gage outputs can be easily related to the applied force. The load cell is usually connected to a bridge circuit to yield a voltage proportional to the load. Two commercial load cells, which are used to measure uniaxial force, are shown in Figure 9.33. An example of the application of load cells is in commercially available laboratory materials testing machines for measuring forces applied to a test specimen. Load cells are also used in weight scales, and they are sometimes included as integral parts of mechanical structures to monitor forces in the structures.

A Strain Gage Load Cell for an Exteriorized Skeletal Fixator

DESIGN EXAMPLE 9.1

Orthopedic biomechanics is a field that deals with the analysis of the loading of the skeletal system, engineering approaches to understanding the mechanical properties of biological tissues, and choosing appropriate systems to replace the tissues when they fail. It is a major growth industry in the health care field and offers many opportunities to engineers interested

(continued)

(continued)

in problems that have a biological or medical flavor. Many interesting engineering design problems are associated with the choice of implantable materials, the long-term strength of materials in the hostile environment of the body, and the specification of engineered materials that must attach securely to biological tissues. Replacement joints are among the most successful designs to date, and you probably know someone who has had a joint replaced with a stainless steel or titanium prosthesis.

We have been involved in a number of exciting bioengineering research projects. One of the most interesting involved the analysis of loading in the limbs in patients who had suffered severely broken bones. When one suffers a multiple fracture in a leg bone, the application of a simple plaster or fiberglass cast is not sufficient to allow the bone to heal. Bone is a living tissue constantly being naturally remodeled or replaced. Interestingly, stress is important in initiating the healing process and in maintaining healthy bones. As a case in point, astronauts in a weightless space environment for extended periods exhibit bone mass loss due to the lack of a gravitational stress.

With a severely broken bone, one that has been fractured into a number of pieces, an interesting biomechanical invention is often used to help the healing process. It is known as an exteriorized skeletal fixator. As illustrated in the following figure, it consists of a structural bar outside the body fastened to stainless steel pins that pierce and hold large bone fragments in place until the bone heals. The engineered structure therefore carries most of the load of the body when the animal or person walks.

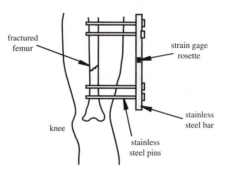

It is important to understand how the fixator is loaded while the patient walks. This information is necessary in order to size the fixator so that it does not deform too much and prevent bone healing. One subtle feature of the healing process is that a very small amount of bone stress is necessary for healing. If all the stress is removed from the bone and no relative motion occurs between the fragments, the bone will not heal. But too much relative motion would hinder healing.

To measure the magnitude of loading for a verification study, we need to design a load cell as part of a skeletal fixator to monitor the complex loading profiles that occur during walking. We are interested in measuring the load profiles in the cylindrical stainless steel bar that supports the broken leg. Bone loads during walking are fairly complex and consist of axial stress, bending stress, and torsional stress.

Our objective is to design a load cell that allows us to easily and reliably determine the axial, bending, and torsional loads while the subject is walking. We can do this by mounting three rectangular strain gage rosettes on the bar 90° apart. The center (*b*) gage of each rosette (see Figure 9.31) will be aligned with the axis of the bar.

If the principal stresses are calculated at each rosette, the axial, bending, and torsion loads can be determined. The principal stresses in the axial direction can be used to determine the axial load and bending loads in two directions. The maximum shear stresses at each rosette can be averaged to determine the torsional load. Thus, the nine strain gages (from three rosettes) yield simultaneous measurements of axial load, bending load, and torsional load.

To obtain accurate measurements, the nine strain gages are now connected to strain gage bridge circuits. A bridge console such as the Measurements Group 2100 provides up to 10 bridges for simultaneous measurement of strain. The analog outputs of the bridge console can be digitized at an appropriate sampling frequency to yield the real-time stress profiles.

■ **CLASS DISCUSSION ITEM 9.12**
Sampling Rate Fixator Strain Gages

What is an appropriate sampling rate for the fixator strain gages in Design Example 9.1 assuming a typical human walking gait?

9.4 TEMPERATURE MEASUREMENT

Because temperature is such an important variable in many engineering systems, an engineer should be familiar with the basic methods of measuring it. Temperature sensors appear in buildings, chemical process plants, engines, transportation vehicles, appliances, computers, and many other devices that require monitoring and control of temperature.

Because many physical phenomena depend on temperature, we can use this dependence to indirectly measure temperature by measuring quantities such as pressure, volume, electrical resistance, and strain and then convert the values using the physical relationship between the quantity and temperature.

The temperature scales used to express temperature are

■ Celsius (°C): Common SI unit of relative temperature.

■ Kelvin (K): Standard SI unit of absolute thermodynamic temperature. Note the absence of the degree symbol.

■ Fahrenheit (°F): English system unit of relative temperature.

■ Rankine (°R): English system unit of absolute thermodynamic temperature.

The relationships between these scales are summarized here:

$$T_C = T_K - 273.15 \qquad (9.43)$$

$$T_F = (9/5)T_C + 32 \qquad (9.44)$$

$$T_R = T_F + 459.67 \qquad (9.45)$$

where T_C is temperature in degrees Celsius, T_K is temperature in Kelvin, T_F is temperature in degrees Fahrenheit, and T_R is temperature in degrees Rankine.

The following subsections introduce common devices used for measuring temperature. They include the liquid-in-gas thermometer, bimetallic strip, resistance temperature device, thermistor, and thermocouple.

9.4.1 Liquid-in-Glass Thermometer

A simple nonelectrical temperature-measuring device is the liquid-in-glass thermometer. It typically uses alcohol or mercury as the working fluid, which expands and contracts relative to the glass container. The upper range is usually on the order of 600°F. When making measurements in a liquid, the depth of immersion is important, as it can result in different measurements. Because readings are made visually, and there can be a meniscus at the top of the working fluid, measurements must be made carefully and consistently.

9.4.2 Bimetallic Strip

Video Demo

9.4 Thermostat with bimetallic strip and mercury switch

Another nonelectrical temperature-measuring device used in simple control systems is the **bimetallic strip.** As illustrated in Figure 9.34, it is composed of two or more metal layers having different coefficients of thermal expansion. The strip can be straight, as shown in the figure, or coiled for a more compact design (e.g., in older thermostats such as that shown in Video Demo 9.4). Because these layers are permanently bonded together, the structure will deform when the temperature changes. This is due to the difference in the thermal expansions of the two metal layers. The deflection δ can be related to the temperature of the strip. Bimetallic strips are used in household and industrial thermostats where the mechanical motion of the strip makes or breaks an electrical contact to turn a heating or cooling system on or off.

9.4.3 Electrical Resistance Thermometer

A **resistance temperature device (RTD)** is constructed of metallic wire wound around a ceramic or glass core and hermetically sealed. The resistance of the metallic wire increases with temperature. The resistance-temperature relationship is usually approximated by the following linear expression:

$$R = R_0[1 + \alpha(T - T_0)] \tag{9.46}$$

where T_0 is a reference temperature, R_0 is the resistance at the reference temperature, and α is a calibration constant. The sensitivity (dR/dT) is $R_0\alpha$. The reference

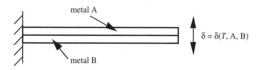

metal A

$\delta = \delta(T, A, B)$

metal B

Figure 9.34 Bimetallic strip.

temperature is usually the ice point of water (0°C). The most common metal used in RTDs is platinum because of its high melting point, resistance to oxidation, predictable temperature characteristics, and stable calibration values. The operating range for a typical platinum RTD is $-220°$ C to $750°$ C. Lower cost nickel and copper types are also available, but they have narrower operating ranges.

A **thermistor** is a semiconductor device, available in probes of different shapes and sizes, whose resistance changes exponentially with temperature. Its resistance-temperature relationship is usually expressed in the form

$$R = R_0 \, e^{\left[\beta\left(\frac{1}{T} - \frac{1}{T_0}\right)\right]} \tag{9.47}$$

where T_0 is a reference temperature, R_0 is the resistance at the reference temperature, and β is a calibration constant called the **characteristic temperature** of the material. A well-calibrated thermistor can be accurate to within 0.01°C or better, which is better than typical RTD accuracies. However, thermistors have much narrower operating ranges than RTDs. Note in Equation 9.47 that a thermistor's resistance actually decreases with increasing temperature. This is very different from metal conductors that experience increasing resistance with increasing temperature (see Section 2.2.1).

9.4.4 Thermocouple

Two dissimilar metals in contact (see Figure 9.35) form a thermoelectric junction that produces a voltage proportional to the temperature of the junction. This is known as the **Seebeck effect.**

Because an electrical circuit must form a closed loop, thermoelectric junctions occur in pairs, resulting in what is called a **thermocouple.** We can represent a thermoelectric circuit containing two junctions as illustrated in Figure 9.36. Here we have wires of metals A and B forming junctions at different temperatures T_1 and T_2, resulting in a potential V that can be measured. The thermocouple voltage V depends on the metal properties of A and B and the difference between the junction

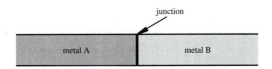

Figure 9.35 Thermoelectric junction.

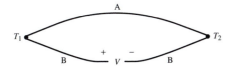

Figure 9.36 Thermocouple circuit.

temperatures T_1 and T_2. The thermocouple voltage is directly proportional to the junction temperature difference:

$$V = \alpha(T_1 - T_2) \qquad (9.48)$$

where α is called the **Seebeck coefficient.** As we will see later in this section, the relationship between voltage and temperature difference is not exactly linear. However, over a small temperature range, α is nearly constant.

Secondary thermoelectric effects, known as the Peltier and Thompson effects, are associated with current flow in the thermocouple circuit, but these are usually negligible in measurement systems when compared with the Seebeck effect. However, when the current is large in a thermocouple circuit, these other effects become significant. The Peltier effect relates the current flow to heat flow into one junction and out of the other. This effect forms the basis of a **thermoelectric cooler (TEC)** or refrigerator.

To properly design systems using thermocouples for temperature measurement, it is necessary to understand the basic laws that govern their application. The five basic laws of thermocouple behavior follow.

1. *Law of leadwire temperatures.*   The thermoelectric voltage due to two junctions in a circuit consisting of two different conducting metals depends only on the junction temperatures T_1 and T_2. As illustrated in Figure 9.37, the temperature environment of the leads away from the junctions (T_3, T_4, T_5) does not influence the measured voltage. Therefore, we need not be concerned about shielding the leadwires from environmental conditions.

2. *Law of intermediate leadwire metals.*   As illustrated in Figure 9.38, a third metal C introduced in the circuit constituting the thermocouple has no influence on the resulting voltage as long as the temperatures of the two new junctions (A-C and C-A) are the same ($T_3 = T_4$). As a consequence of this law, a voltage measurement device that creates two new junctions can be inserted into the thermocouple circuit without altering the resulting voltage.

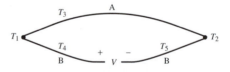

Figure 9.37 Law of leadwire temperatures.

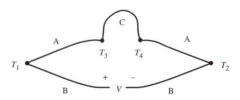

Figure 9.38 Law of intermediate leadwire metals.

3. *Law of intermediate junction metals.* As illustrated in Figure 9.39, if a third metal is introduced within a junction creating two new junctions (A-C and C-B), the measured voltage will not be affected as long as the two new junctions are at the same temperature ($T_1 = T_3$). Therefore, although soldered or brazed joints introduce thermojunctions, they have no resulting effect on the measured voltage. If $T_1 \neq T_3$, the effective temperature at C is the average of the two temperatures ($(T_1 + T_3)/2$).

4. *Law of intermediate temperatures.* Junction pairs at T_1 and T_3 produce the same voltage as two sets of junction pairs spanning the same temperature range (T_1 to T_2 and T_2 to T_3); therefore, as illustrated in Figure 9.40,

$$V_{1/3} = V_{1/2} + V_{2/3} \tag{9.49}$$

This equation can be read: The voltage resulting from measuring temperature T_1 relative to T_3 is the same as the sum of the voltages resulting from T_1 relative to T_2 and T_2 relative to T_3. This result supports the use of a reference junction to allow accurate measurement of an unknown temperature based on a fixed reference temperature (described below).

5. *Law of intermediate metals.* As illustrated in Figure 9.41, the voltage produced by two metals A and B is the same as the sum of the voltages produced by each metal (A and B) relative to a third metal C:

$$V_{A/B} = V_{A/C} + V_{B/C} \tag{9.50}$$

This result supports the use of a standard reference metal (e.g., platinum) to be used as a basis to calibrate all other metals.

A standard configuration for thermocouple measurements is shown in Figure 9.42. It consists of wires of two metals, A and B, attached to a voltage-measuring

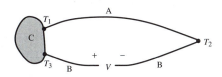

Figure 9.39 Law of intermediate junction metals.

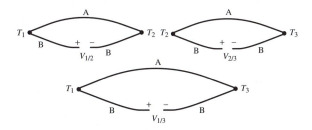

Figure 9.40 Law of intermediate temperatures.

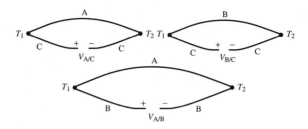

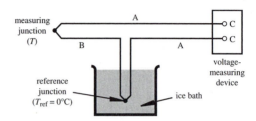

Figure 9.41 Law of intermediate metals.

Figure 9.42 Standard thermocouple configuration.

device with terminals made of metal C. The reference junction is used to establish a temperature reference for one of the junctions to ensure accurate temperature measurements at the other junction relative to the reference. A convenient reference temperature is 0°C, because this temperature can be accurately established and maintained with a distilled water ice bath (i.e., an ice-water mixture). If the terminals of the voltage measuring device are at the same temperature, the law of intermediate leadwire metals ensures that the measuring device terminal metal C has no effect on the measurement. For a given pair of thermocouple metals and a reference temperature, a standard reference table can be compiled for converting voltage measurements to temperatures.

An important alternative to using an ice bath is a semiconductor reference (e.g., a thermistor), which electrically establishes the reference temperature based on solid state physics principles. These reference devices are usually included in thermocouple instrumentation to eliminate the need for an external reference temperature. Video Demo 9.12 describes a commercially available digital thermometer, which can be used for accurate thermocouple temperature measurements.

Figure 9.43 illustrates a two-reference junction configuration, which allows independent choice of the leadwire metal. Copper is a good choice, because copper leadwires are inexpensive and no junctions are introduced at the voltmeter connections, which are usually copper.

Figure 9.44 illustrates the configuration for a **thermopile,** which combines N pairs of junctions, resulting in a voltage N times that of a single pair. In the example shown in the figure, the multiplication factor would be 3. If the measuring junctions (at T) are at different temperatures, the output would represent the average of these temperatures.

Video Demo

9.12 Thermocouple with a digital thermometer

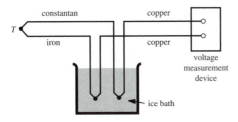

Figure 9.43 Attaching leadwires of selected metal.

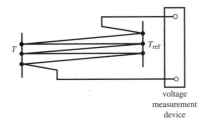

Figure 9.44 Thermopile.

Thermocouple Configuration with Nonstandard Reference

EXAMPLE 9.2

A standard two-junction thermocouple configuration is being used to measure the temperature in a wind tunnel. The reference junction is held at a constant temperature of 10°C. We have only a thermocouple table referenced to 0°C. A portion of the table follows. We want to determine the output voltage when the measuring junction is exposed to an air temperature of 100°C.

Junction Temperature (°C)	Output Voltage (mV)
0	0
10	0.507
20	1.019
30	1.536
40	2.058
50	2.585
60	3.115
70	3.649
80	4.186
90	4.725
100	5.268

By applying the law of intermediate temperature for this example, we can write

$$V_{100/0} = V_{100/10} + V_{10/0}$$

(continued)

(continued)

We wish to find $V_{100/10}$, the voltage measured for a temperature of 100°C relative to a reference junction at 10°C. We can get the other voltages in the equation, $V_{100/0}$ and $V_{10/0}$, from the table because both are referenced to 0°C. Therefore,

$$V_{100/10} = V_{100/0} - V_{10/0} = (5.268 - 0.507)\ \text{mV} = 4.761\ \text{mV}$$

The six most commonly used thermocouple metal pairs are denoted by the letters E, J, K, R, S, and T. The 0°C reference junction calibration for each of the types is nonlinear and can be approximated with a polynomial. The metals in the junction pair, the thermoelectric polarity, the commonly used color code, the operating range, the accuracy, and the polynomial order and coefficients are shown for each type in Table 9.2. The general form for the polynomial using the coefficients in the table is

$$T = \sum_{i=0}^{9} c_i V^i = c_0 + c_1 V + c_2 V^2 + c_3 V^3 + c_4 V^4 + c_5 V^5 + c_6 V^6 + c_7 V^7 + c_8 V^8 + c_9 V^9$$

(9.51)

MathCAD Example

9.1 Thermocouple calculations

where V is the thermoelectric voltage measured in volts and T is the measuring junction temperature in °C, assuming a 0°C reference junction. Figure 9.45 shows the sensitivity curves for some commercially available thermocouple pairs. Even though we use a ninth-order polynomial to represent the temperature voltage relation, providing an extremely close fit, the relationship is close to linear as predicted by the Seebeck effect. MathCAD Example 9.1 shows examples of how the polynomial coefficients in Table 9.2 can be used to make calculations for thermocouple voltage measurements.

9.5 VIBRATION AND ACCELERATION MEASUREMENT

Video Demo

9.13 Bouncing ball accelerometer

An **accelerometer** is a sensor designed to measure acceleration, or rate of change of speed, due to motion (e.g., in a video game controller), vibration (e.g., from rotating equipment), and impact events (e.g., to deploy an automobile airbag). Accelerometers are normally mechanically attached or bonded to an object or structure for which acceleration is to be measured. The accelerometer detects acceleration along one axis and is insensitive to motion in orthogonal directions. Strain gages or piezoelectric elements (described in Section 9.5.1) constitute the sensing element of an accelerometer, converting acceleration into a voltage signal. As a simple example of acceleration sensing, Video Demo 9.13 shows a toy ball with blinking lights that can sense acceleration associated with a bounce. Its sensing element is simply a spring surrounding a metal post.

Table 9.2 Thermocouple data

	Type E	Type J	Type K	Type R	Type S	Type T
Metal pair	Chromel (+) and constantan (−)	Iron (+) and constantan (−)	Chromel (+) and alumel (−)	87% platinum, 13% rhodium (+) and platinum (−)	90% platinum, 10% rhodium (+) and platinum (−)	Copper (+) and constantan (−)
Color code	Purple	Black	Yellow	Green	Green	Blue
Operating range	−100°C to 1000°C	0°C to 760°C	0°C to 1370°C	0°C to 1000°C	0°C to 1750°C	−160°C to 400°C
Accuracy	±0.5°C	±0.1°C	±0.7°C	±0.5°C	±0.1°C	±0.5°C
Approximate sensitivity (mV/°C)	0.079	0.054	0.042	0.012	0.011	0.049
Polynomial order	9	5	8	8	9	7
c_0	0.104967	−0.0488683	0.226585	0.263633	0.927763	0.100861
c_1	17,189.5	19,873.1	24,152.1	179,075.	169,527.	25,727.9
c_2	−282,639.	−218,615.	67,233.4	-4.88403×10^7	-3.15684×10^7	−767,346.
c_3	1.26953×10^7	1.15692×10^7	2.21034×10^6	1.90002×10^{10}	8.99073×10^9	7.80256×10^7
c_4	-4.48703×10^8	-2.64918×10^8	-8.60964×10^8	-4.82704×10^{12}	-1.63565×10^{12}	-9.24749×10^9
c_5	1.10866×10^{10}	2.01844×10^9	4.83506×10^{10}	7.62091×10^{14}	1.88027×10^{14}	6.97688×10^{11}
c_6	-1.76807×10^{11}	—	-1.18452×10^{12}	-7.20026×10^{16}	-1.37241×10^{16}	-2.66192×10^{13}
c_7	1.71842×10^{12}	—	1.38690×10^{13}	3.71496×10^{18}	6.17501×10^{17}	3.94078×10^{14}
c_8	-9.19278×10^{12}	—	-6.33708×10^{13}	-8.03104×10^{19}	-1.56105×10^{19}	—
c_9	2.06132×10^{13}	—	—	—	1.69535×10^{20}	—

Source: G. Burns, M. Scroger, and G. Strouse, "Temperature-Electromotive Force Reference Functions and Tables for the Letter-Designated Thermocouple Types Based on the ITS-90," NIST Monograph 175, April 1993.

ANSI DESIGNATION	ALLOY (Generic or Trade Names)
JP	Iron
JN, EN, or TN	Constantan, Cupron, Advance
KP or EP	Chromega, Tophel, T$_1$, Thermokanthal KP
KN	Alomega, Nial, T$_2$, Thermokanthal KN
TP	Copper
RN or SN	Pure Platinum
RP	Platinum 13% Rhodium
SP	Platinum 10% Rhodium

Trade Names: Advance T—Driver Harris Co., Chromega and Alomega—OMEGA Engineering, Inc., Cupron, Nial and Tophel—Wilbur B. Driver Co., Thermokanthal KP and Thermokanthal KN—The Kanthal Corporation.

ANSI LETTER DESIGNATIONS—Currently thermocouple and extension wire is ordered and specified by an ANSI letter designation. Popular generic and trade name examples are Chromega/Alomega—ANSI Type K: Iron/Constantan— ANSI Type J: Copper/Constantan—ANSI Type T: Chromega/Constantan—ANSI Type E: Platinum/Platinum 10% Rhodium— ANSI Type S: Platinum/Platinum 13% Rhodium—ANSI Type R. The positive and negative legs are identified by letter suffixes P and N, respectively, as listed in the tables.

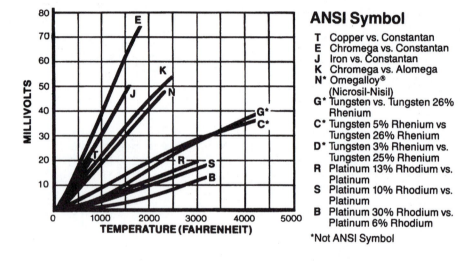

ANSI Symbol

T Copper vs. Constantan
E Chromega vs. Constantan
J Iron vs. Constantan
K Chromega vs. Alomega
N* Omegalloy®
(Nicrosil-Nisil)
G* Tungsten vs. Tungsten 26%
Rhenium
C* Tungsten 5% Rhenium vs
Tungsten 26% Rhenium
D* Tungsten 3% Rhenium vs.
Tungsten 25% Rhenium
R Platinum 13% Rhodium vs.
Platinum
S Platinum 10% Rhodium vs.
Platinum
B Platinum 30% Rhodium vs.
Platinum 6% Rhodium

*Not ANSI Symbol

Figure 9.45 Thermocouple types and characteristics. *(Courtesy of OMEGA Engineering Inc., Stamford, CT)*

The design of an accelerometer is based on the inertial effects associated with a mass connected to a moving object through a spring, damper, and displacement sensor. Figure 9.46 illustrates the components of an accelerometer along with the displacement references, terminology, and free-body diagram. When the object accelerates, there is relative motion between the housing and the seismic mass. A displacement transducer senses the relative motion. Through a frequency response analysis of the second-order system modeling the accelerometer, we can relate the

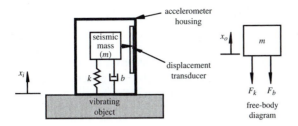

Figure 9.46 Accelerometer displacement references and free-body diagram.

displacement transducer output to either the absolute position or acceleration of the object.

To determine the frequency response of the accelerometer, we first express the forces shown in the free-body diagram. To do this, we define the relative displacement x_r between the seismic mass and the object as

$$x_r = x_o - x_i \tag{9.52}$$

It is measured by a position transducer between the seismic mass and the housing. Therefore, the spring force can be expressed as

$$F_k = k(x_o - x_i) = kx_r \tag{9.53}$$

and the damper force can be expressed as

$$F_b = b(\dot{x}_o - \dot{x}_i) = b\dot{x}_r \tag{9.54}$$

where the overdots represent time derivatives. Applying Newton's second law, the equation of motion for the seismic mass is

$$\sum F_{\text{ext}} = m\ddot{x}_o \tag{9.55}$$

or

$$-F_k - F_b = m\ddot{x}_o \tag{9.56}$$

The forces have negative signs in this equation because they are in the opposite direction from the reference direction x_o in the free-body diagram. Substituting Equations 9.53 and 9.54, the result is

$$-kx_r - b\dot{x}_r = m\ddot{x}_o \tag{9.57}$$

Because the relative displacement x_r is

$$x_r = x_o - x_i \tag{9.58}$$

we can replace $\ddot{x}_o$ with

$$\ddot{x}_o = \ddot{x}_r + \ddot{x}_i \tag{9.59}$$

Therefore, we can write Equation 9.57 as

$$-kx_r - b\dot{x}_r = m(\ddot{x}_r + \ddot{x}_i) \tag{9.60}$$

which can be rearranged as

$$m\ddot{x}_r + b\dot{x}_r + kx_r = -m\ddot{x}_i \tag{9.61}$$

This second-order differential equation relates the measured relative displacement x_r to the input displacement x_i. As in the analysis of a second-order system presented in Section 4.10, we can rewrite this equation as

$$\ddot{x}_r + 2\zeta\omega_n\dot{x}_r + \omega_n^2 x_r = -\ddot{x}_i \tag{9.62}$$

where the natural frequency is

$$\omega_n = \sqrt{\frac{k}{m}} \tag{9.63}$$

and the damping ratio is

$$\zeta = \frac{b}{2\sqrt{km}} \tag{9.64}$$

■ CLASS DISCUSSION ITEM 9.13
Effects of Gravity on an Accelerometer

The free-body diagram and resulting equation of motion for the accelerometer do not show a gravitational force explicitly. Explain why.

For a frequency response analysis, the input displacement is assumed to be composed of sinusoidal terms of the form:

$$x_i(t) = X_i \sin(\omega t) \tag{9.65}$$

Because the system is linear, the resulting relative output displacement is also sinusoidal of the same frequency but different phase:

$$x_r(t) = X_r \sin(\omega t + \phi) \tag{9.66}$$

Using the procedure presented in Section 4.10.2, the frequency response analysis results in the amplitude ratio (Question 9.27)

$$\frac{X_r}{X_i} = \frac{(\omega/\omega_n)^2}{\left(\left[1 - \left(\frac{\omega}{\omega_n}\right)^2\right]^2 + 4\zeta^2\left(\frac{\omega}{\omega_n}\right)^2\right)^{1/2}} \tag{9.67}$$

and the phase angle

$$\phi = -\tan^{-1}\left(\frac{2\zeta(\omega/\omega_n)}{1 - \left(\frac{\omega}{\omega_n}\right)^2}\right) \tag{9.68}$$

To relate the relative output displacement signal x_r to the input acceleration $\ddot{x}_i$, we differentiate Equation 9.65, resulting in

$$\ddot{x}_i(t) = -X_i\omega^2\sin(\omega t) \tag{9.69}$$

Note that the amplitude of the input acceleration is

$$X_i\omega^2 \tag{9.70}$$

Rearranging Equation 9.67, we have

$$H_a(\omega) = \frac{X_r\omega_n^2}{X_i\omega^2} = \frac{1}{\left(\left[1-\left(\dfrac{\omega}{\omega_n}\right)^2\right]^2 + 4\zeta^2\left(\dfrac{\omega}{\omega_n}\right)^2\right)^{1/2}} \tag{9.71}$$

where $H_a(\omega)$ is used to represent the ratio $(X_r\omega_n^2)/(X_i\omega^2)$ as a function of frequency ω. Figures 9.47 and 9.48 illustrate this term and the phase angle relationship (Equation 9.68) graphically for different values of damping ratio.

The denominator of $H_a(\omega)$ is the input acceleration amplitude $X_i\omega^2$, and the numerator is the product of the output displacement amplitude X_r and the square of the natural frequency ω_n^2. Therefore, we can relate the measured output displacement amplitude to the input acceleration amplitude as

$$X_r = \left(\frac{1}{\omega_n^2}\right) H_a(\omega)(X_i\omega^2) \tag{9.72}$$

so the input acceleration amplitude can be expressed as

$$(X_i\omega^2) = \frac{X_r\omega_n^2}{H_a(\omega)} \tag{9.73}$$

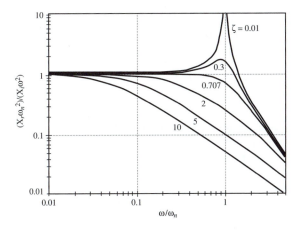

Figure 9.47 Ideal accelerometer amplitude response.

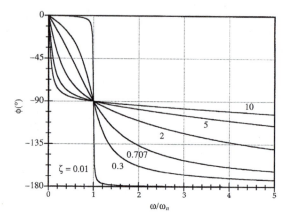

Figure 9.48 Ideal accelerometer phase response.

If we design the accelerometer so that $H_a(\omega)$ is nearly 1 over a large frequency range, then the input acceleration amplitude is given directly in terms of the relative displacement amplitude scaled by a constant factor ω_n^2:

$$(X_i\omega^2) = (\omega_n^2)X_r \tag{9.74}$$

As can be seen in Figure 9.47, the largest frequency range resulting in a unity amplitude ratio occurs when the damping ratio ζ is 0.707 and the natural frequency ω_n is as large as possible. Also, as is clear in Figure 9.48, a ζ of 0.707 results in the best phase linearity for the system. We can make the natural frequency large by choosing a small seismic mass and a large spring constant. This is easy to accomplish in the very small package common to commercial accelerometers.

Equation 9.74 applies to every frequency component ω lying within the bandwidth of the sensor. If we have an arbitrary input composed of a number of frequencies that lie within the bandwidth, each frequency contributes to the signal according to Equation 9.74. Therefore, the total acceleration, due to all frequency components, is also directly related to the total measured relative displacement:

$$\ddot{x}_i(t) = \omega_n^2 x_r(t) \tag{9.75}$$

The same spring-mass-damper configuration used to measure acceleration can be designed to measure displacement instead. This type of device is called a **vibrometer.** The amplitude ratio given in Equation 9.67 provides us with the necessary relationship between the input and output displacements. As we did in the accelerometer analysis, we can now define a displacement ratio as

$$H_d(\omega) = \frac{X_r}{X_i} \tag{9.76}$$

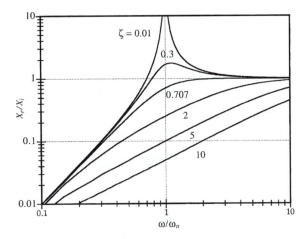

Figure 9.49 Vibrometer amplitude response.

Figure 9.49 illustrates the amplitude ratio to frequency relationship for different values of the damping ratio. The phase angle relationship is the same as for the accelerometer (see Figure 9.48). The input displacement amplitude X_i is related to the measured relative displacement amplitude X_r as

$$X_i = \frac{X_r}{H_d(\omega)} \tag{9.77}$$

If we design the vibrometer so that $H_d(\omega)$ is nearly 1 over a large frequency range, then the input displacement amplitude is given directly by the relative displacement amplitude:

$$X_i = X_r \tag{9.78}$$

As can be seen in Figure 9.49, the largest frequency range resulting in a unity amplitude ratio occurs when the damping ratio ζ is 0.707 and the natural frequency ω_n is as small as possible (so ω/ω_n is large). We can make the natural frequency small by choosing a large seismic mass and a small spring constant. This explains the large size of seismographs used to measure motion due to an earthquake.

9.5.1 Piezoelectric Accelerometer

The highest quality accelerometers are constructed using a **piezoelectric crystal,** a material whose deformation results in charge polarization across the crystal. In a reciprocal manner, application of an electric field to a piezoelectric material results in deformation. As illustrated in Figure 9.50a, a piezoelectric accelerometer consists of a crystal in contact with a mass, supported in a housing by a spring.

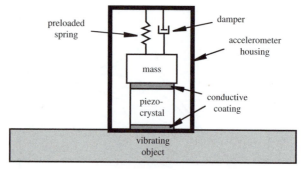

(*a*) schematic illustration

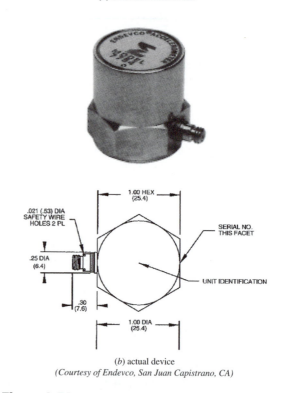

(*b*) actual device
(*Courtesy of Endevco, San Juan Capistrano, CA*)

Figure 9.50 Piezoelectric accelerometer construction.

The purpose for the preload spring is to help keep the mass in contact with the crystal and to keep the crystal in compression, which can help prolong its life. The crystal itself also has stiffness that contributes to the overall stiffness of the system. Figure 9.50b shows a commercially available piezoelectric accelerometer. In addition to the natural damping properties inherent in the crystal, additional damping is sometimes incorporated (e.g., by filling the housing with oil). When the supporting object experiences acceleration, relative displacement occurs between the case and the mass due to the inertia of the mass. The resulting strain in the piezoelectric crystal causes a displacement charge between the crystal conductive coatings as a result of the piezoelectric effect. Note that an accelerometer using a piezoelectric crystal requires no external power supply. Also, it is important to recognize that the accelerometer measures acceleration only in the direction in which it is mounted (i.e., along the axis of the spring, mass, and crystal).

A piezoelectric material produces a large output for its size. Naturally occurring piezoelectric materials are Rochelle salt, tourmaline, and quartz. Some crystalline materials can be artificially polarized to take on piezoelectric characteristics by heating and then slowly cooling them in a strong electric field. Such materials are barium titanate, lead zirconate (PZT), lead titanate, and lead metaniobate. These ferroelectric ceramics are more often used in accelerometers because the sensitivity can be controlled during manufacturing.

A simple equivalent circuit for a piezoelectric crystal is shown in Figure 9.51, implying that the crystal is effectively a capacitor and a charge source that generates a charge q across the capacitor plates proportional to the deformation of the crystal. Representing the accelerometer by a Thevenin equivalent circuit (see Figure 9.52), the open circuit voltage V is equal to the charge, typically in the picocoulomb range, divided by the capacitance, typically in the picofarad range:

$$V = \frac{q}{C_p}$$

(9.79)

The sensitivity of the accelerometer is the ratio of the charge output to the acceleration of the housing expressed in pC/g, (rms pC)/g, or (peak pC)/g, where g is the acceleration due to gravity. The output of the accelerometer is attached to a **charge amplifier,** which converts the displacement charge on the crystal to a voltage

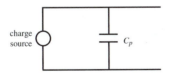

Figure 9.51 Equivalent circuit for piezoelectric crystal.

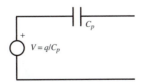

Figure 9.52 Thevenin equivalent of piezoelectric crystal.

that can be measured. Most accelerometers are calibrated in millivolts per *g* for a specified charge amplifier.

In general, piezoelectric accelerometers cannot measure constant or slowly changing accelerations, because the crystals can measure a change in force only by sensing a change in strain. But they are excellent for dynamic measurements such as vibration and impacts. The response at low frequencies is further limited by the low-frequency cutoff of the charge amplifier, which may be on the order of a few hertz. The response at high frequencies is a function of the mechanical characteristics of the accelerometer. The dynamic range of an accelerometer ranges from a few hertz to a fraction of the resonant frequency given by

$$f_n = \frac{1}{2\pi}\sqrt{\frac{k}{m}} \tag{9.80}$$

usually in the kHz range. A typical frequency response curve for a piezoelectric accelerometer is shown in Figure 9.53.

Lab Exercise 14 explores how to use an accelerometer to monitor vibrations from rotating equipment. The health of bearings in a system can be determined by monitoring the spectrum of the vibration signal. This is called bearing signature analysis. Video Demo 9.14 shows a demonstration of the experiment, and Internet Links 9.5 and 9.6 point to some example results. Scratches and wear in roller bearings create more high-frequency content in the vibration spectrum.

Lab Exercise

Lab 14 Vibration measurement with an accelerometer

Video Demo

9.14 Accelerometer bearing signature analysis experiment

Internet Link

9.5 Sample bearing signature analysis result for a good bearing

9.6 Sample bearing signature analysis result for a bad bearing

---

■ **CLASS DISCUSSION ITEM 9.14**
Piezoelectric Sound

How does a piezoelectric microphone work? How about a piezoelectric buzzer?

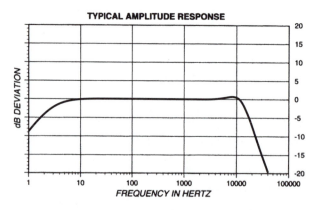

Figure 9.53 Piezoelectric accelerometer frequency response.
(Courtesy of Endevco, San Juan Capistrano, CA)

9.6 PRESSURE AND FLOW MEASUREMENT

Most techniques for measuring pressure involve sensing a displacement or deflection and relating it to pressure through calibration or theoretical relations. One type of pressure sensor is a manometer, which measures a static pressure or pressure difference by detecting fluid displacements in a gravitational field. Another type is the elastic diaphragm, bellows, or tube where deflection of an elastic member is measured and related to pressure. Another type is the piezoelectric pressure transducer that can measure dynamic pressures when the piezoelectric crystal deforms in response to the applied pressure. References that cover these and other techniques are included in the bibliography at the end of the chapter. Internet Link 9.7 is also an excellent resource.

Internet Link

9.7 Pressure measurement techniques and devices

There are also many techniques for measuring gas and liquid flow rates. A pitot tube measures the difference between total and static pressure of a moving fluid. Venturi and orifice meters are based on measuring pressure drops across obstructions to flow. Turbine flow meters detect the rate of flow by measuring the rate of rotation of an impeller in the flow. Coriolis flow meters measure mass flow rate through a U-tube in rotational vibration. Hot-wire anemometers sense the resistance changes in a hot current-carrying wire. The temperature and resistance of the wire depend on the amount of heat transferred to the moving fluid. The heat transfer coefficient is a function of the flow rate. Laser doppler velocimeters (LDVs) sense the frequency shift of laser light scattered from particles suspended in a moving fluid. Most fluid mechanics textbooks present the theoretical foundation of a variety of flow measurement techniques. References that cover these and other techniques are included in the bibliography at the end of the chapter. Internet Link 9.8 is also an excellent resource.

Internet Link

9.8 Flow measurement techniques and devices

9.7 SEMICONDUCTOR SENSORS AND MICROELECTROMECHANICAL DEVICES

The advent of wide-scale semiconductor electronic design and manufacturing changed more than the way we process electronic signals. Many of the techniques developed for producing integrated circuits have been adapted for the design of a new class of semiconductor sensors and actuators called **microelectromechanical (MEM)** devices. In 1980, the first MEM sensor was developed using integrated circuit technology to etch silicon and produce a device that responded to acceleration. It consisted of a tiny silicon cantilever with an integral semiconductor strain gage. Acceleration deflected the cantilever (due to its inertia), and the strain gage sensed the magnitude of acceleration. The strain gage and the bar were etched from a single piece of silicon using the processing methods developed earlier to modify silicon for semiconductor electronics. MEM accelerometers are now used in automobiles to control airbag systems. Other common successful applications of MEM sensors include pressure sensors for automobile tire pressure monitoring systems (TPMS), and accelerometers and gyros for detecting orientation and motion of video game and TV controllers and portable electronic devices (e.g., smart phones and cameras).

Video Demo

5.3 Chip manufacturing process
5.4 How a CPU is made

ICs are made by a series of processes consisting of photoresist lithographic layering, light exposure, controlled chemical etching, vapor deposition, and doping (see Video Demos 5.3 and 5.4). The chemical etching process is important in that tiny mechanical devices can be created by a technique known as micromachining. Using carefully designed masks and timed immersion in chemical baths, microminiature versions of accelerometers, static electric motors, and hydraulic or gas-driven motors can be formed.

Semiconductor sensor designs are based on different electromagnetic properties of doped silicon and gallium arsenide and the variety of ways that semiconductors function in different physical environments. The following list summarizes some of the semiconductor properties that are the basis for different classes of semiconductor MEMs:

- The piezoresistive characteristic of doped silicon, the coupling between resistance change and deformation, is the basis for semiconductor strain gages and pressure sensors.

- The magnetic characteristics of doped silicon, principally the Hall effect, are the basis of semiconductor magnetic transistors where the collector current can be modulated by an external magnetic field.

- Electromagnetic waves and nuclear radiation induce electrical effects in semiconductors, forming the basis of light color sensors and other radiation detectors.

- The thermal properties of semiconductors are the basis for thermistors, thermal conductivity sensors, humidity sensors, and temperature sensor ICs.

Surface acoustic wave (SAW) devices are an important class of MEM sensors. A SAW device consists of a flat piezoelectric substrate with metallic patterns lithographically deposited on the surface. These patterns form interdigital transducers (IDTs) and reflection coupler gratings as shown in Figure 9.54. An input signal applied to an interdigital transducer excites a deformation in the piezoelectric

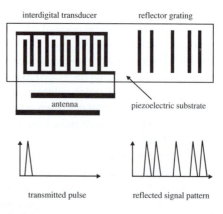

Figure 9.54 Surface acoustic wave transponder device.

substrate generating an acoustic wave that propagates on the surface. Conversely, a SAW wave can induce a voltage in an IDT resulting in an output signal.

Figure 9.54 illustrates a common application of a SAW device: a wireless identification system. A transmitter sends out a pulse that is received by the passive SAW device via an antenna. The resulting SAW wave is reflected as a pattern of pulses unique to the spacings within the reflector grating. The pulses are retransimitted through the same antenna back to a receiver allowing identification of the SAW device. This device is commonly used in automatic highway toll booths to identify vehicles as they pass under a transceiver.

Other applications of SAW devices include delay lines, frequency filters, and a variety of sensors. Their function depends on the substrate SAW propagation and resonance and the patterns and spacings of IDTs and reflector gratings.

With the development of so many different semiconductor sensors, engineers have now begun to integrate sensors and signal processing circuits together in a hybrid circuit that contains transducers, A/D converters, programmable memory, and a microprocessor. The complete package provides a **micromeasurement system (MMS)** that is small, accurate, and less expensive than a discrete measuring system. Distributed micromeasurement systems will be used more and more in future mechatronic system design.

QUESTIONS AND EXERCISES

Section 9.2   Position and Speed Measurement

9.1. Draw schematics for each switch in Figure 9.4.

9.2. Show how an SPDT switch can be wired, with a 5 V supply and resistor(s), to provide a digital signal that is low when the switch is open and high when it is closed.

9.3. Show how an NO pushbutton switch can be wired, with a 5 V supply and resistor(s), to provide a negative logic (active low) reset signal to a microprocessor. The signal should be low only while the button is held down.

9.4. Draw a schematic for a DPDT switch and show how it can be wired to switch two separate circuits on and off.

9.5. Document a complete and thorough answer to Class Discussion Item 9.1.

9.6. Document a complete and thorough answer to Class Discussion Item 9.2.

9.7. Document a complete and thorough answer to Class Discussion Item 9.3.

9.8. You are using a poorly constructed 4-bit natural binary-coded absolute encoder and observe that, as the encoder rotates through the single step from code 3 to code 4, the encoder outputs several different and erroneous values. If misalignment between the photosensors and the code disk is the problem, what possible codes could result during the code 3-to-4 transition?

9.9. Document a complete and thorough answer to Class Discussion Item 9.4.

9.10. What is the angular resolution of a two-channel incremental encoder with a 2X quadrature decoder circuit if the code disk track has 1000 radial lines?

9.11. Write a PicBasic program to perform the functionality in Figure 9.18.

9.12. Document a complete and thorough answer to Class Discussion Item 9.5.

Section 9.3   Stress and Strain Measurement

9.13. Derive Equation 9.9 for a circular conductor instead of a rectangular conductor.

9.14. A new experimental strain gage is mounted on a 0.25 inch diameter steel bar in the axial direction. The gage has a measured resistance of 120 Ω, and when the bar is loaded with 500 lb in tension, the gage resistance increases by 0.01 Ω. What is the gage factor of the gage?

9.15. Document a complete and thorough answer to Class Discussion Item 9.8.

9.16. A steel bar with modulus of elasticity 200 GPa and diameter 10 mm is loaded in tension with an axial load of 50 kN. If a strain gage of gage factor 2.115 and resistance 120 Ω is mounted on the bar in the axial direction, what is the change in resistance of the gage from the unloaded state to the strained state? If the strain gage is placed in one branch of a Wheatstone bridge (R_1) with the other three legs having the same base resistance ($R_2 = R_3 = R_4 = 120\ \Omega$), what is the output voltage of the bridge (V_{out}) in the strained state, as a function of the excitation voltage? What is the stress in the bar?

9.17. Document a complete and thorough answer to Class Discussion Item 9.10.

9.18. Draw a schematic diagram illustrating the wiring of a Wheatstone bridge circuit that takes advantage of both three-wire lead connections and dummy gage temperature compensation.

9.19. A strain gage bridge used in a load cell dissipates energy. Why? Compare the power dissipated in a bridge circuit with equal resistance arms for gages of 350 Ω and 120 Ω when the excitation voltage is 10 V. What strategies can one employ if heating becomes a problem? Do these strategies have any deficiencies?

9.20. Even when we use three-wire connections to a strain gage to reduce the effects of leadwire resistance changes with temperature, we can have a phenomenon called leadwire desensitization. If the magnitude of the leadwire resistance exceeds 0.1% of the nominal gage resistance, significant error can result. Assuming a 22AWG leadwire (0.050 Ω/m) and a standard 120 Ω gage, how long can the leads be before there is leadwire desensitization?

Section 9.4   Temperature Measurement

9.21. Find the approximate sensitivity (in mV/°C) of a J-type thermocouple by evaluating or plotting its polynomial function.

9.22. Find the approximate sensitivity (in mV/°C) of a T-type thermocouple by evaluating or plotting its polynomial function.

9.23. If a J-type thermocouple is used in a standard two-junction thermocouple configuration (see Figure 9.42) with a 0°C reference temperature, what voltage would result for an input temperature of 200°C?

9.24. If a J-type thermocouple is used in a standard two-junction thermocouple configuration (see Figure 9.42) with a 100°C reference temperature, what measurement temperature would correspond to a measured voltage of 30 mV?

9.25. Solve Question 9.24 if the reference temperature were 11°C instead of 100°C with everything else the same.

9.26. Using a J-type thermocouple with a fixed 0°C reference, how much would the voltage reading change if the measurement temperature changes from 10°C to 120°C?

Section 9.5 Vibration and Acceleration Measurement

9.27. Derive Equation 9.67.

9.28. An accelerometer is designed with a seismic mass of 50 g, a spring constant of 5000 N/m, and a damping constant of 30 Ns/m. If the accelerometer is mounted to an object experiencing displacement, $x_{in}(t) = 5 \sin(100t)$ mm, find each of the following:

   a. the actual acceleration amplitude of the object.

   b. the amplitude of the steady state relative displacement between the seismic mass and the housing of the accelerometer.

   c. the acceleration amplitude, as measured by the accelerometer.

   d. an expression for the steady state relative displacement of the seismic mass relative to the housing as a function of time [$x_r(t)$].

9.29. A spring-mass-damper vibrometer is designed with a seismic mass of 1 kg, a coil spring of spring constant 2 N/m, and a dashpot with damping constant 2 Ns/m. Determine an expression for the steady state displacement of the seismic mass $x_{out}(t)$ given an object input displacement of $x_{in}(t) = 10 \sin(1.25t)$ mm.

BIBLIOGRAPHY

Beckwith, T., Marangoni, R., and Lienhard, J., *Mechanical Measurements,* 5th Edition, Addison-Wesley, Reading, MA, 1993.

Burns, G., Scroger, M., and Strouse, G., "Temperature-Electromotive Force Reference Functions and Tables for the Letter-Designated Thermocouple Types Based on the ITS-90," NIST Monograph 175, April 1993.

Complete Temperature Measurement Handbook and Encyclopedia, vol. 28, Omega Engineering, Stamford, CT, 1992.

Dally, J. and Riley, W., *Experimental Stress Analysis,* 3rd Edition, McGraw-Hill, New York, 1991.

Doeblin, E., *Measurement Systems Application and Design,* 4th Edition, McGraw-Hill, New York, 1990.

Figliola, R. and Beasley, D., *Theory and Design for Mechanical Measurements,* 2nd Edition, John Wiley, New York, 1995.

Gardner, J., *Microsensors: Principles and Applications,* John Wiley, New York, 1994.

Hauptmann, P., *Sensors, Principles and Applications,* Carl Hanser Verlag, 1991.

Holman, J., *Experimental Methods for Engineers,* 6th Edition, McGraw-Hill, New York, 1994.

Janna, W., *Introduction to Fluid Mechanics,* Brooks/Cole Engineering Division, Monterey, CA, 1983.

Kovacs, G., *Micromachined Transducers Sourcebook,* WCB/McGraw-Hill, New York, 1998.

Measurements Group Education Division, "Strain Gage Based Transducers: Their Design and Construction," Measurements Group, Raleigh, NC, 1988.

Measurements Group Education Division, "Student Manual for Strain Gage Technology," Measurements Group, Raleigh, NC, 1991.

Miu, D., *Mechatronics: Electromechanics and Contromechanics,* Springer-Verlag, New York, 1993.

Pallas-Areny, R. and Webster, J., *Sensors and Signal Conditioning,* John Wiley, New York, 1991.

Proceedings of the Sixth UK Mechatronics Forum International Conference, C5—Novel Sensors and Actuators session, Skovde, Sweden, 1998.

Sze, S., *Semiconductor Sensors,* John Wiley, New York, 1994.

Walton, J., *Engineering Design: From Art to Practice,* pp. 117–119, West Publishing, St. Paul, MN, 1991.

White, F., *Fluid Mechanics,* 7th Edition, McGraw-Hill, New York, 2010.

Actuators

T his chapter describes various actuators important in mechatronic system
design. ∎

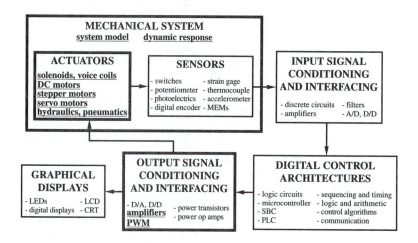

After you read, discuss, study, and apply ideas in this chapter, you will:

1. Be able to identify different classes of actuators, including solenoids, DC
motors, AC motors, hydraulics, and pneumatics

2. Understand the differences between series, shunt, compound, permanent magnet, and stepper DC motors

3. Understand how to design electronics to control a stepper motor

4. Be able to select a motor for a mechatronics application

5. Be able to identify and describe the components used in hydraulic and pneumatic systems

10.1 INTRODUCTION

Most mechatronic systems involve motion or action of some sort. This motion or action can be applied to anything from a single atom to a large articulated structure. It is created by a force or torque that results in acceleration and displacement. **Actuators** are the devices used to produce this motion or action.

Up to this point in the book, we have focused on electronic components and sensors and associated signals and signal processing, all of which are required to produce a specific mechanical action or action sequence. Sensor input measures how well a mechatronic system produces its action, open loop or feedback control helps regulate the specific action, and much of the electronics we learned about is required to manipulate and communicate this information. Actuators produce physical changes such as linear and angular displacement. They also modulate the rate and power associated with these changes. An important aspect of mechatronic system design is selecting the appropriate type of actuator. This chapter covers some of the most important actuators: solenoids, electric motors, hydraulic cylinders and rotary motors, and pneumatic cylinders. Putting it poetically, this chapter is "where the rubber meets the road." Internet Link 10.1 provides links to vendors and online resources for various commercially available actuators and support equipment.

Internet Link

10.1 Actuator online resources and vendors

10.2 ELECTROMAGNETIC PRINCIPLES

Many actuators rely on electromagnetic forces to create their action. When a current-carrying conductor is moved in a magnetic field, a force is produced in a direction perpendicular to the current and magnetic field directions. **Lorentz's force law,** which relates force on a conductor to the current in the conductor and the external magnetic field, in vector form is

$$\vec{F} = \vec{I} \times \vec{B} \tag{10.1}$$

where $\vec{F}$ is the force vector (per unit length of conductor), $\vec{I}$ is the current vector, and $\vec{B}$ is the magnetic field vector. Figure 10.1 illustrates the relationship between these vectors and indicates the **right-hand rule** analogy, which states that if your right-hand index finger points in the direction of the current and your middle finger is aligned with the field direction, then your extended thumb (perpendicular to the index and middle fingers) will point in the direction of the force. Another way to apply the right-hand rule is to align your extended fingers in the direction of the $\vec{I}$ vector and orient your palm so you can curl (flex) your fingers toward the direction of the $\vec{B}$ vector. Your hand will then be positioned such that your extended thumb points in the direction of $\vec{F}$.

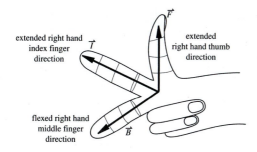

extended right hand
index finger
direction

$\vec{I}$

$\vec{F}$

extended
right hand thumb
direction

flexed right hand
middle finger
direction

$\vec{B}$

Figure 10.1 Right-hand rule for magnetic force.

Another electromagnetic effect important to actuator design is field intensification within a coil. Recall that, when discussing inductors in Chapter 2, we stated that the magnetic flux through a coil is proportional to the current through the coil and the number of windings. The proportionality constant is a function of the permeability of the material within the coil. The permeability of a material characterizes how easily magnetic flux penetrates the material. Iron has a permeability a few hundred times that of air; therefore, a coil wound around an iron core can produce a magnetic flux a few hundred times that of the same coil with no core. Most electromagnetic devices we will present use iron cores of one form or another to enhance magnetic flux. Cores are usually laminated (made up of insulated layers of iron stacked parallel to the coil-axis direction) to reduce the eddy currents induced when the cores experience changing magnetic fields. Eddy currents, which are a result of Faraday's law of induction, result in inefficiencies and undesirable core heating.

10.3 SOLENOIDS AND RELAYS

As illustrated in Figure 10.2, a **solenoid** consists of a coil and a movable iron core called the **armature.** When the coil is energized with current, the core moves to increase the flux linkage by closing the air gap between the cores. The movable core is usually spring-loaded to allow the core to retract when the current is switched off. The force generated is approximately proportional to the square of the current and inversely proportional to the square of the width of the air gap. Solenoids are inexpensive, and their use is limited primarily to on-off applications such as latching, locking, and triggering. They are frequently used in home appliances (e.g., washing machine valves), automobiles (e.g., door latches and the starter solenoid), pinball machines (e.g., plungers and bumpers), and factory automation. Video Demos 10.1 through 10.3 show examples of interesting student projects using solenoids in creative ways.

An electromechanical **relay** is a solenoid used to make or break mechanical contact between electrical leads. A small voltage input to the solenoid controls a potentially large current through the relay contacts. Applications include power switches and electromechanical control elements. A relay performs a function similar to a power transistor switch circuit but has the capability to switch much larger currents.

Video Demo

10.1 Magic piano
10.2 Automated melodica
10.3 LED fountain system

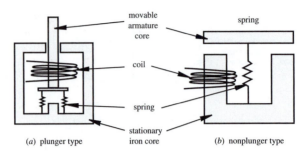

(a) plunger type   *(b)* nonplunger type

Figure 10.2 Solenoids.

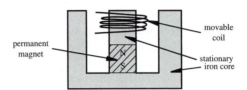

Figure 10.3 Voice coil.

Video Demo

10.4 Relay
and transistor
switching circuit
comparison

10.5 Computer
hard-drive with
voice coil

10.6 Computer
hard-drive
track seeking
demonstration

10.7 Computer
hard-drive super-
slow-motion video
of track finding

Relays, because they create a mechanical connection and don't require voltage bias-ing, can be used to switch either DC or AC power. Also, the input circuit of a relay is electrically isolated from the output circuit, unlike the common-emitter transistor circuit, where there is a common ground between the input and output. Because the relay is electrically isolated, noise, induced voltages, and ground faults occur-ring in the output circuit have minimal impact on the input circuit. One disadvan-tage of relays is that they have slower switching times than transistors. And because they contain contacts and mechanical components, they wear out much faster. Video Demo 10.4 demonstrates how relays and transistors respond to different switching speeds.

As illustrated in Figure 10.3, a **voice coil** consists of a coil that moves in a magnetic field produced by a permanent magnet and intensified by an iron core. Figure 10.4 shows the coil and iron core of a commercially available voice coil, which can be used as either a sensor or an actuator. When used as an actuator, the force on the coil is directly proportional to the current in the coil. The coil is usually attached to a movable load such as the diaphragm of an audio speaker, the spool of a hydraulic proportional valve, or the read-write head of a computer disk drive. The linear response, small mass of the moving coil, and bidirectional capability make voice coils more attractive than solenoids for control applications.

Video Demos 10.5 and 10.6 show how a computer disk drive functions, where a voice coil is used to provide the pivoting motion of the read-write head. Video Demo 10.7 shows a super-slow-motion clip, filmed with a special high-speed camera, which dramatically demonstrates the accuracy and speed of the voice coil motion. The read-write head comes to a complete stop on one track before moving to another. In real-time (e.g., in Video Demo 10.6), this motion is a total blur.

■ **CLASS DISCUSSION ITEM 10.1**
Examples of Solenoids, Voice Coils, and Relays

Make a list of common household and automobile devices that contain solenoids, voice coils, and relays. Describe why you think the particular component was selected for each of the devices you cite.

Figure 10.4 Photograph of a voice coil iron core and coil.

10.4 ELECTRIC MOTORS

Electric motors are by far the most ubiquitous of the actuators, occurring in virtually all electromechanical systems. Electric motors can be classified either by function or by electrical configuration. In the functional classification, motors are given names suggesting how the motor is to be used. Examples of functional classifications include torque, gear, servo, instrument servo, and stepping. However, it is usually necessary to know something about the electrical design of the motor to make judgments about its application for delivering power and controlling position. Figure 10.5 provides a configuration classification of electrical motors found in mechatronics applications. The differences are due to motor winding and rotor designs, resulting in a large variety of operating characteristics. The price-performance ratio of electric motors continues to improve, making them important additions to all sorts of mechatronic systems from appliances to automobiles. AC induction motors are particularly important in industrial and large consumer appliance applications. In fact, the AC induction motor is sometimes called the workhorse of industry. Video Demos 10.8 through 10.10 show some examples and describe how the motors function.

Video Demo

10.8 AC induction motor (single phase)
10.9 AC induction motor with a soft start for a water pump
10.10 AC induction motor variable frequency drive for a building air handler unit

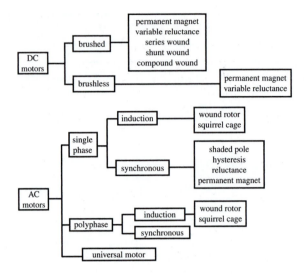

Figure 10.5 Configuration classification of electric motors.

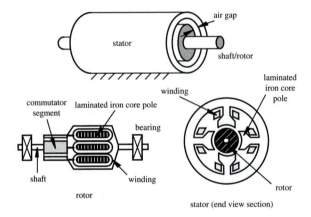

Figure 10.6 Motor construction and terminology.

Figure 10.6 illustrates the construction and components of a typical electric motor. The stationary outer housing, called the **stator,** supports radial magnetized poles. These poles consist of either permanent magnets or wire coils, called **field coils,** wrapped around laminated iron cores. The purpose of the stator poles is to provide radial magnetic fields. The iron core intensifies the magnetic field inside the coil due to its permeability. The purpose for laminating the core is to reduce the effects of eddy currents, which are induced in a conducting material (see Class Discussion Item 10.2). The **rotor** is the part of the motor that rotates. It consists of a rotating shaft supported by bearings, conducting coils usually referred to as the **armature** windings, and an iron core that intensifies the fields created by the windings. There is a small **air gap** between the rotor and the stator where the magnetic fields interact. In many **DC motors,** the rotor also includes a **commutator** that delivers and controls the direction of current through the armature windings. For motors with a commutator, **"brushes"** provide stationary electrical contact to the moving commutator conducting segments. Brushes in early motors consisted of bristles of copper wire flexed against the commutator, hence the term brush; but now they are usually made of graphite, which provides a larger contact area and is self-lubricating. The brushes are usually spring-loaded to ensure continual contact with the commutator. Video Demo 10.11 shows a small, brushed, permanent-magnet DC motor disassembled so you can see the various components and how they function.

A **brushless** DC motor has permanent magnets on the rotor and a rotating field in the stator. The permanent magnets on the rotor eliminate the need for a commutator. Instead, the DC currents in the stator coils are switched in response to proximity sensors that are triggered as the shaft rotates. Video Demos 10.12 and 10.13 show two examples of brushless DC motors. One advantage of a brushless motor is that it does not require maintenance to replace worn brushes. Also, because there are no rotor windings or iron core, the rotor inertia is much smaller, sometimes making control easier. There are also no rotor heat dissipation problems, because there are no rotor windings and hence no I^2R heating. Another advantage of not having brushes is that there is no arcing associated with mechanical commutation. Therefore, brushless motors create less EMI and are more suitable in environments where explosive gases might be present. One disadvantage of brushless motors is that they can cost more due to the sensors and control circuitry required.

Figure 10.7 shows examples of commercially available assembled motors. In the top figure, the motor on the left is an AC induction motor with a gearhead speed reduction unit attached. The motor on the right is a two-phase stepper motor. Motors come in standard sizes with standard mounting brackets, and they usually include nameplates listing some of the motor's specifications. The bottom figure shows the internal construction of a permanent-magnet-rotor stepper motor. Video Demo 10.14 shows other examples of commercially available regular DC motors and stepper motors.

Video Demo

10.11 DC motor components

10.12 Brushless DC motor from a computer fan

10.13 Brushless DC motor gear pump

10.14 DC and stepper motor examples

■ **CLASS DISCUSSION ITEM 10.2**
Eddy Currents

Describe the causes of eddy currents that are induced in a conducting material experiencing a changing magnetic field. The iron core in a motor rotor is usually laminated. Explain why. What is the best orientation for the laminations?

Torque is produced by an electric motor through the interaction of either stator fields and armature currents or stator fields and armature fields. We illustrate both principles starting with the first. Figure 10.8 illustrates a DC motor with six armature windings. The direction of current flow in the windings is illustrated in the figure.

(a) AC induction and stepper motor

Bracket Rotor   Stator   Flange

(b) exploded view of stepper motor with a
permanent magnet rotor

Figure 10.7 Examples of commercial motors. *(Courtesy of Oriental Motor, Torrance, CA)*

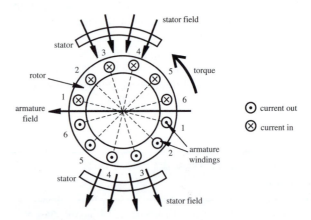

Figure 10.8 Electric motor field-current interaction.

As a result of Equation 10.1, the interaction of the fixed stator field and the currents in the armature windings produce a torque in the counterclockwise direction. You can verify this torque direction by applying the right-hand rule to the armature current and stator field directions. To maintain the torque as the rotor rotates, the spatial arrangement of the armature currents relative to the stator field must remain fixed.

A commutator accomplishes this by switching the currents in the armature windings in the correct sequence as the rotor turns.

Figure 10.9 illustrates an example commutator. It consists of a ring of alternating conductive and insulating materials connected to the rotor windings. Current is directed through the windings via the brushes, which slide on the surface of the commutator as it rotates. In the position shown, the current flows through windings A, B, and C in the clockwise direction and through F, E, and D in the counterclockwise direction.

When the rotor turns clockwise one sixth of a full rotation from the position shown, the currents in windings C and F switch directions. As the brushes slide over the rotating commutator, this process continues in sequence. With appropriate winding configurations, the commutator maintains a consistent spatial arrangement of the currents relative to the fixed stator fields. This continually maintains the torque in the desired direction.

Another method by which electric motors can create torque is through the interaction of stator and rotor magnetic fields. The torque is produced by the fact that like field poles attract and unlike poles repel. Figure 10.10 illustrates this principle of operation with a simple two-pole DC motor. The stator poles generate fixed magnetic fields with permanent magnets or coils carrying DC current. The winding in the rotor is commutated to cause changes in direction of its magnetic field. The interaction of the changing rotor field and the fixed stator fields produce a torque on the shaft, causing rotation. With the rotor in position i, the right brush contacts commutator segment A and the left brush contacts segment B, creating a current in the rotor winding, resulting in the magnetic poles as shown. The rotor magnetic poles oppose the stator magnetic poles, creating a torque causing clockwise motion of the rotor. In position ii, the stator poles both oppose and attract the rotor poles to enhance the clockwise rotation. Between positions iii and v the commutator contacts switch, changing the direction of the rotor current and hence the direction of the magnetic field. In position iv, both brushes temporarily lose contact with the commutator, but

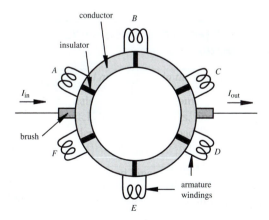

Figure 10.9 Electric motor six-winding commutator.

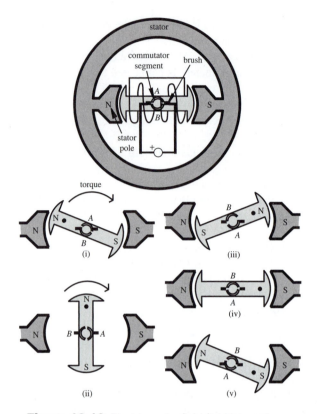

Figure 10.10 Electric motor field-field interaction.

the rotor continues to move due to its momentum. In position *v* reversed magnetic field in the rotor again opposes the stator field, continuing the clockwise torque and motion.

Video Demo

10.11 DC motor components

■ CLASS DISCUSSION ITEM 10.3
Field-Field Interaction in a Motor

Does the armature field in Figure 10.8 have any effect on the torque produced by the motor?

A problem with the simple two-pole design illustrated in Figure 10.10 is that starting would not occur if the motor happens to be in position iv, where the brushes are located over the commutator gaps. This problem can be avoided by designing the motor with more poles and more commutator segments with overlapping switching. This allows the brushes to always contact two active segments, even while switching (see Video Demo 10.11 and Class Discussion Item 10.4).

Other problems not discussed with these simple models are a back electromotive force (emf) and induction. As the rotor windings cut through the stator magnetic field, a **back emf** is induced opposing the voltage applied to the rotor. Also, when the commutator switches the direction of current, a voltage is induced to oppose the change in current direction.

The principles of operation of **AC motors** are similar regarding interaction of the magnetic fields, but commutation is not required. This is because the magnetic field rotates around the stator as a result of the AC voltages and the arrangement of the coils around the stator housing. The rotor windings of **asynchronous** AC motors have no external voltage applied; rather, voltages are induced in the rotor windings due to the rotating fields around the stator. The rotor rotates at slower speeds than the rotating stator fields (this is called **slip**), making the induction possible, hence the term asynchronous. Because of this action, asynchronous motors are sometimes referred to as **induction machines.** With **synchronous** AC motors, the rotor windings are energized but through **slip rings** instead of a commutator. Brushes provide constant uninterrupted contact with the slip rings, causing fields to rotate around the rotor windings at the same rate as the fields rotate around the stator. Due to the interaction of these fields, the rotor rotates at the same speed as the stator fields, hence the term synchronous.

■ **CLASS DISCUSSION ITEM 10.4**
Dissection of Radio Shack Motor

Purchase an inexpensive 1.5–3-V DC motor (e.g., Radio Shack Catalog No. 273-223) and disassemble it. Identify the brushes, the commutator segments, the armature windings, the laminated rotor poles, and the stator permanent magnets. Sketch the magnetic field produced by the stator permanent magnets. For different commutator positions, determine the direction of current flow (and the resulting field direction) in the armature windings. Determine the direction of torque produced by the field-field and field-current interactions. Which effect do you think is stronger in this motor?

Internet Link 10.2 provides excellent illustrations and animations showing the fundamentals of how various motors, generators, and transformers function.

10.5 DC MOTORS

Direct current (DC) motors are used in a large number of mechatronic designs because of the torque-speed characteristics achievable with different electrical configurations. DC motor speeds can be smoothly controlled and in most cases are reversible. Since DC motors have a high ratio of torque to rotor inertia, they can respond quickly. Also, **dynamic braking,** where motor-generated energy is fed to a

Internet Link

10.2 Electric motor illustrations and animations

resistor dissipater, and **regenerative braking,** where motor-generated energy is fed back to the DC power supply, can be implemented in applications where quick stops and high efficiency are desired.

Figure 10.11 illustrates a typical **torque-speed curve** that displays the torques a motor can provide at different speeds at rated voltage. For a given torque provided by the motor, the **current-torque curve** can be used to determine the amount of current required when rated voltage is applied. As a general rule of thumb, motors deliver large torques at low speeds, and large torques imply large motor currents. The **starting torque** or **stall torque** T_s is the maximum torque the motor can produce, at zero speed, associated with starting or overloading the motor. The **no-load speed** ω_{max} is the maximum sustained speed the motor can attain. This speed can be reached only when no load or torque is applied to the motor (i.e., only when it is free running).

Based on how the stator magnetic fields are created, DC motors are classified into four categories: permanent magnet, shunt wound, series wound, and compound wound. The electrical schematics, torque-speed curves, and current-torque curves for each configuration are illustrated in Figures 10.12 through 10.15. In the figures, V is the DC voltage supply, I_A is the current in the rotor (armature) windings, I_F is the current in the stator (field) windings, and I_L is the total load current delivered by the DC supply.

The stator fields in **permanent magnet (PM) motors** (see Figure 10.12) are provided by permanent magnets, which require no external power source and therefore produce no I^2R heating. A PM motor is lighter and smaller than other, equivalent DC motors because the field strength of permanent magnets is high. PM motors are easily reversed by switching the direction of the applied voltage, because the current and field change direction only in the rotor. The PM motor is ideal in control applications because of the linearity of its torque-speed relation. The design of a controller is always easier when the actuator is linear since the system analysis is greatly simplified. When a motor is used in a position or speed control application with sensor feedback to a controller, it is referred to as a **servomotor.** PM motors are used only in low-power applications since their rated power is usually limited to 5 hp (3728 W) or less, with fractional horsepower ratings being more common. PM DC motors can be brushed, brushless, or stepper motors.

Shunt motors (see Figure 10.13) have armature and field windings connected in parallel, which are powered by the same supply. The total load current is the sum of the armature and field currents. Shunt motors exhibit nearly constant speed over a large range of loading, have starting torques about 1.5 times the rated operating torque, have the lowest starting torque of any of the DC motors, and can be economically converted to allow adjustable speed by placing a potentiometer in series with the field windings.

Series motors (see Figure 10.14) have armature and field windings connected in series so the armature and field currents are equal. Series motors exhibit very high starting torques, highly variable speed depending on load, and very high speed when the load is small. In fact, large series motors can fail catastrophically when they are suddenly unloaded (e.g., in a belt drive application when the belt fails) due

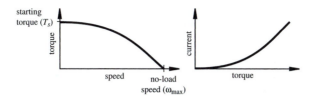

Figure 10.11 Motor torque-speed curve.

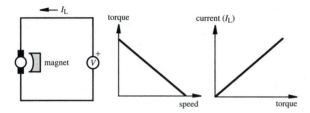

Figure 10.12 DC permanent magnet motor schematic and torque-speed curve.

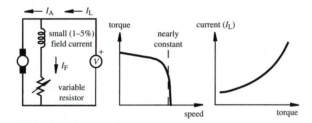

Figure 10.13 DC shunt motor schematic and torque-speed curve.

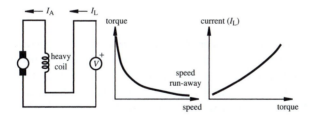

Figure 10.14 DC series motor schematic and torque-speed curve.

to dynamic forces at high speeds. This is called **run-away.** As long as the motor remains loaded, this poses no problem. The torque-speed curve for a series motor is hyperbolic in shape, implying an inverse relationship between torque and speed and nearly constant power over a wide range.

Compound motors (see Figure 10.15) include both shunt and series field windings, resulting in combined characteristics of both shunt and series motors. Part of the load current passes through both the armature and series windings, and the remaining load current passes through the shunt windings only. The maximum speed

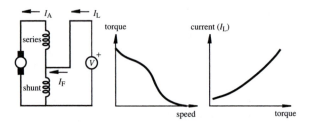

Figure 10.15 DC compound motor schematic and torque-speed curve.

of a compound motor is limited, unlike a series motor, but its speed regulation is not as good as with a shunt motor. The torque produced by compound motors is somewhat lower than that of series motors of similar size.

Note that, unlike the permanent magnet motor, when voltage polarity for a shunt, series, or compound DC motor is changed, the direction of rotation does not change (i.e., these motors are not reversible). The reason for this is that the polarity of both the stator and rotor changes because the field and armature windings are excited by the same source.

10.5.1 DC Motor Electrical Equations

As illustrated in Figure 10.16, the impedance of a DC motor's armature can be modeled as a resistance R in series with the parallel combination of an inductance L and a second resistance R_L. However, as the conducting armature begins to rotate in the magnetic field produced by the stator, a voltage called the **back emf** V_{emf} is induced in the armature windings opposing the applied voltage. The back emf is proportional to the speed of the motor ω in rad/sec:

$$V_{emf} = k_e \omega \tag{10.2}$$

where the proportionality constant k_e is called the **electrical constant** of the motor.

R_L, the equivalent resistive loss in the magnetic circuit, is usually an order of magnitude larger than R, the resistance of the windings, and can be neglected. If the voltage applied to the armature is V_{in}, the current through the armature is I_{in}, and if we assume $R_L \approx 0$, the electrical equation for the motor is

$$V_{in} = L\frac{dI_{in}}{dt} + RI_{in} + k_e \omega \tag{10.3}$$

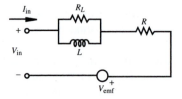

Figure 10.16 Motor armature equivalent circuit.

10.5.2 Permanent Magnet DC Motor Dynamic Equations

Since the permanent magnet DC motor is so important and is the easiest to understand and analyze, we look at its governing equations in more detail. Due to the interaction between the stator field and the armature current, the torque generated by a PM DC motor is directly proportional to armature current:

$$T = k_t I_{in} \tag{10.4}$$

where k_t is defined as the **torque constant** of the motor. A PM motor's electrical constant k_e and torque constant k_t are very important parameters, and they are often reported in manufacturer specifications. When the dynamics of a motor and its load are considered, the generated motor torque T is given by

$$T = (J_a + J_L)\frac{d\omega}{dt} + T_f + T_L \tag{10.5}$$

where J_a and J_L are the polar moments of inertia of the armature and the attached load, T_f is the frictional torque opposing armature rotation (including drag created by air resistance), and T_L is the torque dissipated by the load.

When voltage is applied to a PM DC motor, the rotor accelerates until a steady state operating condition is attained. At steady state, Equation 10.3 becomes

$$V_{in} = RI_{in} + k_e\omega \tag{10.6}$$

Note that, at steady state, from Equation 10.5, the motor torque balances the friction and load torques.

By solving for I_{in} in Equation 10.4 and substituting into Equation 10.6, we get

$$V_{in} = \left(\frac{R}{k_t}\right)T + k_e\omega \tag{10.7}$$

and by solving for the motor torque in this equation, we get

$$T = \left(\frac{k_t}{R}\right)V_{in} - \left(\frac{k_e k_t}{R}\right)\omega \tag{10.8}$$

This equation predicts the linear torque-speed relation for a PM DC motor with a fixed applied voltage.

Figure 10.17 shows the linear torque-speed curve and the power-speed curve for a permanent magnet DC motor with a fixed supply voltage V_{in}. The linear relation in Equation 10.8 can also be expressed in terms of the starting torque T_s and the no-load speed ω_{max} as

$$T(\omega) = T_s\left(1 - \frac{\omega}{\omega_{max}}\right) \tag{10.9}$$

The no-load speed is the steady-state speed at which the motor will run if there is no load torque ($T_L = 0$). At this speed, the motor torque balances the friction torque (i.e., $T = T_f$).

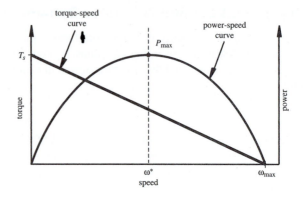

Figure 10.17 Permanent magnet DC motor characteristics.

By comparing Equations 10.8 and 10.9, you can see that the stall torque and no-load speed are related to the motor parameters according to

$$T_s = \left(\frac{k_t}{R}\right) V_{in} \tag{10.10}$$

and

$$\omega_{max} = \frac{T_s R}{k_e k_t} \tag{10.11}$$

Using Equation 10.9, the power delivered by the motor at different speeds can be expressed as

$$P(\omega) = T\omega = \omega T_s \left(1 - \frac{\omega}{\omega_{max}}\right) \tag{10.12}$$

The maximum power output of the motor occurs at the speed where

$$\frac{dP}{d\omega} = T_s \left(1 - \frac{2\omega}{\omega_{max}}\right) = 0 \tag{10.13}$$

Solving for the speed gives

$$\omega^* = \frac{1}{2}\omega_{max} \tag{10.14}$$

Therefore, the best speed to run a permanent magnet motor to achieve maximum power output is half the no-load speed.

In addition to the electrical and torque constants, manufacturers often specify the armature resistance R. Using Equations 10.4 and 10.10, the **stall current** I_s can be found in terms of the armature resistance and supply voltage:

$$I_s = \frac{V_{in}}{R} \tag{10.15}$$

This equation for current is valid only when the motor rotor is not turning; otherwise, the rotor current is affected by the back emf induced in the rotor windings. The stall current is the maximum current through the motor for a given supply voltage.

10.5.3 Electronic Control of a Permanent Magnet DC Motor

The simplest form of motor control is open loop control, where one simply sets the drive voltage value, and the motor characteristics and load determine the operating speed and torque. But most interesting problems require some sort of automatic control, where the voltage is automatically varied to produce a desired motion. This is called **closed-loop** or **feedback control,** and it requires an output speed and/or torque sensor to feed back output values to continuously compare the actual output to a desired value, called the **set point.** The controller then actively changes the motor output to move closer to the set point. Electronic speed controllers are of two types: linear amplifiers and pulse width modulators. Pulse width modulation controllers have the advantage that they either drive bipolar power transistors rapidly between cutoff and saturation or they turn FETs on and off. In either case, power dissipation is small. Servo amplifiers using linear power amplification are satisfactory but produce a lot of heat, because they function in the transistor linear region. You will find commercial servo controllers using linear amplifiers, but because of lower power requirements, ease of design, smaller size, and lower cost, we focus on the switched amplifier designs, which are generally called **pulse-width modulation (PWM)** amplifiers.

The principle of a PWM amplifier is shown in Figure 10.18. A DC power supply voltage is rapidly switched at a fixed frequency f between two values (e.g., ON and OFF). This frequency is often in excess of 1 kHz. The high value is held during a variable pulse width t within the fixed period T where

$$T = \frac{1}{f} \tag{10.16}$$

The resulting asymmetric waveform has a **duty cycle** defined as the ratio between the ON time and the period of the waveform, usually specified as a percentage:

$$\text{duty cycle} = \frac{t}{T}100\% \tag{10.17}$$

As the duty cycle is changed (by the controller), the average current through the motor changes, causing changes in speed and torque at the output. It is primarily the duty cycle, and not the value of the power supply voltage, that is used to control the speed of the motor. Video Demo 10.15 explains how PWM signals are used to control standard radio-controlled (RC) servomotors. Lab Exercise 11 explores how to use a microcontroller to generate a PWM signal to vary the speed of a DC motor. The end result of the exercise is demonstrated in Video Demo 7.5.

The block diagram of a PWM speed feedback control system for a DC motor is shown in Figure 10.19. A voltage tachometer produces an output linearly related

Video Demo

10.15 RC servo with pulse-width-modulation input

7.5 Pulse-width-modulation speed control of a DC motor, with keypad input

Lab Exercise

Lab 11 Pulse-width-modulation motor speed control with a PIC

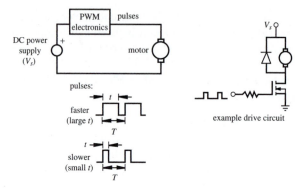

Figure 10.18 Pulse width modulation of a DC motor.

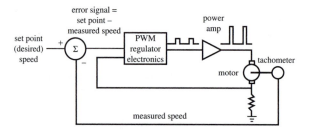

Figure 10.19 PWM velocity feedback control.

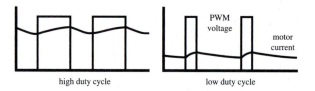

Figure 10.20 PWM voltage and motor current.

to the motor speed. This is compared to the desired speed set point (another voltage that can be manually set or computer controlled). The error and the motor current are sensed by a pulse-width-modulation regulator that produces a width-modulated square wave as an output. This signal is amplified to a level appropriate to drive the motor.

In a PWM motor controller, the armature voltage switches rapidly, and the current through the motor is affected by the motor inductance and resistance. Since the switching speed is high, the resulting current through the motor has a small fluctuation around an average value, as illustrated in Figure 10.20. As the duty cycle grows larger, the average current grows larger and the motor speed increases.

H-Bridge Drive for a DC Motor

Although it is possible to design and build a drive circuit for a servomotor with discrete control and power components, several integrated circuit designs are available that save enormously on time and money in mechatronic design. Consider the very ordinary problem of controlling a DC motor. Your ultimate intent may be to control speed, direction of rotation, angle, and/or torque.

To control the speed of a DC motor, we must be able to change the current supplied to the motor. To control the direction of rotation, the direction of current supplied to the motor must be reversible. This requires a current amplifier and some means to switch the current direction. The concept of an **H-bridge** meets these requirements. It uses four switches (relays or transistors) arranged in an H configuration around the DC motor (see the following figure). The switches are turned on one pair at a time for the desired direction of motion.

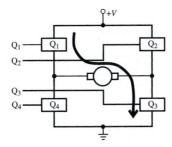

If switches Q_1 and Q_3 are on and Q_2 and Q_4 are off, current flows through the motor in the direction shown and the motor rotates in one direction. Alternatively, if switches Q_2 and Q_4 are on and Q_1 and Q_3 are off, the motor rotates in the other direction.

An H-bridge can be constructed easily with relays (four SPSTs or two DPDTs); however, relays cannot be switched very fast, so this limits our PWM options. Also, because relays are mechanical devices, they can wear out and fail after many switching cycles. However, for applications where you need only reversibility and not PWM speed control, relays can be a good option. For applications where we want PWM speed control, transistors or solid-state relays are better choices. An excellent resource for information and advice on how to build H-bridges using a variety of components can be found at Internet Link 10.3.

One can also build a discrete H-bridge with power BJTs or MOSFETs, but it may be difficult to properly choose and bias the transistors. Therefore, we utilize a monolithic solution using National Semiconductor's line of motion control ICs, which can be conveniently adapted for driving DC motors. Consider the LMD15200, a 3 A, 55 V H-bridge specifically designed to drive DC and stepper motors. It allows for current direction control, and also offers features for overcurrent and overtemperature detection, pulse width modulation, and dynamic braking. The functional diagram follows.

Internet Link

10.3 H-bridge resource

(continued)

(continued)

Functional Diagram

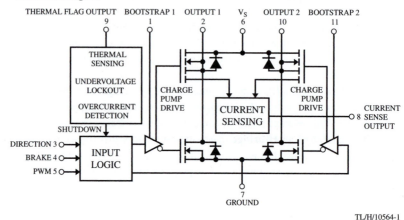

TL/H/10564-1

Function Diagram and Ordering Information

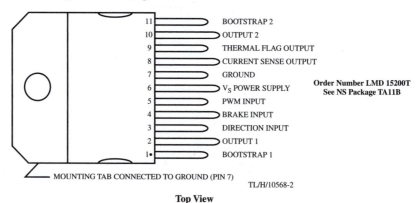

**Order Number LMD 15200T
See NS Package TA11B**

MOUNTING TAB CONNECTED TO GROUND (PIN 7)

TL/H/10568-2

Top View

(Courtesy of National Semiconductor Inc., Santa Clara, CA)

    This design uses power MOSFETs with flyback protection diodes that suppress large reverse transients across the transistors. The motor poles are connected between Output 1 and Output 2. The voltage supply can be up to 55 V. External digital signals control direction, braking, and pulse width modulation. A thermal sensor shuts down the outputs when the device temperature exceeds 170°C.

    The complete block diagram for our speed controller design is shown in the next figure. The speed set point is controlled with a potentiometer or input voltage value. A tachometer is added to the motor as the sensor to provide a measure of the motor speed. The National Semiconductor pulse width modulation IC LM3524D is conveniently used here to drive the H-bridge motor controller input.

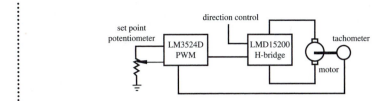

DC motor position and speed controller—H-bridge driver and PWM speed control   C.5

The figure below shows the functional diagram for Threaded Design Example C (see Section 1.3 and Video Demo 1.8), with the portion described here highlighted.

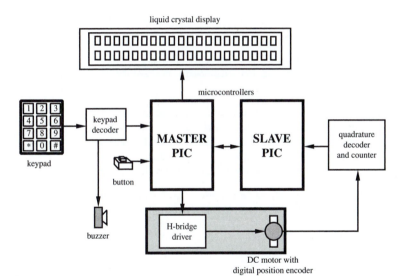

Video Demo

1.8 DC motor position and speed controller

The complete circuit required for this part of the design is shown below. A commercially available S17-3A-LV H-bridge was selected because of its ample current capacity and compatibility with the selected motor. It also exhibits a near linear relationship between PWM duty cycle and motor speed over a fairly large range. Details about the H-bridge and motor can be found at Internet Link 10.4. The two inputs to the H-bridge are a direction line to indicate whether the motor turns clockwise or counterclockwise, and a PWM line to control the speed of the motor via an appropriate duty cycle.

Internet Link

10.4 S17-3A-LV H-bridge

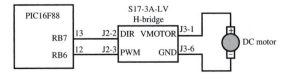

(continued)

(continued) The code used to rotate the motor at a desired speed is presented below. The subroutine *pwm_periods* calculates the pulse segment widths to create a duty cycle proportional to the desired motor speed. The subroutine *pwm_pulse* is called continuously by *run_motor* to keep the motor rotating at the selected speed. The comments in the code explain some of the details involved with the solution.

```
' Define I/O pin names
motor_dir Var PORTB.7     ' motor H-bridge direction line
motor_pwm Var PORTB.6     ' motor H-bridge pulse-width-modulation line

' Declare Variables
motor_speed Var BYTE     ' motor speed as percentage of maximum (0 to 100)
motion_dir Var BIT       ' motor direction (1:CW/Forward 0:CCW/Reverse)
on_time Var WORD         ' PWM ON pulse width
off_time Var WORD        ' PWM OFF pulse width
pwm_cycles Var BYTE      ' # of PWM pulses sent during the position control loop

' Define constants
pwm_period Con 50        ' period of each motor PWM signal cycle (in microsec)
                         '   (50 microsec corresponds to 20kHz)

' Initialize I/O and variables
TRISB.6 = 0              ' configure H-bridge DIR pin as an output
TRISB.7 = 0              ' configure H-bridge PWM pin as an output
motion_dir = CW          ' starting motor direction: CW (forward)
motor_speed = 50         ' starting motor speed = 50% duty cycle
Low motor_pwm            ' make sure the motor is off to begin with

' Subroutine to run the motor at the desired speed and direction until the
'    stop button is pressed. The duty cycle of the PWM signal is the
'    motor_speed percentage
run_motor:
    ' Set the motor direction
    motor_dir = motion_dir

    ' Output the PWM signal
    Gosub pwm_periods           ' calculate the on and off pulse widths
    While (stop_button == 0)  ' until the stop button is pressed
        Gosub pwm_pulse ' send out a full PWM pulse
    Wend
Return

'Subroutine to calculate the PWM on and off pulse widths based on the desired
' motor speed (motor_speed)
pwm_periods:
    ' Be careful to avoid integer arithmetic and
    '   WORD overflow [max=65535] problems
    If (pwm_period >= 655) Then
        on_time = pwm_period/100 * motor_speed
        off_time = pwm_period/100 * (100-motor_speed)
```

```
    Else
        on_time = pwm_period * motor_speed / 100
        off_time = pwm_period *(100-motor_speed) / 100
    Endif
Return

' Subroutine to output a full PWM pulse based on the data from pwm_periods
pwm_pulse:
    ' Send the ON pulse
    High motor_pwm
    Pauseus on_time

    ' Send the OFF pulse
    Low motor_pwm
    Pauseus off_time
Return
```

10.6 STEPPER MOTORS

A special type of DC motor, known as a **stepper motor,** is a permanent magnet or variable reluctance DC motor that has the following performance characteristics: it can rotate in both directions, move in precise angular increments, sustain a holding torque at zero speed, and be controlled with digital circuits. It moves in accurate angular increments, known as **steps,** in response to digital pulses sent to an electric drive circuit. The number and rate of the pulses control the position and speed of the motor shaft. Generally, stepper motors are manufactured with steps per revolution of 12, 24, 72, 144, 180, and 200, resulting in shaft increments of 30°, 15°, 5°, 2.5°, 2°, and 1.8° per step. Special **micro-stepping** circuitry can be designed to allow many more steps per revolution, often 10,000 steps/rev or more.

Stepper motors are either **bipolar,** requiring two power sources or a switchable polarity power source, or **unipolar,** requiring only one power source. They are powered by DC sources and require digital circuitry to produce coil energizing sequences for rotation of the motor. Feedback is not always required for control, but the use of an encoder or other position sensor can ensure accuracy when exact position control is critical. The advantage of operating without feedback (i.e., in open-loop mode) is that a closed-loop control system is not required. Generally, stepper motors produce less than 1 hp (746 W) and are therefore used only in low-power position control applications. Video Demos 9.2 and 10.16 show interesting student project examples of stepper motor applications.

A commercial stepper motor has a large number of poles that define a large number of equilibrium positions of the rotor. In the case of a permanent magnet stepper motor, the stator consists of wound poles, and the rotor poles are permanent magnets. Exciting different stator winding combinations moves and holds the rotor in different positions. The **variable reluctance** stepper motor has a ferromagnetic rotor rather than a permanent magnet rotor. Motion and holding result from the attraction of stator and rotor poles to positions with minimum magnetic reluctance (magnetic field resistance) that allow for maximum magnetic flux. A variable reluctance motor

Video Demo

9.2 Automated laboratory rat exercise machine with IR sensor and stepper motor
10.16 Multifunction slot machine

has the advantage of a lower rotor inertia and therefore a faster dynamic response. The permanent magnet stepper motor has the advantage of a small residual holding torque, called the **detent torque,** even when the stator is not energized.

To understand how the rotor moves in an incremental fashion, consider a simple design consisting of four stator poles and a permanent magnet rotor as illustrated in Figure 10.21. In step 0, the rotor is in equilibrium, because opposite poles on the stator and rotor are adjacent to and attract each other. Unless the magnet polarities of the stator poles are changed, the rotor remains in this position and can withstand an opposing torque up to a value called the **holding torque.** When the stator polarities are changed as shown (step 0 to step 1), a torque is applied to the rotor, causing it to move 90° in the clockwise direction to a new equilibrium position shown as step 1. When the stator polarities are again changed as shown (step 1 to step 2), the rotor experiences a torque driving it to step 2. By successively changing the stator polarities in this manner, the rotor can move to successive equilibrium positions in the clockwise direction. The sequencing of the pole excitations is the means by which the direction of rotation occurs. Counterclockwise motion can be achieved by applying the polarity sequence in the opposite direction. The motor torque is directly related to the magnetic field strength of the poles and the rotor.

The dynamic response of the rotor and attached load must be carefully considered in applications that involve starting or stopping quickly, changing or ramping speeds quickly, or driving large or changing loads. Due to the inertia of the rotor and attached load, rotation can exceed the desired number of steps. Also, as illustrated in Figure 10.22, a stepper motor driving a typical mechanical system through one step will exhibit an underdamped response. If damping is increased in the system, for example, with mechanical, frictional, or viscous damping, the response can be modified to reduce oscillation, as shown in the figure. Note, however, that even with an ideal choice for damping, the motor requires time to totally settle into a given position, and this settling time varies with the step size and the amount of damping. It is also important to note that the torque required from the motor increases with added damping. Video Demo 10.17 shows an example of a typical underdamped second-order system response of a stepper motor with a fairly large step size. It also shows how the response changes as the step rate increases. Video Demo 10.18 shows

Video Demo

10.17 Stepper motor step response and acceleration through resonance

10.18 High-speed video of medium speed response

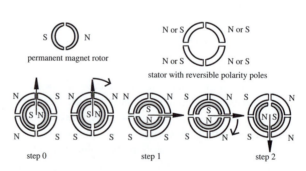

step 0   step 1   step 2

Figure 10.21 Stepper motor step sequence.

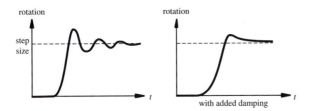

Figure 10.22 Dynamic response of a single step.

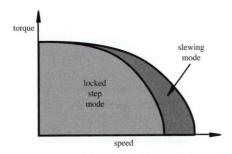

Figure 10.23 Stepper motor torque-speed curves.

slow-motion footage of the motor turning at a medium speed to show how the step response effects are less pronounced at higher step rates.

The torque-speed characteristics for a stepper motor are usually divided into two regions as illustrated in Figure 10.23. In the **locked step mode,** the rotor decelerates and may even come to rest between each step. Within this region, the motor can be instantaneously started, stopped, or reversed without losing step integrity. In the **slewing mode,** the speed is too fast to allow instantaneous starting, stopping, or reversing. The rotor must be gradually accelerated to enter this mode and gradually decelerated to leave the mode. While in slewing mode, the rotor is in synch with the stator field rotation and does not settle between steps. The curve between the regions in the figure indicates the maximum torques that the stepper can provide at different speeds without slewing. The curve bordering the outside of the slewing mode region represents the absolute maximum torques the stepper can provide at different speeds.

Figure 10.24 illustrates a unipolar stepper motor field coil schematic with external power transistors that must be switched on and off to produce the controlled sequence of stator polarities to cause rotation. The schematic in Figure 10.24 shows six wires connected to the motor. Because the second and fifth wires are usually connected externally as shown, manufacturers sometimes connect them inside the motor, in which case the motor only has five external wires. The wires are usually color coded by the manufacturers to help the user make a correspondence to the schematic. Figure 10.24 includes a common color scheme used for a six-wire unipolar stepper motor: yellow (coil 1), red (1/2 common), orange (coil 2), black (coil 3), green (3/4 common), brown (coil 4). Another common six-wire color scheme is

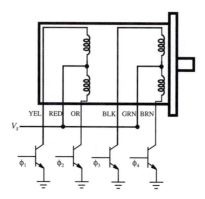

Figure 10.24 Standard unipolar stepper motor field coil schematic.

Internet Link

10.5 How to identify the different wires of a stepper motor

green (coil 1), white (1/2 common), blue (coil 2), red (coil 3), white (3/4 common), black (coil 4). A common color scheme for a five-wire unipolar stepper is red (coil 1), green (coil 2), black (common), brown (coil 3), white (coil 4). If you come across a motor for which you have no documentation and the color scheme is unknown, there is a testing procedure you can follow to determine the wire identities (e.g., see Internet Link 10.5 , being aware that the coils are numbered differently in the linked Web page. Coils 3-2-1-4 in the Web page correspond to coils 1-2-3-4 or ϕ_1-ϕ_2-ϕ_3-ϕ_4 here).

Figure 10.25 illustrates the construction of and stepping sequence for a four-phase unipolar stepper motor. It consists of a two-pole permanent magnet rotor and a four-pole stator, with each pole wound by two complementary windings (e.g., ϕ_1 and ϕ_2 wound in opposite directions on the top left pole). Table 10.1 lists the phase sequence required to step the motor in full steps, where two of the four phases are energized (ON) and each stator pole is polarized. Table 10.2 lists the phase sequence for half-stepping, where between each full step only one phase is energized (ON) and

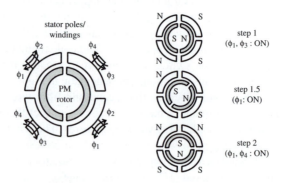

Figure 10.25 Example of a unipolar stepper motor.

Table 10.1 Unipolar full-step phase sequence

	Step	ϕ_1	ϕ_2	ϕ_3	ϕ_4
CW	1	ON	OFF	ON	OFF
↓	2	ON	OFF	OFF	ON
CCW	3	OFF	ON	OFF	ON
↑	4	OFF	ON	ON	OFF

Table 10.2 Unipolar half-step phase sequence

	Step	ϕ_1	ϕ_2	ϕ_3	ϕ_4
CW	1	ON	OFF	ON	OFF
↓	1.5	ON	OFF	OFF	OFF
	2	ON	OFF	OFF	ON
	2.5	OFF	OFF	OFF	ON
CCW	3	OFF	ON	OFF	ON
↑	3.5	OFF	ON	OFF	OFF
	4	OFF	ON	ON	OFF
	4.5	OFF	OFF	ON	OFF

only two stator poles are polarized. The resolution or number of steps of the motor is twice as large in the half-step mode (8 steps/rev at 45°) than in the full-step mode (4 steps/rev at 90°), but the holding torque and drive torque change between two values on alternate cycles in half-step mode. Another technique for increasing the number of steps is called **micro-stepping,** where the phase currents are controlled by fractional amounts, rather than just ON and OFF, resulting in more magnetic equilibrium positions between the poles. In effect, discretized sine waves are applied to the phases instead of square waves. The most common commercially available stepper motors have 200 steps/rev in full-step mode and are sometimes referred to as 1.8° (360°/200) steppers. In micro-stepping mode, 10,000 or more steps per revolution can be achieved.

Figure 10.26 illustrates a bipolar stepper motor schematic with external power transistors that must be switched on and off in ordered combinations to produce the controlled sequence of stator polarities to cause rotation. There are four wires connected to the motor, which easily distinguishes it from a unipolar stepper, which has either five or six wires. The wires are usually color coded by the manufacturer to help make a correspondence to the schematic. Figure 10.26 includes a common color scheme used for a bipolar stepper motor: red (+) and grey (−) for the coil A, and yellow (+) and black (−) for coil B.

Table 10.3 lists the phase sequence required to step a bipolar stepper motor in full steps. Note that the table is the same as for a unipolar stepper (see Table 10.1), but here sets of two transistors are switched together for each ON or OFF state

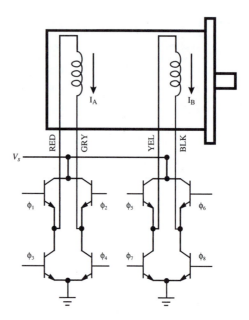

Figure 10.26 Standard bipolar stepper motor field coil schematic.

Table 10.3 Bipolar full-step phase sequence

	Step	ϕ_1 and ϕ_4	ϕ_2 and ϕ_3	ϕ_5 and ϕ_8	ϕ_6 and ϕ_7
CW	1	ON	OFF	ON	OFF
↓	2	ON	OFF	OFF	ON
CCW	3	OFF	ON	OFF	ON
↑	4	OFF	ON	ON	OFF

Internet Link

10.3 H-bridge resource

instead of just one with the unipolar sequence. The configuration of transistors is called an H-bridge configuration (see Design Example 10.1 and Internet Link 10.3 for more information). By switching the transistors in pairs, the current direction can be reversed. For example, in step 1, transistors 1 and 4 are ON and transistors 2 and 3 are OFF. This applies power to coil A in the forward direction, creating a positive current ($I_A > 0$). Likewise, transistors 5 and 8 are ON and 6 and 7 are OFF in step 1, creating a positive current in coil B ($I_B > 0$). In step 2, transistors 5 and 8 are OFF and 6 and 7 are ON, reversing the direction of the current in coil B ($I_B < 0$). Table 10.4 lists the phase sequence for half-stepping. As with the unipolar motor, only half of the holding torque is available in the half-step positions, because only one coil is energized.

Figure 10.27 illustrates the structure, pole geometry, and coil connections of an actual stepper motor in more detail. This particular stepper motor can be wired as

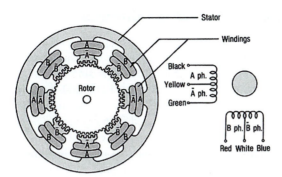

Figure 10.27 Typical stepper motor rotor and stator configuration.
(Courtesy of Oriental Motor, Torrance, CA)

Figure 10.28 Actual stepper motor rotor. *(Courtesy of Oriental Motor, Torrance, CA)*

Table 10.4 Bipolar half-step phase sequence

	Step	ϕ_1 and ϕ_4	ϕ_2 and ϕ_3	ϕ_5 and ϕ_8	ϕ_6 and ϕ_7
CW	1	ON	OFF	ON	OFF
↓	1.5	ON	OFF	OFF	OFF
	2	ON	OFF	OFF	ON
	2.5	OFF	OFF	OFF	ON
CCW	3	OFF	ON	OFF	OFF
↑	3.5	OFF	ON	OFF	OFF
	4	OFF	ON	ON	OFF
	4.5	OFF	OFF	ON	OFF

a four-phase unipolar motor or a two-phase bipolar motor. Figure 10.28 shows the 50-tooth split rotor with one side having north polarity and the other having south polarity, creating a large number of effective poles.

See Internet Link 10.6 for links to many resources and manufacturers for various stepper motor products.

Internet Link

10.6 Stepper motor online resource

10.6.1 Stepper Motor Drive Circuits

A drive circuit for properly phasing the signals applied to the poles of the unipolar stepper motor for rotation in full-step mode is easily and economically produced using the components illustrated in Figure 10.29. A similar drive circuit can be purchased as a single monolithic IC (e.g., E-Lab's EDE1200, Signetics' SAA1027, or Allegro Microsystems' UCN 5804B). The discrete circuit includes 7414 Schmitt trigger buffers, a 74191 up-down counter, and 7486 Exclusive OR gates. The Schmitt triggers (see Section 6.12.2) produce well-defined control signals with sharp rise and fall times in the presence of noise or fluctuations on the direction (CW/ CCW), initialization (RESET), and single-step (STEP) inputs. The up-down counter and the XOR gates in turn create four properly phased motor drive signals. These four digital signals (ϕ_1, ϕ_2, ϕ_3, ϕ_4) are coupled to the bases of power transistors that sequentially energize the respective motor coils connected to the DC motor supply, resulting in shaft rotation. Each square-wave pulse received at the STEP input causes the motor to rotate a full step in the direction determined by the CW/CCW input.

The timing diagram for the two least significant output bits B_0 and B_1 of the counter and the phase control signals is shown in Figure 10.30. Compare the signals

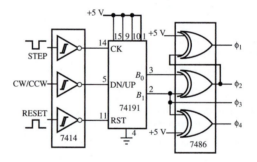

Figure 10.29 Unipolar stepper motor full-step drive circuit.

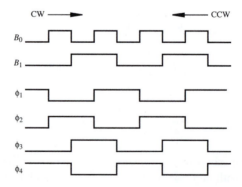

Figure 10.30 Timing diagram for full-step unipolar stepper motor drive circuit.

to the sequence in Table 10.1. They are in agreement. Boolean expressions that produce the four desired phased outputs from the two counter bits can be represented in both AND-OR-NOT and XOR forms:

$$\phi_1 = \overline{\phi_2} = \phi_2 \oplus 1$$
$$\phi_2 = (B_0 \cdot \overline{B_1}) + (\overline{B_0} \cdot B_1) = B_0 \oplus B_1$$
$$\phi_3 = B_1$$
$$\phi_4 = \overline{B_1} = B_1 \oplus 1$$

(10.18)

These expressions can be verified (Class Discussion Item 10.5) by checking the signal values at different times in the timing diagram shown in Figure 10.30. The purpose for representing the Boolean expressions in XOR form is to allow the logic to be executed using a single IC (the quad XOR 7486); otherwise, three ICs would be required for the AND, OR, and NOT representation.

■ **CLASS DISCUSSION ITEM 10.5**
Stepper Motor Logic

Construct a truth table for the timing diagram in Figure 10.30 and verify that Equations 10.18 are correct. Also, show that the sum-of-product and product-of-sum results for ϕ_2 are equivalent.

THREADED DESIGN EXAMPLE

Stepper motor position and speed controller—Stepper motor driver **B.3**

The figure below shows the functional diagram for Threaded Design Example B (see Section 1.3 and Video Demo 1.7), with the portion described here highlighted.

Video Demo

1.7 Stepper motor position and speed controller

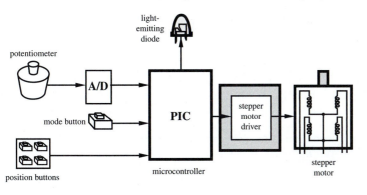

The figure below shows all components and interconnections required to drive a stepper motor from a PIC. A commercially available stepper motor driver IC, the E-Lab EDE1200, is the main component in the design. Detailed information about this component can be found in the data sheet at Internet Link 7.16. Only two signals from the PIC are required to drive the motor: a direction line and a step line. Each time a pulse is sent out on the step line, the stepper motor rotates a single step either clockwise or counterclockwise, as indicated by the direction line. The ULN2003A is required with the EDE1200 to provide enough current to drive typical stepper motor coils. Refer to the EDE1200 data sheet for more information.

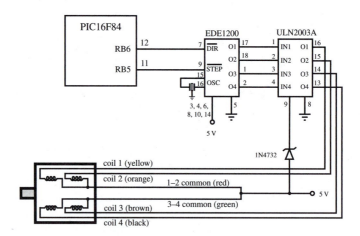

The code required to move the motor follows. The *move* subroutine first determines the required direction and magnitude of the motion, based on the user-selected *new_ motor_ pos* value. This value is compared to the current motor position (*motor_ pos*) to determine and set the motion direction and to calculate the required number of steps. The subroutine *move_ steps* (with the help of subroutine *step_motor*) then sends pulses to the motor causing rotation. The rotational speed is a function of the previously set *step_ period.*

```
' Define I/O pin names
motor_dir Var PORTB.6      ' stepper motor direction bit (0:CW 1:CCW)
motor_step Var PORTB.5     ' stepper motor step driver (1 pulse = 1 step)

' Define Constants
CW Con 0                   ' clockwise motor direction
CCW Con 1                  ' counterclockwise motor direction

' Subroutine to move the stepper motor to the position indicated by motor_pos
'   (the motor step size is 7.5 degrees)
move:
    ' Set the correct motor direction and determine the required displacement
    If (new_motor_pos > motor_pos) Then
        motor_dir = CW
        delta = new_motor_pos - motor_pos
```

```
    Else
        motor_dir = CCW
        delta = motor_pos − new_motor_pos
    EndIf

    ' Determine the required number of steps (given 7.5 degrees per step)
    num_steps = 10* delta / 75

    ' Step the motor the appropriate number of steps
    Gosub move_steps

    ' Update the current motor position
    motor_pos = new_motor_pos
Return

' Subroutine to move the motor a given number of steps (indicated by num_steps)
move_steps:
    For i = 1 to num_steps
        Gosub step_motor
    Next
Return

' Subroutine to step the motor a single step (7.5 degrees) in the motor_dir
'    direction
step_motor:
    Pulsout motor_step, 100 *step_period ' (100 * 10microsec = 1 millisec)
    Pause step_period
    ' Equivalent code:
    ' High motor_step
    ' Pause step_period
    ' Low motor_step
    ' Pause step_period
Return
```

10.7 SELECTING A MOTOR

When selecting a motor for a specific mechatronics application, the designer must consider many factors and specifications, including speed range, torque-speed variations, reversibility, operating duty cycle, starting torque, and power required. These and other factors are described here in a list of questions that a designer must consider when selecting and sizing a motor in consultation with a motor manufacturer. As we will see, the torque-speed curve provides important information, helping to answer many questions about a motor's performance. Recall that the torque-speed curve displays the torques the motor can deliver at different speeds at rated voltage. Figure 10.31 shows an example of a torque-speed curve for a stepper motor, and Figure 10.32 shows an example of a torque-speed curve for a servomotor. These figures are examples from motor manufacturer specification sheets.

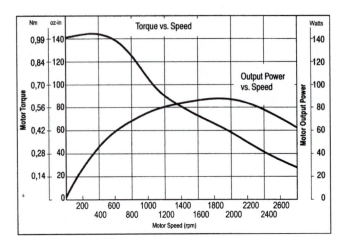

Figure 10.31 Typical stepper motor performance curves.
(Courtesy of Aerotech, Pittsburgh, PA)

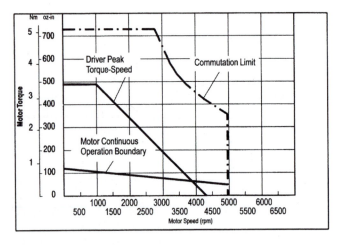

Figure 10.32 Typical servomotor performance curves.
(Courtesy of Aerotech, Pittsburgh, PA)

Some of the salient questions a designer may need to consider when choosing a motor for an application include the following:

■ *Will the motor start and will it accelerate fast enough?*   The torque at zero speed, called the *starting torque,* is the torque the motor can deliver when rotation begins. For the system to be self-starting, the motor must generate torque sufficient to overcome friction and any load torques.

The acceleration of the motor and load at any instant is given by

$$\alpha = (T_{\mathrm{motor}} - T_{\mathrm{load}})/J \qquad\qquad (10.19)$$

where α is the angular acceleration in rad/sec^2, T_{motor} is the torque produced by the motor, T_{load} is the torque dissipated by the load, and J is the total polar moment of inertia of the motor rotor and the load. The difference between motor and load torques determines the acceleration of the system. When the motor torque is equal to the load torque, the system is at a steady state operating speed.

■ *What is the maximum speed the motor can produce?* The zero torque point on the torque-speed curve determines the maximum speed a motor can reach. Note that the motor cannot deliver any torque to the load at this speed. When the motor is loaded, the maximum no-load speed cannot be achieved.

■ *What is the operating duty cycle?* When a motor is not operated continuously, one must consider the operating cycle of the system. The **duty cycle** is defined as the ratio of the time the motor is on with respect to the total elapsed time. If a load requires a low duty cycle, a lower-power motor may be selected that can operate above rated levels but still perform adequately without overheating during repeated on-off cycles.

■ *How much power does the load require?* The power rating is a very important specification for a motor. Knowing the power requirements of the load, a designer should choose a motor with adequate power based on the duty cycle.

■ *What power source is available?* Whether the motor is AC or DC might be a critical decision. Also, if battery power is to be used, the battery characteristics must match the load requirements.

■ *What is the load inertia?* As Equation 10.19 implies, for fast dynamic response, it is desirable to have low motor rotor and load inertia J. When the load inertia is large, the only way to achieve high acceleration is to size the motor so it can produce much larger torques than the load requires under steady state conditions.

■ *Is the load to be driven at constant speed?* The simplest method to achieve constant speed is to select an AC synchronous motor or a DC shunt motor which runs at a relatively constant speed over a significant range of load torques. Stepper motors and servomotors can also be driven at constant and accurate speeds, but these alternatives can involve more cost and might not be available in larger sizes (e. g., as required in industrial applications).

■ *Is accurate position or speed control required?* In the cases of angular positioning at discrete locations and incremental motion, a stepper motor is a good choice. A stepper motor is easily rotated to and held at discrete positions. It also can rotate at a wide range of speeds by controlling the step rate. The stepper motor can be operated with open-loop control, where no sensor feedback is required. However, if you attempt to drive a stepper motor at too fast a step rate or if the load torque is too large, the stepper motor may slip and not execute the number of steps expected. Therefore, a feedback sensor such as an encoder might be included with a stepper motor to check if the motor has achieved the desired motion.

For some complex motion requirements, where precise position or speed profiles are required (e.g., in automation applications where machines need to perform prescribed programmed motion), a servomotor may be the best choice. A **servomotor** is a DC, AC, or brushless DC motor combined with a position sensing device (e.g., a digital encoder). The servomotor is driven by a programmable controller that processes the sensor input and generates amplified voltages and currents to the motor to achieve specified motion profiles. This is called **closed-loop control,** since it includes sensor feedback. A servomotor is typically more expensive than a stepper motor, but it can have a much faster and smoother response. Video Demo 10.19 shows the typical components in a commercial servomotor system, and Internet Link 10.7 provides links to resources and vendors of motion control products.

For small-scale robotics and hobby projects, an RC servo is a good option. An RC servo is a small DC motor with an integral potentiometer (to sense shaft angle) and feedback electronics to provide position control directed by a PWM input signal. Originally developed and used mostly for radio or remote-controlled (RC) model planes, cars, and boats, they are versatile components for small projects requiring precise motion. The pulse width of the PWM signal dictates the motor position within a limited range of motion. When pulses of a certain width are sent to the motor, the shaft turns to and holds the corresponding position until the pulse width is changed. RC servos can also be modified to operate in continuous rotation in either direction at varying speeds. An excellent resource dealing with selecting, using, controlling, and modifying RC servos can be found at Internet Link 10.8. Video Demos 10.20 and 10.21 show interesting examples of student projects utilizing RC servos.

■ *Is a transmission or gearbox required?*   Often loads require low speeds and large torques. Since motors usually have better performance at high speed and low torque, a speed-reducing transmission (gear box or belt drive) is often needed to match the motor output to the load requirements. The term **gear motor** is used to refer to a motor-gearbox assembly sold as a single package.

When a transmission is used, the effective inertia of the load is

$$J_{\text{eff}} = J_{\text{load}}\left(\frac{\omega_{\text{load}}}{\omega_{\text{motor}}}\right)^2 \tag{10.20}$$

where J_{eff} is the effective polar moment of inertia of the load as seen by the motor. The sum of this inertia and the motor rotor inertia can be used in Equation 10.19 (i.e., $J = J_{motor} + J_{eff}$) to calculate acceleration. The speed ratio in Equation 10.20 is called the **gear ratio** of the transmission. It is often specified as a ratio of two numbers, where one or both numbers are integers (which is always the case when using meshing gears, which have integer numbers of teeth). So a gear ratio is sometimes written N : $M,$ which can be read as: N to M gear reduction. This means N turns of the motor are required to create M turns of the load, so for an N : M gear ratio, the speeds are related by

$$\omega_{\text{load}} = (M/N)\omega_{\text{motor}} \tag{10.21}$$

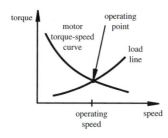

Figure 10.33 Motor operating speed.

■ *Is the motor torque-speed curve well matched to the load?* If the load has a well-defined torque-speed relation, called a **load line,** it is wise to select a motor with a similar torque-speed characteristic. If this is the case, the motor torque can match the load torque over a large range of speeds, and the speed can be controlled easily by making small changes in voltage to the motor.

■ *For a given motor torque-speed curve and load line, what will the operating speed be?* As Figure 10.33 illustrates, for a given motor torque-speed curve and a well-defined load line, the system settles at a fixed speed operating point. Furthermore, the operating point is self-regulating. At lower speeds, the motor torque exceeds the load torque and the system accelerates toward the operating point, but at higher speeds, the load torque exceeds the motor torque, reducing the speed toward the operating point. The operating speed can be actively changed by adjusting the voltage supplied to the motor, which in turn changes the torque-speed characteristic of the motor.

■ *Is it necessary to reverse the motor?* Some motors are not reversible due to their construction and control electronics, and care must be exercised when selecting a motor for an application that requires rotation in two directions.

■ *Are there any size and weight restrictions?* Motors can be large and heavy, and designers need to be aware of this early in the design phase.

■ **CLASS DISCUSSION ITEM 10.6**
Motor Sizing

Why is it important not to oversize a motor for a particular application?

■ **CLASS DISCUSSION ITEM 10.7**
Examples of Electric Motors

Make a list of the different types of electric motors found in household devices and automobiles. Describe the reasons why you think the particular type of motor is used for each example you cite.

10.8 HYDRAULICS

Hydraulic systems are designed to move large loads by controlling a high-pressure fluid in distribution lines and pistons with mechanical or electromechanical valves. A hydraulic system, illustrated in Figure 10.34, consists of a pump to deliver high-pressure fluid, a pressure regulator to limit the pressure in the system, valves to control flow rates and pressures, a distribution system composed of hoses or pipes, and linear or rotary actuators. The infrastructure, which consists of the elements contained in the dashed box in the figure, is typically used to power many hydraulic valve-actuator subsystems.

A hydraulic **pump** is usually driven by an electric motor (e.g., a large AC induction motor) or an internal combustion engine. Typical fluid pressures generated by pumps used in heavy equipment (e.g., construction equipment and large industrial machines) are in the 1000 psi (6.89 MPa) to 3000 psi (20.7 MPa) range. The hydraulic fluid is selected to have the following characteristics: good lubrication to prevent wear in moving components (e.g., between pistons and cylinders), corrosion resistance, and incompressibility to provide rapid response. Most hydraulic pumps act by **positive displacement,** which means they deliver a fixed volume of fluid with each cycle or rotation of the pump. The three main types of positive displacement pumps used in hydraulic systems are gear pumps, vane pumps, and piston pumps. An example of a **gear pump,** which displaces the fluid around a housing between teeth of meshing gears, is shown in Figure 10.35. Note that the meshing teeth provide a seal,

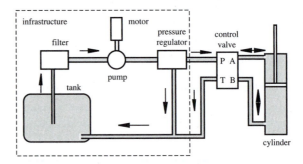

Figure 10.34 Hydraulic system components.

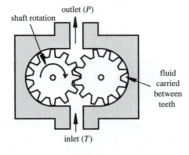

Figure 10.35 Gear pump.

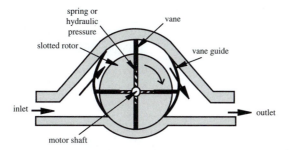

Figure 10.36 Vane pump.

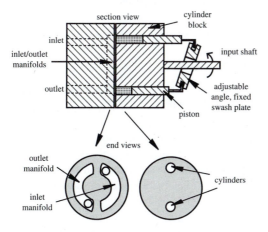

Figure 10.37 Swash plate piston pump.

and the fluid is displaced from the inlet to the outlet along the nonmeshing side of the gears. Video Demos 10.22 and 10.23 show and describe various types of gear pumps.

Figure 10.36 illustrates a **vane pump,** which displaces the fluid between vanes guided in rotor slots riding against the housing and vane guide. The vane guide supports the vanes from one side of the housing to the next and is constructed to allow the fluid to pass. The output displacement can be varied (with a constant motor speed) by moving the shaft vertically relative to the housing.

Figure 10.37 illustrates a **piston pump.** The cylinder block is rotated by the input shaft, and the piston ends are driven in and out as they ride in the fixed **swash plate** slot, which is angled with respect to the axis of the shaft. A piston draws fluid from an inlet manifold over half the swash plate and expels fluid into the outlet manifold during the other half. The displacement of the pump can be changed simply by changing the angle of the fixed swash plate. Table 10.5 lists and compares the general characteristics of the different pump types.

Since positive displacement hydraulic pumps provide a fixed volumetric flow rate, it is necessary to include a pressure relief valve, called a **pressure regulator,**

Video Demo

10.22 Gear pumps
10.23 Hydraulic gear pumps

Table 10.5 Comparison of pump characteristics

Pump type	Displacement	Typical pressure (psi)	Cost
Gear	Fixed	2000	Low
Vane	Variable	3000	Medium
Piston	Variable	6000	High

to prevent the pressure from exceeding design limits. The simplest pressure regulator is the spring-ball arrangement illustrated in Figure 10.38. When the pressure force exceeds the spring force, fluid is vented back to the tank, preventing a further increase in pressure. The threshold pressure, or **cracking pressure,** is usually adjusted by changing the spring's compressed length and therefore its resisting force.

10.8.1 Hydraulic Valves

There are two types of hydraulic valves: the **infinite position valve** that allows any position between open and closed to modulate flow or pressure, and the **finite position valve** that has discrete positions, usually just open and closed, each providing a different pressure and flow condition. Inlet and outlet connections to a valve are called **ports.** Finite position valves are commonly described by an x/y designation, where x is the number of ports and y is the number of positions. As an example, a 4/3 valve, with 4 ports and 3 positions, is illustrated in schematic form in Figure 10.39. In position 1, system pressure is vented to tank; in position 2, output port A is pressurized and port B is vented to tank; and in position 3, output port B is pressurized and port A is vented to tank. As illustrated in Figure 10.40, this particular valve is useful in controlling a double-acting hydraulic cylinder where ports A and B connect

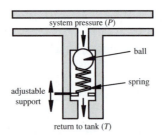

Figure 10.38 Pressure regulator.

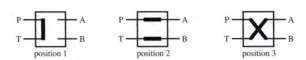

Figure 10.39 4/3 valve schematic.

to opposite ends of the cylinder, applying or venting pressure on opposite sides of the piston. In position 1, the cylinder does not move, because the pressure is vented to the tank. In position 2, the cylinder moves to the right, since pressure is applied to the left side of the piston. In position 3, the cylinder moves to the left, because pressure is applied to the right side of the piston.

Common types of fixed position valves are check valves, poppet valves, spool valves, and rotary valves. Figure 10.41 illustrates check and poppet valves. The **check valve** allows flow in one direction only. The **poppet valve** is a check valve that can be forced open to allow reverse flow.

As illustrated in Figure 10.42, a **spool valve** consists of a cylindrical spool with multiple lobes moving within a cylindrical casing containing multiple ports. The spool can be moved back and forth to align spaces between the spool lobes with input and output ports in the housing to direct high-pressure flow to different conduits in the system. The spool is balanced in each position because the static pressure is the same on opposing internal faces of the lobes. Therefore, no force is required to hold a position. In the left position, port A is pressurized, and port B is vented to the tank; and in the right position, port B is pressurized and port A is vented. To move the spool between positions, an axial force is required, from an actuator or manual control lever, to overcome the hydrodynamic forces associated with changing the momentum of the flow.

In the design of a spool valve where large hydrodynamic forces occur, a **pilot valve** is added, as shown in Figure 10.43. The pilot valve operates at a lower pressure, called **pilot pressure,** and at much lower flow rates and therefore requires less force to actuate. The pilot valve directs pilot pressure to one side of the main spool, and the force generated by the pressure acting over the main spool lobe face is large enough to actuate the main valve. The effect of the pilot valve is to amplify force provided by the solenoid or mechanical lever acting on the pilot spool. In the figure, the pilot spool is in the full left position, causing pilot pressure to be applied to the

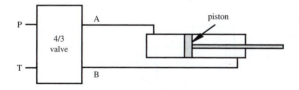

Figure 10.40 Double-acting hydraulic cylinder.

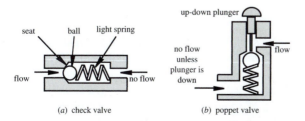

Figure 10.41 Check and poppet valves.

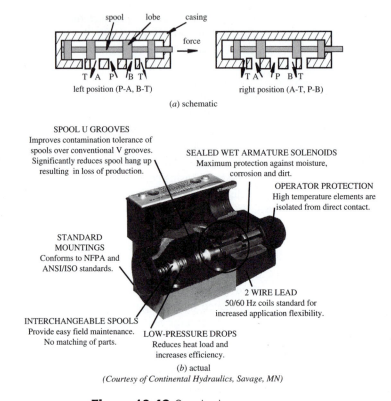

left position (P-A, B-T)   right position (A-T, P-B)

(*a*) schematic

SPOOL U GROOVES
Improves contamination tolerance of
spools over conventional V grooves.
Significantly reduces spool hang up
resulting in loss of production.

SEALED WET ARMATURE SOLENOIDS
Maximum protection against moisture,
corrosion and dirt.

OPERATOR PROTECTION
High temperature elements are
isolated from direct contact.

STANDARD
MOUNTINGS
Conforms to NFPA and
ANSI/ISO standards.

2 WIRE LEAD
50/60 Hz coils standard for
increased application flexibility.

INTERCHANGEABLE SPOOLS
Provide easy field maintenance.
No matching of parts.

LOW-PRESSURE DROPS
Reduces heat load and
increases efficiency.

(*b*) actual
(Courtesy of Continental Hydraulics, Savage, MN)

Figure 10.42 Spool valve.

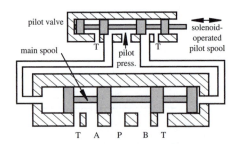

Figure 10.43 Pilot-operated spool valve.

Video Demo

10.24 Pilot valve
hydraulic amplifier
cut-away

left side of the main spool and venting fluid from the right side of the main spool to
tank, thus driving the main spool to the full right position. This applies main pressure
to port B and vents port A. Video Demo 10.24 illustrates how a pilot valve can be
used to create an amplified hydraulic force.

The discussion of spool valves so far has been limited to operation between
two positions only: on and off. Continuous operation can be achieved by using a
proportional valve, one whose spool moves a distance proportional to a mechanical

or electrical input (e.g., a lever or an adjustable-current solenoid), thus changing the rate of flow and varying the speed and force of the actuator. When the spool position is controlled by electrical solenoids, the proportional valve is called an **electrohydraulic valve.** These valves may be used in open-loop control situations with no feedback, but they often include sensors to monitor spool position or actuator output. Proportional valves equipped with sensor and control circuitry are often called **servo valves.** Electrohydraulic valves are often pilot operated where the solenoids drive the pilot spool, which in turn controls the position of the primary spool. The pilot spool can be driven by a single solenoid with a spring return or a set of opposing solenoids. The solenoid currents, and therefore displacements, can be controlled by amplifiers linked to analog or digital controllers.

10.8.2 Hydraulic Actuators

The most common hydraulic actuator is a simple **cylinder** with a piston driven by the pressurized fluid. As illustrated in Figure 10.44, a cylinder can be **single acting,** where it is driven to and held in one position by pressure and returned to the other position by a spring or by the weight of the load, or **double acting,** where pressure is used to drive the piston in both directions. As illustrated in Figure 10.45, the linear actuator can be very versatile in achieving a variety of motions. Cylinder motion in the hydraulic elevator drives the elevator directly. The scissor jack converts small linear motion in the horizontal direction to larger linear motion in the vertical direction. Linear motion of the cylinder in the crane results in rotary motion of its pivoted boom.

Rotary motion can also be achieved with hydraulic systems directly with a rotary actuator. One type of rotary actuator, called a **gear motor,** is simply a gear pump (see Figure 10.35) driven in reverse where pressure is applied, resulting in rotation of a shaft.

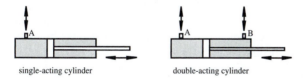

Figure 10.44 Single-acting and double-acting cylinders.

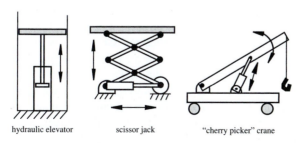

Figure 10.45 Example mechanisms driven by a hydraulic cylinder.

Internet Link

10.9 Hydraulic
Valves.org
(valve, pump,
motor, cylinder
manufactures and
information)

> ■ **CLASS DISCUSSION ITEM 10.8**
> *Force Generated by a Double-Acting Cylinder*
>
> For a given system pressure, is the force generated by a double-acting cylinder different depending on the direction of actuation? How is the force determined for each direction of motion?

Hydraulic systems have the advantage of generating extremely large forces from very compact actuators. They also can provide precise control at low speeds and have built-in travel limits defined by the cylinder stroke. The drawbacks of hydraulic systems include the need for a large infrastructure (high-pressure pump, tank, and distribution lines); potential for fluid leaks, which are undesirable in a clean environment; possible hazards associated with high pressures (e.g., a ruptured line); noisy operation; vibration; and maintenance requirements. Because of these disadvantages, electric motor drives are often the preferred choice. However, in large systems, which require extremely large forces, hydraulics often provide the only alternative. For more information, Internet Link 10.9 provides links to online resources and manufacturers of hydraulic components and systems.

10.9 PNEUMATICS

Pneumatic systems are similar to hydraulic systems, but they use compressed air as the working fluid rather than hydraulic liquid. The components in a pneumatic system are illustrated in Figure 10.46. A compressor is used to provide pressurized air, usually on the order of 70 to 150 psi (482 kPA to 1.03 MPa), which is much lower than hydraulic system pressures. As a result of the lower operating pressures, pneumatic actuators generate much lower forces than hydraulic actuators.

After the inlet air is compressed, excess moisture and heat are removed from the air with an air treatment unit (see Figure 10.46). Unlike hydraulic pumps, which provide positive displacement of fluid at high pressure on demand, compressors cannot provide high volume of pressurized air responsively; therefore, a large volume of high-pressure compressed air is stored in a reservoir or tank. The working pressure delivered to the system can be controlled by a pressure regulator to be much lower than the reservoir pressure. The reservoir is equipped with a pressure-sensitive switch that activates the compressor when the pressure starts to fall below the desired level. Control valves and actuators act in much the same way as in hydraulic systems, but instead of returning fluid to a tank, the air is simply returned (exhausted) to the atmosphere. Pneumatic systems are open systems, always processing new air, and hydraulic systems are closed systems, always recirculating the same oil. This eliminates the need for a network of return lines in pneumatic systems. Another advantage of pneumatic systems is that air is "cleaner" than oil, although air does not have the self-lubricating features of hydraulic oil.

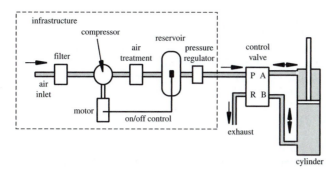

Figure 10.46 Pneumatic system components.

When sources of compressed air are readily available, as they often are in engineering-related facilities, pneumatic actuators may be a good choice. Double-acting or single-acting pneumatic cylinders are ideal for providing low-force linear motion between two well-defined endpoints. Since air is compressible, pneumatic cylinders are not typically used for applications requiring accurate motion between the endpoints, especially in the presence of a varying load. Video Demo 10.25 demonstrates various types of pneumatic cylinders, and Video Demo 10.26 shows an interesting example of an apparatus driven by a pneumatic system.

Another advantage of pneumatic systems is the possibility of replacing the infrastructure with a high-pressure storage tank and regulator. The tank serves a function analogous to a battery in an electrical system. This makes possible light, mobile pneumatic systems (e.g., a pneumatically actuated walking robot). In these applications, the capacity of the tank limits the range or operating time of the system.

Video Demo

10.25 Pneumatic cylinders of various types and sizes

10.26 Pneumatic biomechanics exercise apparatus overview

QUESTIONS AND EXERCISES

Section 10.3   Solenoids and Relays

10.1. A solenoid can be modeled as an inductor in series with a resistor. Design a circuit to use a digital output to control a 24 V solenoid.

Section 10.5   DC Motors

10.2. Why can the presence of electric motors or solenoids affect the function of nearby electronic circuits?

10.3. If a manufacturer's specifications for a PM DC motor are as follows, what are the motor's no-load speed, stall current, starting torque, and maximum power for an applied voltage of 10 V?

- ■ Torque constant = 0.12 Nm/A
- ■ Electrical constant = 12 V/1000 RPM
- ■ Armature resistance = 1.5 Ω

10.4. Recognizing that the H-bridge IC in Design Example 10.1 includes a current sense output, draw the block diagram for torque control of a brushed DC motor using the IC discussed. (Hint: The current sense output pin from the H-bridge yields 377 mA of current for each amp that is output to the motor. Convert this output to a voltage and use it as the input to a PWM chip like the LM3524D. Include a torque adjust potentiometer on the LM3524D.)

Section 10.6   Stepper Motors

10.5. For the full-step drive circuit in Figure 10.29, complete the following timing diagram by adding signals B_0, B_1, ϕ_1, ϕ_2, ϕ_3, and ϕ_4. Using Table 10.1, verify that the intended motion results.

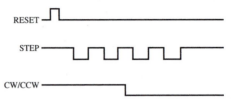

10.6. Document a complete and thorough answer to Class Discussion Item 10.5.

10.7. A unipolar stepper motor full-step drive circuit is available as a single component from some manufacturers. Explore the Internet to find a stepper motor driver, show a labeled pin-out diagram, and indicate how to connect it to a stepper motor properly.

10.8. A designer wishes to use a stepper motor coupled to a gearbox to drive an indexed conveyor belt to achieve a linear resolution of 1 mm and a maximum speed of 10 cm/sec. The gearbox is a speed reducer with a gear ratio of 3 to 1, and the conveyor is driven by a 10 cm drum attached to the output shaft of the gearbox. What is the minimum resolution required for the stepper motor? Also, what step rate would be required to achieve the maximum speed at this resolution?

Section 10.7   Selecting a Motor

10.9. For each of the following applications, what is a good choice for the type of electric motor used? Justify your choice.

   a.   robot arm joint

   b.   ceiling fan

   c.   electric trolley

   d.   circular saw

   e.   NC milling machine

   f.   electric crane

   g.   disk drive head actuator

   h.   disk drive motor

   i.   windshield wiper motor

   j.   industrial conveyor motor

   k.   washing machine

   l.   clothes dryer

10.10. A permanent magnet DC motor is coupled to a load through a gearbox. If the polar moments of inertia of the rotor and load are J_r and J_l, the gearbox has a $N{:}M$ reduction (where $N > M$) from the motor to the load, the motor has a starting torque T_s and a no-load speed ω_{max}, and the load torque is proportional to its speed ($T_l = k\omega$),

    a. What is the maximum acceleration that the motor can produce in the load?

    b. What is the steady state speed of the motor and the load?

    c. How long will it take for the system to reach 95% of its steady state speed?

Section 10.8   Hydraulics

10.11. Document a complete and thorough answer to Class Discussion Item 10.8.

10.12. What is the maximum force a 1 inch inner diameter single acting hydraulic cylinder can generate with a fluid pressure of 1000 psi?

10.13. If a machine requires that a 1 cm inner diameter single-acting hydraulic cylinder produce 2000 N of force, what is the minimum required system pressure in MPa?

Section 10.9   Pneumatics

10.14. You are designing a pneumatic system that requires use of a pneumatic valve. On the Internet, find the specifications for a pneumatic valve capable of handling 100 psi, and draw a schematic showing how you would interface it to a digital system.

10.15. You are designing a pneumatic actuator using a dual-acting cylinder to produce 100 lbf using a 2000-psi scuba tank as a pressure source. Specify the elements required in the system in block diagram form. Be as specific as possible about component specifications.

BIBLIOGRAPHY

Kenjo, T., *Electric Motors and Their Controls,* Oxford Science Publications, Oxford, England, 1994.

Khol, R., editor, "Electrical & Electronics Reference Issue," *Machine Design,* v. 57, n. 12, May 30, 1985.

McPherson, G., *An Introduction to Electrical Machines and Transformers,* John Wiley, New York, 1981.

National Semiconductor Corporation, "National Power ICs Databook," 2900 Semiconductor Drive, P.O. Box 58090, Santa Clara, CA 95052.

Norton, R. L., *Design of Machinery,* 3rd ed., McGraw-Hill, New York, 2003.

Shultz, G. P., *Transformers and Motors,* Macmillan, Carmel, IN, 1989.

Westbrook, M. H. and Turner, J. D., *Automotive Sensors,* Institute of Physics Publishing, Philadelphia, 1994.

Williamson, L., "What You Always Wanted to Know About Solenoids," *Hydraulics and Pneumatics,* September, 1980.

11 CHAPTER

Mechatronic Systems—Control Architectures and Case Studies

This chapter provides a summary of mechatronic system digital control architectures, introduces the concepts of PID control, and presents three case studies of mechatronic system design. ■

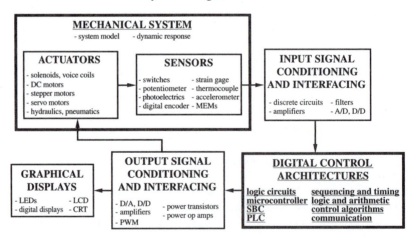

CHAPTER OBJECTIVES

After you read, discuss, study, and apply ideas in this chapter, you will:

1. Know why many engineering designs today can be classified as mechatronic systems

2. Be aware of the various types of control architectures possible in mechatronic system design

3. Understand the basic principles of feedback control systems

4. Appreciate the variety of designs that can result from limited specifications of a mechatronic system

5. Understand the basics of programmable logic controller (PLC) programming using ladder logic

6. Recognize that many aerospace, automotive, and consumer products are mechatronic systems

11.1 INTRODUCTION

In the previous chapters of this book, we developed the foundations for the integration of mechanical devices, sensors, and signal and power electronics into mechatronic systems. As we progressed through the chapters, we included increasingly challenging mechatronic design examples to help you connect the theory and analysis with actual applications to expand your knowledge and design skills. To obtain completeness in the integration of mechanical devices, sensors, and signal and power electronics into the most advanced mechatronic systems, we must include microprocessor-based control systems. Chapter 7 presented the basics of microcontroller programming and interfacing. Here, we briefly introduce other control architectures also useful in mechatronic system design. Then we present three case studies of complete mechatronic systems: a myoelectrically controlled robotic arm, an articulated walking machine for movement in unusual environments, and a coin counter. These projects have been used as mechatronic projects in coursework at Colorado State University. To provide a final perspective, we list examples of mechatronic systems in many industries.

11.2 CONTROL ARCHITECTURES

Many mechatronic systems have multiple inputs and outputs related by deterministic relationships that result in some form of control of the outputs. A designer can choose from a wide spectrum of control architectures, ranging from simple open-loop control to complex feedback control. Implementation of the control can be as simple as using a single op amp or as complicated as programming massively parallel microprocessors. Here, we describe a hierarchy of basic control approaches you may consider in the design of a mechatronic system.

11.2.1 Analog Circuits

Many simple mechatronic designs require a specific actuator output based on an analog input signal. In some cases, analog signal processing circuits consisting of op amps and transistors can be employed to effect the desired control. Op amps can be used to perform comparisons and mathematical operations such as analog addition, subtraction, integration, and differentiation. They can also be used in amplifiers for linear control of actuators. Analog controllers are often simple to design and easy to implement and can be less expensive than microprocessor-based systems.

11.2.2 Digital Circuits

If the input signals are digital or can be converted to a finite set of states, then combinational or sequential logic controllers may be easy to implement in mechatronic design. The simplest designs use a few digital chips to create a digital controller. To generate complex Boolean functions on a single IC, specialized digital devices such as **programmable array logic (PAL)** controllers and **programmable logic arrays (PLAs)** can be used to reduce design complexity. PALs and PLAs contain many gates and a gridwork of conductors that can be custom connected using a programming device. Once programmed, the ICs implement the designed Boolean function between the inputs and outputs. PALs and PLAs may offer an alternative to complex sequential and combinational logic circuits that require many ICs.

Video Demo

11.1 Table tennis assistant controlled by an FPGA

Another type of programmable logic-gate-based device is the **field-programmable gate array (FPGA)**. Like PALs and PLAs, an FPGA contains a large number of reconfigurable gates that can be programmed to create a wide range of logic functions. FPGAs are different from PALs and PLAs because they also can include memory, I/O ports, arithmetic functions, and other functionality found in microcontrollers. Furthermore, FPGAs are usually programmed with a high-level software language (e.g., VHDL) that allows for fairly sophisticated functionality. Video Demo 11.1 shows an example of an FPGA development system used to control a simple device.

Sometimes, it may be economically feasible to design an **application-specific integrated circuit (ASIC)** that provides unique functionality on a single IC. Logic functions, memory, computation, signal processing, and other digital and analog features can be custom built onto a single ASIC. Design and setup for manufacturing can be expensive, but in high-volume manufacturing applications, an ASIC solution can be cheaper in the long run. ASICs are also attractive because the integrated solution will usually be smaller in size and consume less power.

11.2.3 Programmable Logic Controller

Programmable logic controllers (PLCs) are specialized industrial devices for interfacing to and controlling analog and digital devices. They are designed with a small instruction set suitable for industrial control applications. They are usually programmed with **ladder logic,** which is a graphical method of laying out the connectivity and logic between system inputs and outputs. PLCs are designed with industrial control and industrial environments specifically in mind. Therefore, in addition to being flexible and easy to program, they are robust and relatively immune to external interference.

The symbols, notation, and basic constructs used to define and create ladder logic diagrams are shown in Figure 11.1. Programming a PLC is a simple matter of constructing a diagram like this by interactively dragging and dropping components in a graphical user interface provided by the manufacturers. Referring to the figure,

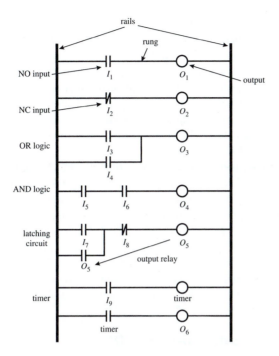

Figure 11.1 Ladder logic symbols and basic functions.

the **rails** on the ladder represent power (on the left) and ground (on the right), and the **rungs** represent current-flow paths (from left to right) to output devices. An **input** is typically a switch or proximity sensor that is represented by a **normally open (NO)** or **normally closed (NC)** relay symbol. Inputs can be arranged in series and parallel configurations in a rung to represent various logic functions. An **output** is included on the right side of a rung, and it is energized when the input relays on the left side of the rung result in a closed circuit. An output can be a motor, heater, lamp, solenoid, or other controlled device. An output state can also be used as an input relay to control other rungs in the ladder. Special features such as timers and counters are also available to create sequencing, stopping, and restarting of a series of events.

In Figure 11.1 output O_1 is energized only while No input I_1 is closed. I_2 is an NC input, and output O_2 is off when I_2 is open. Output O_3 is energized only while NO input I_3 is closed *or* NO input I_4 is closed. Output O_4 is energized only while NO input I_5 is closed *and* NO input I_6 is closed. The O_5 rung is called a **latching circuit.** It is triggered (i.e., O_5 is energized) when NO input I_7 is closed. Also, once triggered, the output remains energized, even if input I_7 returns to the open state, until NC input I_8 opens. This is accomplished by using the output signal as an additional input (called an **output relay**) OR'd with the trigger input I_7. The term *latch* is used because the output's ON state is maintained (stored) until it is reset by a second input (I_8). The last two rungs show how a timer is used. When NO input I_9 closes, a timer is started. When the timer reaches a preprogrammed period, the timer input closes, energizing output O_6.

Figure 11.2 shows an example ladder logic diagram for a common assembly-line control system where a motor or double-acting pneumatic or hydraulic cylinder is actuated back and forth based on a trigger event (e.g., a widget being detected by a proximity sensor next to a conveyor belt). Limit switches or sensors detect the desired ends of travel of the actuator in each direction. Here is the sequence of events: when the NO "trigger" input closes, the actuator is driven in the forward direction. The latching circuit maintains this forward motion until the NC "end limit detect" signal changes state. This opens the NC "end limit detect" input in the first rung and closes the NO "end limit detect" input in the second rung, which starts the timer. When the timer reaches its preset time, the third rung is triggered and the actuator reverses until the NC "start limit detect" signal is detected. In summary, when triggered, the actuator moves forward until the end of travel is detected, then the system pauses before returning the actuator back to the start position.

Video Demo 11.2 shows an example PLC development system and demonstrates how ladder logic can be used to create a fluid-level control system, and Video Demo 11.3 shows an interesting student project example using a PLC. Internet Link 11.1 provides links to various online resources for and manufacturers of PLC products, and Internet Link 11.2 provides an excellent set of tutorials on various PLC topics.

11.2.4 Microcontrollers and DSPs

The **microcontroller,** which is a microcomputer on a single IC, provides a small, flexible control platform that can be easily embedded in a mechatronic system. The microcontroller can be programmed to perform a wide range of control tasks. Designing with microcontrollers usually requires knowledge of a high-level programming language (e.g., C or Basic) or assembly language and experience in interfacing digital and analog devices. See Chapter 7 for background information and examples. Internet Link 7.1 provides links to various online resources for and manufacturers of microcontrollers.

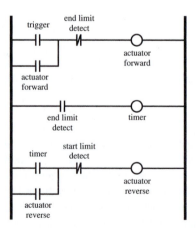

Figure 11.2 Ladder logic motor control cycle example.

Another type of single-IC microcomputer device is a **digital signal processor** (**DSP**). A DSP can have functionality similar to that of a microcontroller, but DSPs are usually better suited to high-speed floating point calculations. DSPs are useful in communication, audio/video, and control applications where fast calculation of digital filters and weighted sums is important for fast cycle times. Internet Link 11.3 provides links to various DSP online resources and manufacutrers.

Internet Link

11.3 Digital signal processor resources and vendors

11.2.5 Single-Board Computer

When an application requires more features or resources than can be found on a typical microcontroller and size is not a major concern, a **single-board computer** offers a good alternative. Most single-board computers have enough RAM and offer compilers to support programming in a high-level language such as C. Single-board computers are also easily interfaced to a personal computer. This is useful in the testing and debugging stages of design development and for downloading software into the single-board computer's memory.

We use the term **minicontroller** to refer to another class of device that falls between a microcontroller and a single-board computer. Examples are the Handyboard, Basic Stamp, and Arduino. These boards contain microcontrollers and other peripheral components that make it easier to interface to external components. Internet Link 11.4 provides links to online resources for single-board computer and mini-controller products.

Internet Link

11.4 Single-board computer and minicontroller online resources and vendors

11.2.6 Personal Computer

In the case of large sophisticated mechatronic systems, a desktop or laptop **personal computer (PC)** may serve as an appropriate control platform. Also, for those not experienced with microcontrollers and single-board computers, the PC may be an attractive alternative. The PC can be easily interfaced to sensors and actuators using commercially available plug-in data acquisition cards or modules (see Section 8.3). These devices typically include software drivers that enable programming with standard high-level language compilers and development environments. Due to the ease and convenience of this approach, PC-controlled mechatronic systems are especially common in R&D testing and product development laboratories, where fast prototyping is required but where large-quantity production and miniaturization are not concerns.

Video Demo

10.3 Computer hard-drive track seeking demonstration

10.4 Computer hard-drive super-slow-motion video of track finding

11.3 INTRODUCTION TO CONTROL THEORY

In mechatronic system design, we often need to accurately control an output (e.g., the position or speed of a motor's shaft) with fast response. An impressive example of position control is shown in Video Demo 10.3, where the read-write head of a computer hard drive is controlled with amazing speed and accuracy (see Video Demo 10.4 for a super-slow-motion playback of the motion). For accurate control, we need to use feedback from sensors (e.g., an encoder or a tachometer). By subtracting a feedback signal from a desired input signal (called the **set point** value), we have a measure of the error in the response. By continually changing the command signal to the system based on the error signal, we can improve the response of the system. This is called **feedback** or **closed-loop control.** The goal of this section is to introduce the basics of feedback control system design.

The theory and practice of control system design can be very complicated and may involve some mathematical techniques and software tools with which you might not be familiar. However, it is important for the reader to understand the approach to control system design, first to appreciate the value of its application and second to develop a desire to learn more about the field of controls (e.g., via follow-on coursework). In the subsections that follow, we explore the concepts of feedback control through a basic but important example of controlling the speed of a DC motor.

Before continuing, you may want to view Video Demos 11.4 and 11.5. They demonstrate two laboratory experiments that illustrate various control system topics. These examples might help you better relate to the material presented in this section. Video Demo 11.3 does a particularly good job of explaining some of the basic principles of proportional-integral-derivative (PID) control, which forms the basis of many control systems.

Video Demo

11.4 PID control of the step response of a mechanical system

11.5 Inverted pendulum uprighting and balancing with linear cart motion

11.3.1 Armature-Controlled DC Motor

An important electromechanical device that is incorporated into many mechatronic systems is a permanent magnet or field-controlled DC motor. In mechatronic system applications, we might need to carefully control the output position, speed, or torque of the motor to match design specifications. This is a good example where feedback control is required. Our first problem, as in every mechatronic control system, is to model the motor. The model of a physical system is often referred to as the **plant.** Then we apply linear feedback analysis to the problem to help in selecting the design parameters for a controller that will reasonably track the desired output (specified input). We must create a good mathematical model of the motor (the system model) before we continue with controller design.

The basic equations governing the response a DC motor were presented in Section 10.5, but we will take the analysis further here. A DC motor has an armature with inductance L and resistance R, such that when it rotates in a magnetic field, the armature will produce an output torque and angular velocity. Due to the motor construction, a back emf (voltage) proportional to angular velocity is caused by the armature coils moving through the stator magnetic fields. Figure 11.3 illustrates the salient elements in the system. The shaft of the armature transmits torque T to a load of inertia I_L and damping c. The motor armature also has inertia, and we will call the total moment of inertia of the armature and load l.

The model for the armature-controlled DC motor assumes that the motor produces a torque T proportional to the armature current i_a:

$$T(t) = k_t i_a(t) \tag{11.1}$$

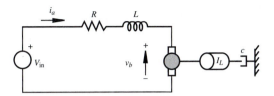

Figure 11.3 Armature-controlled DC motor.

and the back emf or voltage is proportional to the motor speed, ω:

$$v_b(t) = k_e \omega(t) \tag{11.2}$$

where k_t and k_e are parameters resulting from the motor's design that are provided by the motor manufacturer. If we draw a freebody diagram for the mechanical part of the system and apply Kirchhoff's voltage law to the electrical part, we can write two coupled differential equations that define the motor's electrical and mechanical properties:

$$v(t) \;=\; i_a R + L\frac{\mathrm{d}i_a}{\mathrm{d}t} + k_e \omega \tag{11.3}$$

$$I\frac{\mathrm{d}\omega}{\mathrm{d}t} \;=\; T - c\omega \;=\; k_t i_a - c\omega \tag{11.4}$$

The coupling occurs through i_a and ω.

In advanced engineering systems analysis, Laplace transforms are commonly used to simplify the mathematics and help in the interpretation of the results. A mathematical transform converts an equation into a different form that may be easier to handle. The significant advantage of the Laplace transform is that an ordinary differential equation can be transformed into an algebraic equation. As was described in step 1 of the procedure in Section 4.10.2, the variables become functions of the complex variable s instead of time, and derivatives become powers of s. If we take the Laplace transform of the two differential equations 11.3 and 11.4, this converts them into algebraic expressions in the s domain, and we get

$$V(s) = (Ls + R)I_a(s) + k_e\Omega(s) \tag{11.5}$$

$$Is\Omega(s) = k_t I_a(s) - c\Omega(s) \tag{11.6}$$

Capital letters are used to indicate the Laplace transform of the appropriate time-domain function (e.g., $\Omega(s)$ is the Laplace transform of $\omega(t)$).

We can use the Laplace transform form of the system equations (Equations 11.5 and 11.6) to draw a **block diagram** for the motor that illustrates the signal flow in the system (see Figure 11.4). The terms in the equations are shown in the diagram so you can see how the diagram relates to the equations. As an aside, notice that the back emf (the $k_e\Omega(s)$ term) actually occurs in a negative feedback loop, which tends to stabilize the speed of the motor, although we are not explicitly designing this into the system.

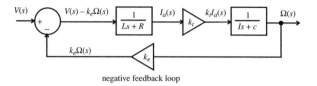

Figure 11.4 Block diagram for the DC motor system (plant).

An important step in controller design is expressing the input-output relationship of the plant, known as the **transfer function** of the system. Using Equations 11.5 and 11.6, we obtain

$$\frac{\Omega(s)}{V(s)} = \frac{k_t}{(Is + c)(Ls + R) + k_e k_t} = G(s) \tag{11.7}$$

This transfer function $G(s)$ relates the motor output speed $\Omega(s)$ to the input voltage $V(s)$. The polynomial in s in the denominator is known as the **characteristic equation.** The order of this polynomial predicts system behavior. Here it is second order, so your knowledge of second-order response will be helpful in understanding the angular velocity response to a step change in input voltage.

11.3.2 Open-Loop Response

In general, the transfer function can tell us a great deal about the response of the system. s is a complex variable, and the transfer function will have **poles** (values of s that make the denominator polynomial zero) that have extremely important consequences in interpreting the response of the DC motor. They have a profound effect on the motor stability, that is, whether any disturbance or input will cause unrestrained growth of the output. In fact, examining the location of the poles in the complex plane tells us whether the system is stable. They also determine whether or not the motor output oscillates when the input voltage command changes. This is where the software tool **Matlab** becomes a valuable asset in the analysis. We can use Matlab to find the poles easily and to determine the responses to various inputs, particularly the step response of the motor. The coefficients of the characteristic equation are determined by the motor parameters only, and they are of importance in determining the system response. We saw this in Chapter 4 where we distinguished among zero-, first-, second-, and higher order systems and identified the characteristics of their responses. Therefore, knowing the order of the system, we can predict the pattern of the response.

Now let us calculate the response of the armature-controlled DC motor as an example. This is very easy to do by creating a **Simulink** model in Matlab, setting the parameters, applying a step input, and plotting the0 response. Excellent tutorials for learning how to use and apply Matlab and Simulink can be found at Internet Links 11.5 and 11.6. In the block diagram model shown in Figure 11.5, the electrical and mechanical transfer functions have been separated. A unit step input of voltage is applied, and the output angular velocity is plotted, as shown in Figure 11.6.

Internet Link

11.5 Matlab and Simulink tutorials and learning resources

11.6 Control tutorials for Matlab and Simulink

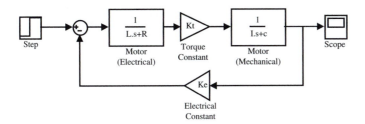

Figure 11.5 Simulink model block diagram.

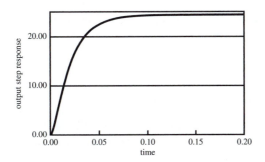

Figure 11.6 The angular velocity output for a step input voltage.

The output angular velocity takes time to respond to the step input and asymptotically reaches a final value due to the second-order nature of the characteristic equation. This overdamped response is what we call the open-loop response of the motor because no feedback has been added to the system (except for the existing internal back emf). It is our objective with a feedback controller to alter the response so that it meets design specifications such as faster rise time, some limited overshoot, and no oscillation.

11.3.3 Feedback Control of a DC Motor

In order to control the motor more effectively, we must add a sensor to monitor the output (in this case, a tachometer to measure speed), feed back the output and compare it to the desired speed input, and apply a control algorithm to the resulting error signal to improve the output response. Figure 11.7 shows a feedback control system in block diagram form. The input is the desired motor speed, and the output is the actual motor speed (e.g., as measured by a tachometer). The error signal is the difference between these signals:

$$\text{error signal} = \text{input} - \text{output} \tag{11.8}$$

The controller changes its command signal (e.g., the motor voltage) to the system or plant (e.g., the motor) in response to the error signal.

There is some art and some engineering involved in designing a controller for a mechatronics system. The general steps for designing a controller include:

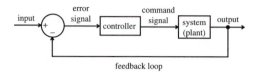

Figure 11.7 General feedback control system.

1. Choose a control action. We will describe that process in the example; however, generally, you start with a proportional control action, where the adjustment is made to the input in proportion to the size of the error signal. Then, integral and differential control actions, which are in proportion to the integral and derivative of the error signal, can be added to help better achieve the desired output specifications.

2. Check to see that the closed-loop system is stable by identifying the pole locations on the complex *s*-plane. Any poles on the right-hand side (with a positive real part) represent system instability.

3. Make sure the steady-state response meets design specifications.

4. Make sure the transient response meets design specifications.

5. Check the overall performance of the mechatronic system through testing with various inputs.

General controllers can take many forms but a majority of industrial applications use **PID** or proportional-integral-derivative controllers. The mathematical form of a PID controller, where the error signal is expressed as $e(t)$, is as follows:

$$\text{command signal} = K_p e(t) + K_d \frac{\mathrm{d}}{\mathrm{d}t} e(t) + K_i \int e(t)\mathrm{d}t \tag{11.9}$$

where K_p is referred to as the proportional gain, K_d is the derivative gain, and K_i is the integral gain. Proportional control is the most intuitive because the control signal is proportional to the error. The larger the error is, the larger the corrective action will be. A large proportional gain creates a fast response, but it can lead to overshoot and oscillation, especially if the system has little damping. Derivative gain responds to the rate of change of the error signal. This allows the controller to anticipate changes in response of the system, which can result in less overshoot and damped oscillation. Integral gain helps eliminate steady-state error by summing up errors over time. The longer the error stays on one side of the desired set point, the larger the corrective action becomes as a result of integral gain. Video Demo 11.4 describes and demonstrates the effect of each of the three control actions, as applied to position control of a DC servo motor system.

Video Demo

11.4 PID control of the step response of a mechanical system

The choice of the gain parameters in a PID controller is a significant part of the design, and there are various analytical and empirical methods that can help one choose the gains. The control design can be specified in terms of important criteria such as settling time, overshoot, steady-state error, and rise time (see Figure 4.18 in Chapter 4 for more information). Please refer to a controls book (e.g., Ogata or

Palm in the bibliography) if you want to explore some of the analytical methods for controller design.

An alternative to model-based analytical design is to interactively adjust the PID gains via simulation. We will illustrate this method with the example of speed control of an armature-controlled DC motor. The complete Simulink model, including the motor model along with a PID feedback controller, is shown in Figure 11.8. The input is a step change in speed. The proportional, derivative, and integral gains of the PID controller are easily adjusted in the Simulink software, and the resulting output step response can be calculated and displayed almost instantaneously. This allows the user to test various gain combinations quickly during design iterations.

A good approach for varying the PID gains during design iterations in simulation is to first gradually increase the proportional gain (K_p) until the rise time is fast enough without too much overshoot and oscillation. Figure 11.9 shows the resulting step responses for various proportional gain values. With low gains, the response is slow, and there is significant steady-state error (i.e., the final value is not very close to the desired final speed of 1.0). With a large proportional gain, the response is fast, but there is significant overshoot and oscillation in the response.

After adjusting the proportional gain, derivative gain can be added to limit the overshoot and oscillation. Sometimes the proportional gain will need to be decreased as more derivative gain is added to limit the overall gain of the system. Figure 11.10 shows the resulting step responses for various derivative gain values. With a low

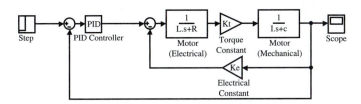

Figure 11.8 Simulink model of an example motor with a PID controller.

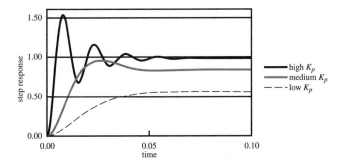

Figure 11.9 The effects of various proportional gains.

gain, the overshoot and oscillation are still present. With very high derivative gain, there is no overshoot or oscillation, but the response is very sluggish.

For most systems, with just proportional and derivative gains, the output response will exhibit a steady-state error in the step response (i.e., the output doesn't quite reach the desired level). For the motor speed control example (see Figure 11.10), the reason for this is clear. The command signal to the motor (see Figure 11.7) must be nonzero to maintain a constant speed in steady state. Because the command signal is a function of the error signal, with just proportional control (note that the derivate term will be zero in steady state) there must be a nonzero error signal for the controller to output a command signal to the motor to keep it turning. Integral gain allows the controller to overcome this limitation and drive the error to zero. The integral term sums the error over time and adds a larger correction to the command signal the longer an error persists. Figure 11.11 shows the resulting step responses for various integral gain values. For a high gain, the error dissipates very quickly, but overshoot will be larger than if there was no integral gain. With a low integral gain, the error diminishes slowly, but it does eventually dissipate to zero. The medium gains curve exhibits a very good step response. There is very little overshoot and almost no oscillation, and the output settles quickly with no steady-state error.

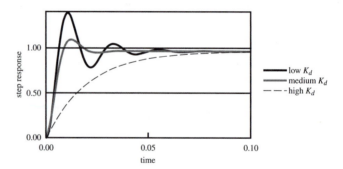

Figure 11.10 The effects of various derivative gains.

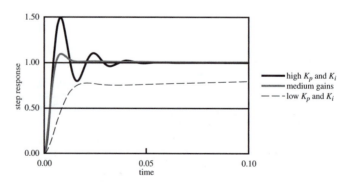

Figure 11.11 The effects of various integral gains.

11.3.4 Controller Empirical Design

Video Demo

8.8 National Instruments DC motor demo

8.9 National Instruments LabVIEW DC motor controller design

In cases where it is difficult or impossible to model a system analytically, there are techniques to empirically design a controller by performing tests on an actual system. The procedure described in the previous section for simulation is one such approach, where the gains are adjusted iteratively. There are also software tools available that can perform system identification automatically, where a model can be approximated by analyzing the system's response to various inputs. A controller can then be designed, possibly with the help of other software tools. Video Demo 8.8 shows a demonstration of how an example set of software tools can be used to develop a speed controller for a simple DC motor system. Video Demo 8.9 provides much more background and detailed demonstrations of the individual steps in the process.

A simple empirical method for tuning a PID controller that is sometimes used in industry is called the Ziegler-Nichols (Z-N) method (see Palm in the bibliography). The Z-N method is applied by observing the step response of the system under controlled conditions. From the observations, PID gains can then be selected to provide a fast system response with minimal overshoot and oscillation.

A PID controller expressed in the s domain can be written as

$$G_{controller}(s) = K_p + \frac{K_i}{s} + K_d s \qquad (11.10)$$

where K_p, K_i, and K_d are the proportional, integral, and derivative gains. Using the Ziegler-Nichols method, the controller is usually expressed as

$$G_{controller}(s) = K_p(1 + \frac{1}{T_i s} + T_d s) \qquad (11.11)$$

where T_i is called the reset time and T_d is the derivative time.

To apply the Z-N method, start by using proportional control action alone ($T_i = $ infinity, $T_d = 0$) and increase K_p from zero until the observed output shows sustained (undamped) oscillation. The resulting period of the oscillations is observed and labeled P_{cr}, and the corresponding gain value is called the critical gain K_{cr}. Then K_p is reduced by a factor, and K_i (T_i) and K_d (T_d) are selected based on the proportions in Table 11.1. Ziegler and Nichols showed that these proportions result in a good system response for the selected type of controller. Note that K_p is lower for a PI controller than a P controller and higher for a PID controller because the I component increases the order of the system and may destabilize it, and the D component tends to stabilize the system, so K_p can be increased a little.

Table 11.1 Ziegler-Nichols recommended gains

Controller	K_p	T_i	T_d
P	$0.5\,K_{cr}$	Infinity	0
PI	$0.45\,K_{cr}$	$P_{cr}/1.2$	0
PID	$0.6\,K_{cr}$	$P_{cr}/2$	$0.125\,P_{cr}$

Often, the Z-N gains will serve only as a starting design, and you may need to tune the gains (i.e., tweak the K_p, K_i, and K_d parameters) to achieve your desired design specifications. This usually requires some trial and error.

11.3.5 Controller Implementation

In previous sections, everything was done in simulation, where the physical system was represented by a mathematical model. To implement a PID controller in an actual physical system, the model is replaced by actual hardware, and the controller must be built as an analog circuit or with a microprocessor-based system running digital software. In Chapter 5, we learned how to construct proportional gain, integrator, differentiator, summer, and difference circuits using operational amplifiers. These circuits can serve as the building blocks for an analog controller. Figure 11.12 shows how the various circuits can be combined, in schematic form, to create an analog PID controller. Each control action has a gain (K_p, K_i, or K_d) created by appropriate choices of component values in the op amp circuits.

An alternative to an op-amp-based analog controller is a digital controller created with software in a microprocessor-based system (e.g., a microcontroller). A digital control system is different from an analog controller because it requires a discrete amount of time to perform control updates. During each update cycle, the sensor signal is acquired, the controller output is calculated, and the controller signal is output. The time delay corresponding to the control loop cycle affects the response of the system. This effect must be accounted for in the mathematical model of the system to be able to predict system response accurately and to choose control parameters intelligently. The concept of the z-transform, where the continuous s-domain is transformed into a discrete representation, enables us to model and analyze such systems. Please refer to a book on modern control theory (e.g., Ogata or Palm in the bibliography at the end of this chapter) if you want to pursue this topic.

To implement a digital controller, the differentiation and integration must be discretized. If successive error signal samples are referred to as e_1, e_2, e_3, . . ., e_{i-1}, e_i, e_{i+1}, . . ., we can accumulate an approximation to the integral with the following equation:

$$I_i = I_{i-1} + \Delta t \, e_i \tag{11.12}$$

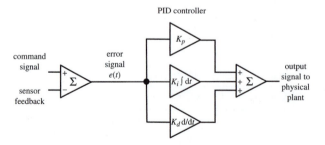

Figure 11.12 Analog PID controller constructed from op amp circuits.

where $I_0 = 0$ and Δt is the cycle time of the control loop. The derivative can be approximated with a finite difference approximation. For example,

$$D_i = (e_i - e_{i-1}) / \Delta t \qquad (11.13)$$

Although, a digital filter usually needs to be applied to this calculation to minimize the undesirable effects of high-frequency noise in the position signal (see Class Discussion Item 11.1). Code for an example control loop cycle might look like the following:

```
error_previous = 0
integral = 0

loop:
    Gosub get_set_point_value    ' acquire set_point value
    Gosub acquire_sensor_        ' acquire sensor_value

    error = sensor - set_point
    integral = integral + error*DT
    derivative = (error - error_previous)/DT

    output = KP*error + KI*integral + KD*derivative
    Gosub send_output_to_system ' update command signal

    error_previous = error
Goto loop
```

Video Demo

8.8 National Instruments DC motor demo

8.9 National Instruments LabVIEW DC motor controller design

■ **CLASS DISCUSSION ITEM 11.1**
Derivative Filtering

As mentioned below Equation 11.13, derivative calculations often need to be filtered. One approach to doing this is to average a set of previous derivative calculations (i.e., use a running average). What is the effect of such an approach, and how can the example code be modified to achieve this?

11.3.6 Conclusion

This section presented a very brief overview of control theory. Although this topic cannot really be covered adequately in such a small amount of space, you at least now have a basic understanding of the main concepts. Anyone interested in pursuing this topic further needs references (and coursework) completely dedicated to this topic.

There are many software tools available to help with modeling, analysis, and controller design. Above, we used Matlab and Simulink for simulation and design. The LabVIEW software introduced in Chapter 8 also provides tools to help with these tasks. Video Demos 8.8 and 8.9 demonstrate the whole process of model-based controller design with LabVIEW. Many additional control systems video demonstrations can be found at Internet Link 11.7. Please review some or all of these videos to help improve your level of understanding of the application of control theory.

Internet Link

11.7 Control system demonstrations

11.4 CASE STUDY 1—MYOELECTRICALLY CONTROLLED ROBOTIC ARM

This case study is an extension of Design Example 5.1, which dealt with myogenic control of a prosthetic limb. Here, more detail is presented, and the control is applied instead to a robotic arm. The problem is presented in steps according to the microcontroller-based system design procedure presented in Section 7.9. This case study is a good example of how to use PIC microcontrollers to interface to and communicate with an assortment of devices including analog circuits, desktop computers, and standard serial interfaces. Video Demo 11.6 demonstrates the finished product in action. You might want to view the video first so you can better relate to the information presented below.

1. Define the problem

The goal of this project is to design a system that uses myoelectric voltages from a person's biceps as a control signal for a robotic arm. As shown in Figure 11.13, this project can be divided into three phases: data acquisition, classification, and actuation.

Myoelectric signals, or **surface electromyograms (sEMG),** are produced during muscle contraction when ions flow in and out of muscle cells. When a nerve sends the signal to initiate muscle contraction, an "action potential" of ions travels along the length of the muscle. This ionic current can be transduced into electronic current with Ag-AgCl electrodes placed on the surface of the skin above the contracting muscle. Typically, the greater the contraction level, the higher the measured amplitude of the sEMG signal. However, even if a contraction level is constant, the sEMG signal can be quite irregular.

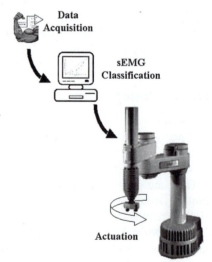

■ Data Acquisition—*Measuring and digitizing the myoelectric signal*

■ Classification—*Estimating the muscle force based on the myoelectric signal*

■ Actuation—*Moving a robotic arm to a position corresponding to the estimated force*

Figure 11.13 Project phases.

Research has shown that a typical sEMG signal has the following characteristics:

Amplitude	0–5 mV
Frequency range	0–500 Hz
Dominant frequency range	50–150 Hz

As can be seen, the sEMG signal is very small, in the millivolt range. In fact, electrical noise on the surface of the skin can be of greater magnitude than the signal of interest. There are three main frequency ranges in which noise may be present:

a. 0–10 Hz: low-frequency motion artifacts (e.g., wire sway)

b. 60 Hz: line noise (e.g., electrical equipment in the room)

c. >500 Hz: high-frequency noise (e.g., random movement between the electrodes relative to the muscle)

2. Draw a functional diagram

Figure 11.14 shows a block diagram depicting the flow of information between the system's required components. Before digitizing the sEMG signal, it must be amplified to take full advantage of the input range of the A/D converter (see Chapter 8 for more details). However, we cannot simply pass the signal into an op amp; if we did, the noise would also be amplified, and it would be impossible to distinguish the sEMG from the noise. Consequently, we need to filter out the noise and amplify the signal prior to A/D conversion. This stage of data acquisition is called **signal conditioning.**

After the signal is amplified and the noise is removed, it is ready for A/D conversion. The digitized signal is then sent to a PC where the muscle force is estimated based on analysis of the sEMG data. The estimated force is sent via an interface circuit to the robotic arm. Then the arm moves to a position that corresponds to the estimated force. For example, at rest the robotic arm will be at zero degrees; at the maximum contraction level the robotic arm will be at the maximum angle; and at intermediate contraction levels the robotic arm will be at corresponding angles in-between. Although we could build a robotic arm from DC or stepper motors, this project used an AdeptOne-MV robotic arm. This type of robot is usually used for industrial purposes (e.g., assembly line work), but it also serves as a good laboratory model of a prosthetic arm.

3. Identify I/O requirements and 4. Select appropriate microcontroller models

Although a single microcontroller could probably be used with some creative programming, two microcontrollers are used to simplify the problem. One microcontroller is dedicated to performing the A/D conversion and sending the digitized signal to a PC. Another microcontroller serves as the interface between the PC and the robotic arm. For the A/D microcontroller, the primary constraint is that it must have an analog input with a sampling rate over 1000 Hz (because the sampling theorem states that we must sample at least two times the highest frequency

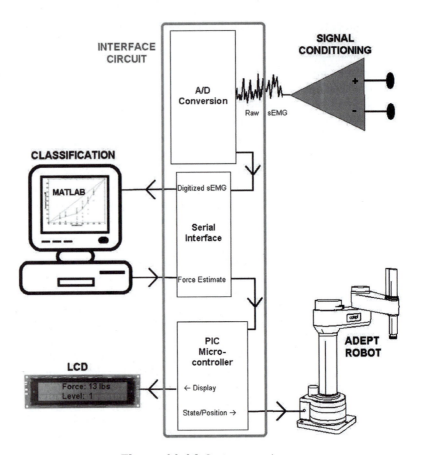

Figure 11.14 System overview.

component of the signal, which is 500 Hz after it is filtered). A PIC16F819 was chosen, although any PIC with A/D capability would meet this criterion. The only salient difference between PICs with analog inputs is their resolution; some are 8 or 12 bit, but most are 10 bit (such as the 16F819). With a 20 MHz oscillator, the PIC can sample 10 bit values at about 50 kHz. Clearly, this is well above the 1 kHz required. However, the limiting factor in this process is not the sampling rate but the time required to transmit the digitized value to a PC. Because of the convenient functions provided in PicBasic for serial communication, this was chosen as the communications protocol. The fastest standard serial baud rate for the PIC is 38400 bits per second. Because each byte of data is limited to 8 bits, each 10 bit value is split up into 2 bytes (plus start and stop bits for each byte). Thus, each 10 bit value requires 20 bits of data to be sent (2 start bits + 2 stop bits + 2 bytes). Consequently, the PIC can only send digitized values at 1920 Hz (38,400 bits per sec / 20 bits). Sending data in a constant stream like this, however, can easily cause the data to be corrupted on the receiving end. If the clocks of the transmitter and receiver are slightly out of sync, then the receiver (i.e., the PC) may lose track of where the bytes of data start and stop. To obviate this problem, a small delay is incorporated in PicBasic before sending each value.

Every 100 msec the PC estimates the muscle force based on the previous 100 msec of data. Also, the estimated force is "binned." For example, an estimated force between 0 and 5 lb would be assigned to bin 0, an estimated force between 5 and 10 lb would be assigned to bin 1, and so on. The bin number directly corresponds to a position of the robotic arm. Every 100 msec, the PC will send two bytes back out the serial port: the estimated force and the bin number. A PIC16F876 was chosen as the microcontroller to interface the PC to the robotic arm and to display information on an LCD. The primary factor considered when choosing a PIC was the number of I/Os. One input is needed for the serial communication, six I/O are required for the LCD interface, five outputs are used to interface to the Adept robot, and one output is for a status LED. Although many PIC models can handle these 13 I/O, the 22 I/O 16F876 is used in case future upgrades are desired.

5. Identify necessary interface circuits

The **signal conditioning circuit** must amplify the small sEMG signal and filter out noise prior to digitization. An instrumentation amplifier is used as the primary amplification component, as well as noise reduction component. As described in Section 5.9, an instrumentation amplifier is essentially a difference amplifier buffered with op amps at each of its two inputs. The buffering op amps provide high input impedance, which improves the difference amplifier's ability to reject noise (i.e., it has a high CMRR). The voltage difference that we will measure is the difference between two electrodes placed on the biceps. As a muscle action potential travels down the biceps, the first electrode will become positive relative to the more distal electrode; conversely, as the action potential continues down the biceps, the second electrode will then become more positive (which, of course, means the first electrode will be negative relative to the second). In theory, ambient noise will reach the electrodes simultaneously and will not be amplified, because the voltage *difference* between the two electrodes, due to noise, will be zero.

To further eliminate noise, high-pass and low-pass filters are implemented. A high-pass filter of 10 Hz is desired to reduce motion artifacts and DC offsets. A low-pass filter of about 500 Hz is desired to reduce high-frequency noise. A low-pass filter is important prior to A/D conversion to prevent aliasing. Unfortunately, 60 Hz line noise is in the middle of the frequency range of the sEMG signal so it would not be a good idea to employ a notch filter to remove this frequency range. Hopefully, the instrumentation amplifier will sufficiently reject the line noise.

A 0–5 V input range is used on the A/D converter. To ensure that the sEMG amplitude is in this range, two more components are incorporated in the signal conditioning circuit: a full wave rectifier and an adjustable gain. The full wave rectifier approximates the absolute value of the signal. Because the sEMG signal is bipolar (i.e., it can be both positive and negative), passing it through a full-wave rectifier will guarantee that the signal is entirely positive. Finally, an adjustable gain is useful to account for differences in electrode size, geometry, and positioning, as well as differences among individuals—all of which affect the amplitude of the signal. By adjusting the gain of the signal conditioning circuit, we can attain a maximum amplitude of about 5 V (i.e., during maximum contraction levels).

The protocol that dictates how most **serial communication** is executed is called **RS-232.** This protocol states, for example, that a logic 1 is between −3 V and −25 V, whereas a logic 0 is between +3 V and +25 V. Because a PIC cannot output negative voltage (and because some PC serial ports have trouble reading voltages less than 5 V), it is a good idea to use an RS-232 level converter chip, such as Maxim's MAX232 (see Figure 11.15). These chips convert TTL/CMOS level signals to RS-232 level signals, and vice versa. It is also important to note that they also invert the signal. For example, when given a +5 V input, it will output about −8 V; when given 0 V, it will output about +8 V. These chips are able to provide these outputs using a 5 V power supply and ground. This is made possible using a technique called **charge pumping,** which uses capacitors to store and boost voltage.

One of the parameters of PicBasic's *Serout* command is the mode. Along with the baud rate, this parameter specifies whether the serial data is driven *true* or *inverted*. Because we are using an RS-232 level converter chip, *true* is used because the chip automatically inverts the signal. The MAX232 chip is used to convert the TTL output of the A/D converter PIC to RS-232 level signals as well as convert the RS-232 output from the PC to TTL level signals for the robot/LCD interface.

DB9 serial ports (see Figure 11.16) are no longer common on PCs because USB and other interfaces have made them obsolete; however, one was available on the computer used for this project. Although the DB9 port has nine pins, only three are useful for our purpose; the other pins are for **handshaking,** a method to help regulate data flow. The only pins necessary for our purposes are listed in Table 11.2. Pin 2 is

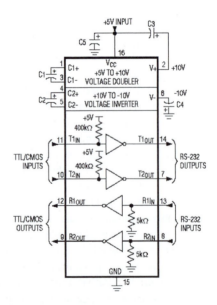

Figure 11.15 MAX232 level converter.

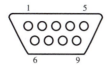

Figure 11.16 Serial port.

Table 11.2 Serial port pins

Pin #	Description
2	Receive data
3	Transmit data
5	Ground

where the PC will receive the digitized sEMG signal, and Pin 3 is where the PC will send the estimated force and bin number to the interface PIC (in both cases, via the MAX232 chip).

6. Decide on a programming language

PicBasic Pro is used. Speed and memory constraints are not primary concerns, so Assembly language is not required. Also, the serial communication and LCD interface commands provided by PicBasic Pro will be quite useful.

7. Draw the schematic

Figure 11.17 shows a circuit diagram for the conditioning circuit. The first stage is an instrumentation amplifier with a gain of 939 with the resistor values shown. The next stage is a simple *RC* filter followed by a buffer op amp. If the buffer op amp were not included, then the resistance of the following stage would **load** the filter (with impedance) and change its behavior. In other words, it would not act as a simple *RC* filter. Next is a **two-pole Sallen-Key** low-pass filter, a type of **active filter** (i.e., it exploits the feedback of an op amp). An entire course can be devoted to analog filter design, but suffice it to say that active filters are more robust than passive filters (such as an RC filter) and that additional "poles" (an *RC* filter is a single-pole filter) are better at attenuating unwanted frequencies. The next stage is an inverting op amp where the feedback resistor is a potentiometer used to adjust the overall gain of the system. Finally, the last stage is a precision full-wave rectifier, which requires no forward bias to turn on the diodes.

After the sEMG signal passes through this circuit, it is amplified by about $1000\times$, depending on the potentiometer setting of the inverting amplifier, bandpass filtered between 8 and 483 Hz, and full-wave rectified. To make this circuit more robust, the design was prototyped on a custom-made **printed circuit board (PCB),** rather than on a breadboard, using *PCB123,* a free PCB layout program available online (see Internet Link 11.8). Figure 11.18 show the traces and soldering pads

Internet Link

11.8 PCB123 software for designing custom printed circuit boards

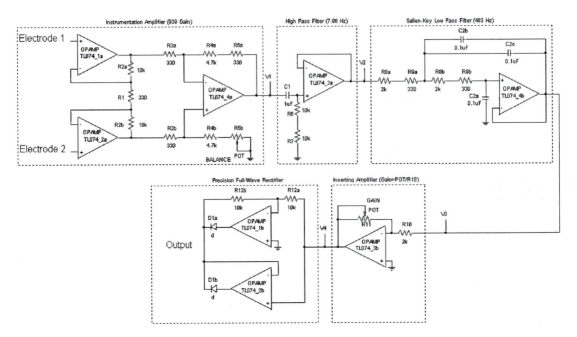

Figure 11.17 Conditioning circuit diagram.

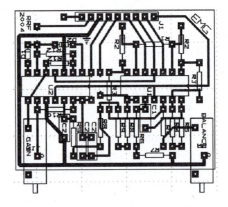

Figure 11.18 Conditioning circuit PCB layout.

Internet Link

11.9 Example of steps in the printed circuit board process

created with the software for the PCB. Figure 11.19 shows the assembled PCB after the components are soldered on. Internet Link 11.9 points to a document that illustrates all of the steps necessary to create a PCB. See Section 2.10.3 for more information about PCBs and soldering.

The only chips used in this example are two TL074s, quad-package JFET op amps. These op amps will operate over a fairly large range, about ±5 to ±18 V. To make the system portable, two 9 V batteries can be used. Furthermore, a voltage regulator can be used to produce the +5 V necessary to operate the PICs.

The output of the conditioning circuit is connected to an analog pin on the PIC16F819. Other than the circuitry required to make this PIC operate, which is the same as the PIC16F84, only this analog input and a wire going from a digital pin to a TTL input of the MAX232 chip are needed. A wire then connects the corresponding RS-232 output of the MAX232 to pin 2 on the PC's serial port. Similarly, a wire connects pin 3 of the serial port to an RS-232 input on the MAX232 chip. The corresponding TTL output is connected to a digital pin on the 16F876 interface PIC. A sample wiring diagram for connecting an LCD to a PIC is shown in Figure 7.13.

At least 9 V is required to register a logic 1 on the Adept robot's input port. On each of the PORT A outputs of the PIC, a transistor and separate power source (such as the positive voltage source used to operate the op amps) are incorporated to help increase the voltage. The schematic in Figure 11.20 shows what is needed to interface each of the PORT A pins to a corresponding Adept input pin. It should be noted that this circuit inverts the logic. If the PIC output is high, the transistor will be saturated, and the Adept input will be at ground potential. Conversely, if the PIC output is low, the transistor will be off and no current will flow through it. With no voltage drop across the 10k resistor, the Adept input will be at 9 V. This negative logic can be accounted for in the Adept code (or, just as easily, in the PIC code).

Figure 11.19 Conditioning circuit photo.

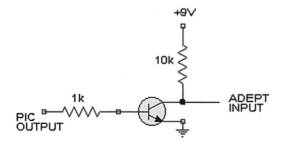

Figure 11.20 PIC to Adept interface circuit.

8.   Draw a program flowchart

The flowchart in Figure 11.21 describes the operation of the A/D converter PIC. After initializing variables and defining how the A/D conversion is performed, the remaining portion is a simple loop that continuously samples the A/D converter and sends the data out through a serial pin.

Figure 11.22 shows the flowchart describing the operation of the interface PIC. This PIC waits to receive 2 bytes on its serial pin (estimated force and bin number), displays those values on an LCD, and outputs the bin number on PORTA. Other than displaying information on an LCD, this PIC is essentially a serial-to-parallel converter. The parallel value is what is read by the Adept robot.

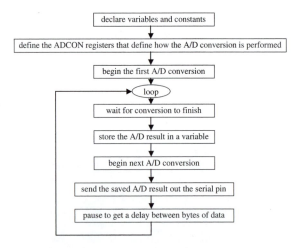

Figure 11.21 A/D converter PIC flowchart.

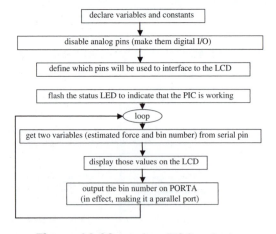

Figure 11.22 Interface PIC flowchart.

9. Write the code

PICs are 8-bit processors, and, as such, any integer value over 8 bits (255+) must be stored in two or more registers. Because the PIC16F819 performs 10 bit A/D conversions, the result is stored in two registers, *ADRESL* and *ADRESH*. One can choose how to "justify" the result: either the eight least significant bits are in one register, or the eight most significant bits are in one register. The other two bits are padded with zeros. Justification, conversion, which pin is being used for the conversion, initiating a conversion, and other A/D parameters are defined by the *ADCON0* and *ADCON1* registers (see the PIC16F819 data sheet for more information). The code for the A/D converter PIC is shown below.

```
'*****************************************************************
'* Name    : ADC.BAS                                            *
'* Version : 2.0                                                *
'* Notes   : PIC16f819                                          *
'*         : Reads analog value, serially sends binary result   *
'*****************************************************************
DEFINE OSC 20 ' 20MHz crystal

' INITIALIZE VARIABLES
ADC_LSB VAR BYTE      ' 8 LSB of the 10 bit A/D conversion
ADC_MSB VAR BYTE      ' 2 MSB of the 10 bit A/D conversion (in bits 0 & 1)
i       VAR BYTE      ' looping variable
serMode CON 6         ' baud rate of serial; 84=9600   bps (driven true)
                      '                      32=19200  bps (driven true)
                      '                       6=38400  bps (driven true)
' SET UP A/D REGISTERS
ADCON0 = %10000000 ' TAD=32Tosc, A/D pin=RA0
ADCON1 = %10001110 ' Right justified (8 LSB are in ADRESL, 2 MSB are in ADRESH)

' DEFINE PORTS
TRISA = %11111111            ' PORTA is all inputs
TRISB = 0                    ' PORTB is all outputs
SYMBOL LED = PORTB.3         ' Status LED
SYMBOL ADC_pin = PORTA.0     ' A/D pin
SYMBOL SERIAL_pin = PORTB.4  ' Serial out pin

' RENAME REGISTERS
SYMBOL ADC_on = ADCON0.0     ' A/D on/off bit (turn off to save power)
SYMBOL ADC_go = ADCON0.2     ' A/D Go/Done bit (manually set to 1 to
                             '   start conversion; automatically is reset
                             '   to 0 when finished with conversion)

initialize:
  ADC_on = 1          ' turn the ADC channel on
  PAUSE 250           ' let things settle a bit (probably not necessary)
  HIGH LED            ' turn on LED to indicate ADC in process
  ADC_go = 1          ' start the 1st A/D conversion
  GOSUB getADC        ' wait until it's done
```

```
main:
  ADC_go = 1              ' start the next A/D conversion
  GOSUB sendADC           ' send the data whilst doing the A/D conversion
  PAUSEUS 126             ' pause 126 us (results in 1500 Hz sampling rate)
  GOSUB getADC            ' get the results of the A/D conversion
GOTO main                 ' do it forever

getADC:
  WHILE (ADC_go ==  1)    ' wait until the conversion is done
  WEND                    ' (should already be done, because SEROUT takes a while
  ADC_MSB = ADRESH        ' save 2 most sig. bits (padded w/ 0s)
  ADC_LSB = ADRESL        ' save 8 least sig. bits
RETURN

sendADC:
  SEROUT2 serial_pin, serMode, [ADC_MSB]      ' send the MSB (padded with six 0s)
  SEROUT2 serial_pin, serMode, [ADC_LSB]      ' send LSB
RETURN

END
```

Most of the code for the interface PIC (see below) is self-evident except, perhaps, for the *DEFINE* statements. *DEFINE* is used to change elements that are predefined in PicBasic Pro. For example, PicBasic Pro assumes a 4 MHz clock and a specific set of pins to be used with an LCD. Because we are using a 20 MHz crystal and a different pin configuration for the LCD, we need to *DEFINE* new parameters. Specifying *DEFINE OSC 20* lets PicBasic know to change all of the time-sensitive commands (e.g., *Pause, Serin, Lcdout*) to account for the faster clock speed. If we did not do this, all of these commands would happen five times too fast, or not work at all.

```
'****************************************************************
'* Name      : PC2PIC.bas                                      *
'* Version   : 2.0                                             *
'* Notes     : PIC16F876                                       *
'*           : Reads two values from serial pin, displays them *
'*           : on an LCD, and converts one of the values to a //*
'*           : value on PORTA                                  *
'****************************************************************

' make sure to set Configure | configuration bits | oscillator - HS
'   before programming (if using 20Mhz osc)

ADCON1 = %11000110        ' turning all of porta to digital
TRISA = %00000000         ' turning all of porta to outputs

DEFINE OSC      20        ' 20 MHz oscillator
DEFINE LCD_DREG PORTB     ' LCD data port
DEFINE LCD_BITS 4         ' 4 parallel data bits
```

```
DEFINE LCD_DBIT        0      ' data bits on PORTB.0 -> PORTB.3
DEFINE LCD_RSREG       PORTC  ' Register Select (RS) port
DEFINE LCD_RSBIT       6      ' RS on PORTC.6
DEFINE LCD_EREG        PORTC  ' Enable (E) port
DEFINE LCD_EBIT        7      ' Enable on PORTC.7
DEFINE LCD_LINES       2      ' 40X2 LCD Display
DEFINE LCD_COMMANDUS 3000     ' command delay time (us) {found
                              ' experimentally 3000}
DEFINE LCD_DATAUS 75          ' data delay time (us) {found experimentally 75}

SYMBOL  LED=PORTB.7           ' Status LED
SYMBOL  serin_pin=PORTB.6     ' serial in pin

i          var    byte       ' looping variable
command    var    byte       ' 1st byte received from serial pin
force      CON    byte       ' 2nd byte received from serial pin
mode       CON    6          ' serial mode (6=38400 baud)

LCDOUT $FE, 1                 ' clear LCD
for i=1 to 8                  ' flash the status LED
    high LED
    pause 100
    low LED
    pause 100
next i

porta = 0
pause 1000
high LED
pause 500


'''''''''''''''''''''''''''''''''''''''''''
'       EMG            |     Force: ## lbs
'Signal Strength       |     Level: #
'''''''''''''''''''''''''''''''''''''''''''

LCDOUT $FE, 128, "      EMG            |     Force:      lbs"
LCDOUT $FE, 192, "Signal Strength      |     Level: "

main:
    SERIN2 serin_pin, mode, [command, force]  ' get 2 serial values

    LCDOUT $FE, 128+31, #force
    LCDOUT " "
    LCDOUT $FE, 192+31, #command
    LCDOUT " "

    'convert serial value to parallel value on portA (for ADEPT robot)
    PortA = command

goto main
```

10. Build and test the system

As with any system, especially more complicated ones, it is vital to test in parts. Break the system down into its smallest functional units and test the inputs and outputs of those units to ensure they are performing as expected. In the signal conditioning circuit, for example, we could test just the full-wave rectifier by passing a sine wave through it and confirming that the output signal is rectified. Also, each of the filters could be tested by inputting a frequency sweep and confirming that the amplitude at the cutoff frequency is about 70.7% of the input amplitude.

With any PIC, it's a good idea to include a status LED that indicates that the PIC is on and functioning to some degree. While troubleshooting, it is also useful to flash the status LED at different points in the code to help you track the execution. LCDs are extremely useful, once they are functioning, in debugging code because they can easily be used to display diagnostic messages and values being stored in variables.

Serial communication between a PIC and PC can be tested using a terminal program. Hyperterminal, the terminal program bundled with Windows, allows you to read from and write to the serial ports. One just needs to be sure to specify the port settings (e.g., baud rate) so that they match the format that the PIC is sending or expecting to receive.

Examples could be given for each element of the system, but the point is to build and test everything incrementally.

Again, a demonstration of the working system can be viewed at Video Demo 11.6.

Video Demo

11.6 Robot controlled by an EMG biosignal

11.5 CASE STUDY 2—MECHATRONIC DESIGN OF A COIN COUNTER

In this section, we present solutions to the following design problem: Design a coin counter that includes an electromechanical device to accept a handful of mixed denomination U.S. coins and that will align and present the coins in serial order to a sensor array capable of acquiring data that can be used to determine the denominations of individual coins. The sensor output is to be interfaced to electronics that can compute both the number of coins presented and the total value of the coins and display those two values in some sort of multiplexed fashion on a single display visible to the user.

We presented this design challenge to a junior-level mechatronics class of 80 students broken down into groups of 4, and more than half the groups created successful solutions given a design period of 6 weeks. All groups were able to present the coins sequentially and display a count of the number of coins. Multiplexing the value of the coins with the count on a single display was more challenging, requiring creative use of digital logic, and not all groups were successful in implementing a design for this part. Given more time to redesign, most groups would have been successful. As you will see, there is no single correct solution to the problem, and the designs are as varied as the people who developed them.

This problem has two significant parts: the design of an electromechanical coin presentation system to align coins sequentially in some fashion so they may be presented to an array of sensors, and the design of an electronic calculator to use the sensor data to display the count and value of the coins. We consider the two parts in succession, realizing that the student design groups often divided their efforts between the mechanical design of the coin presenter and the electronic design of the calculator. Students with significant machine shop experience tackled the electromechanical coin presentation design while students more comfortable with the electronic experience they gained in the course focused on the sensor and counter design. As in all design projects, certain people have affinities for certain parts of the overall design, and it is important to assign responsibilities, communicate progress, document work, and assure compatibility among the various subsystems. We do not want to digress too much on the design process itself here, other than to mention that team coordination and communication in the design process are as critical as the design itself.

The mechanical component of the design requires a chamber to accept a handful of mixed denomination coins and a mechanism to select coins individually and present them to a sensor array that can read their attributes and output digital signals for computation. By observing existing mechanical coin sorters, many students designed inclined rotating disk mechanisms that had circular holes cut in them to accept individual coins. Examples of the mechanical systems designed to present a handful of coins in series are shown in Figure 11.23. Two of the three approaches (Figures 11.23a and 11.23c) use a DC motor to rotate a perforated disk that entrained a single coin. As the disk rotated, the coin would drop into the slot containing the sensor array. Most designs included a DC motor to rotate the disk continuously. Other

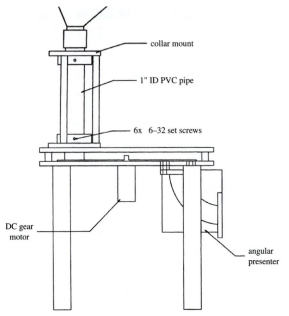

(*a*) horizontal slotted disk design

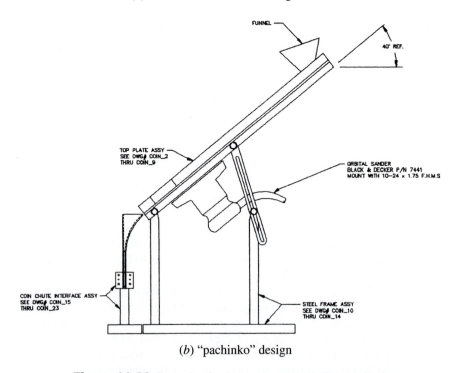

(*b*) "pachinko" design

Figure 11.23 Example of coin counter presentation mechanisms.

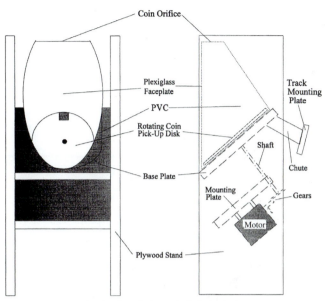

Coin Orifice

Plexiglass Faceplate

PVC

Rotating Coin Pick-Up Disk

Base Plate

Mounting Plate

Plywood Stand

Track Mounting Plate

Shaft

Chute

Gears

Motor

(*c*) inclined slotted disk design

Figure 11.23 *(continued)*

designs (e.g., the one in Figure 11.23b) included vibrating chutes and pachinko-like mechanical arrays. As individual coins were separated, they entered a chute and rolled down a rectangular slot containing the sensor array.

Although people determine the value of coins by sight and feel, automated methods require a well-designed sensing system. The size, weight, and thickness of coins are unique for different denominations. For the sake of simplicity, all groups elected to design a diameter sensor using a set of phototransistor-photodiode pairs. These photo-optic pairs were carefully positioned so that the combination of signals would be unique for each different denomination of coin (see Figure 11.24). For this project, the coin denominations were limited to the U.S. penny, nickel, and quarter to provide significant size differences. Diameter was the primary attribute for assessing each coin's value, assuming that they rolled down the slot smoothly. This meant that the largest coin, the quarter, activated all sensors, and the smallest, the penny, only one. Furthermore, the signals from the sensors would be pulses of different widths affected by the size and speed of the coins. The pulses do not begin or end at the same time (see Figure 11.25), complicating the synchronizing of logic in the design of the evaluator and counter. The sensor outputs were converted to TTL signals using 7404 Schmitt triggers, making the outputs compatible with the computational part of the circuit design.

The sensors produce pulses at different starting times and for different durations depending on the size and speed of the coin. This subtlety requires careful sequential logic design to ensure that the coin is identified correctly. If the TTL signals

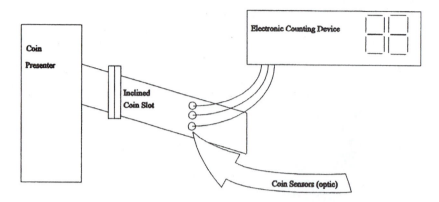

Figure 11.24 Sensor array and chute design.

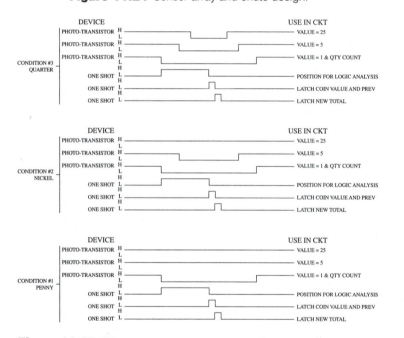

Figure 11.25 TTL outputs corresponding to different coins.

from the sensors were synchronized, then the Boolean expressions for the different denomination coins can be written as follows:

$$X = A \cdot B \cdot C \tag{11.14}$$

$$Y = A \cdot B \cdot \overline{C} \tag{11.15}$$

$$Z = A \cdot \overline{B} \cdot \overline{C} \tag{11.16}$$

where C corresponds to the output of the top sensor, B to the output of the middle sensor, and A to the output of the bottom sensor. X corresponds to the passage of a quarter, Y to the passage of a nickel, and Z to that of a penny.

When this project was assigned, we had not yet started teaching microcontroller programming and interfacing in our course. In lieu of this, we had students develop solutions using basic TTL ICs. Combinational and sequential logic is required to determine the denomination of the coin and increment the displayed output by a value corresponding to the denomination. There were as many solutions for this problem as there were design groups. Figures 11.26 and 11.27 display two student group solutions for the coin counter processing and display circuit. The outputs were transmitted to a digital display driver to multiplex the current number of coins and the accumulated value.

■ **CLASS DISCUSSION ITEM 11.2**
Coin Counter Circuits

Figure 11.26 shows the coin counter electrical schematic from a student group report. The design uses one-shots to latch the value corresponding to the coin and adds that value to the previous sum. There are mistakes in the input portion of the first half of the circuit used to determine whether the coin is a penny, nickel, or quarter. The error is in the combinational logic portion of the circuit. Based on the timing diagrams shown in Figure 11.25, convince yourself that the logic circuit will not identify the coin correctly.

Even if the logic were correct, the design could still have problems due to timing because the one-shot pulse lengths are fixed but the sensor pulse widths depend on how fast the coin rolls past. A sequential logic circuit that would unequivocally identify the coin as a penny, nickel, or quarter follows. Draw a timing diagram that supports the argument that this sequential logic circuit is a robust design. Include signals S_{low} (low sensor), S_{middle} (middle sensor), S_{high} (high sensor), B, C, L, R, X, Y, and Z in your diagram. Refer to Figure 11.25 for the phototransistor sensor signal timing for each type of coin.

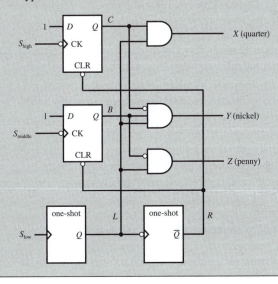

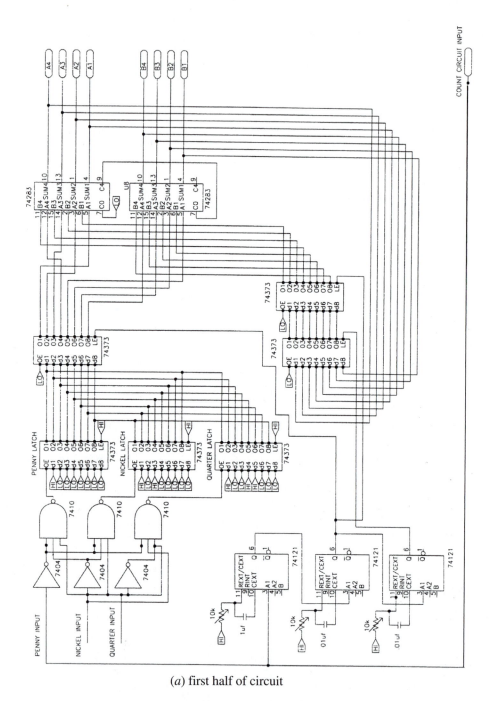

(*a*) first half of circuit

Figure 11.26 Counter design 1.

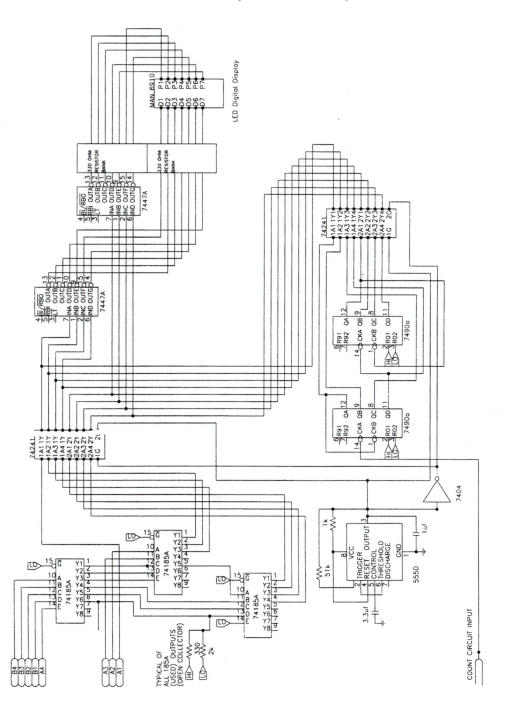

(*b*) second half of circuit

Figure 11.26 (*continued*)

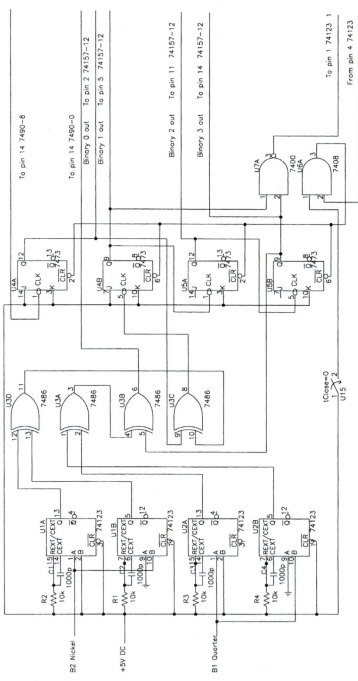

(*a*) first half of circuit

Figure 11.27 Counter design 2.

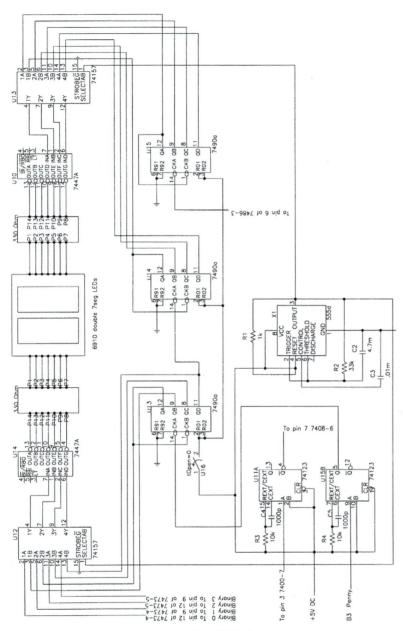

(*b*) second half of circuit

Figure 11.27 *(continued)*

11.6 CASE STUDY 3—MECHATRONIC DESIGN OF A ROBOTIC WALKING MACHINE

In this section, we present a case study of the mechatronic design of an articulated walking machine. It was executed by undergraduate engineering students in 1994, following the completion of our introductory mechatronics course for which this book is used. In 1987, the Society of Automotive Engineers (SAE) began sponsoring an annual Robotic Walking Machine Decathlon, pitting teams from different universities in a challenge to design articulated walking machines that could execute 10 different performance events including a dash, a slalom, obstacle avoidance, and crossing a crevasse. Half of the events included walking motion and obstacle avoidance under tether control, that is, control from a human-operated switch box with an electrical umbilical to the machine. The other half of the events required autonomous control via onboard, preprogrammed systems with no human intervention. Over the years we have seen scores of different walking machine designs, some very simple and capable of completing just a few events, and others of great sophistication and creativity capable of completing all events. The excitement and fun of such competitions is in seeing the fruit of design concepts actually functioning to specifications. Our intent here is not to examine the varieties of walking machines but to focus on a specific design example to illustrate the mechatronic aspects. We begin by displaying three different walking machines, all of which won the national SAE contest (see Figure 11.28).

We now present as a case study the design of the 1994 Colorado State walking machine that the students, applying their ever-present wit, named *Airratic*. This design was a refinement of the first pneumatic design from 1992, which was a dramatic break from the evolving electromechanical designs of the previous seven years. With the air-powered designs, the students had to contend with a whole new set of design constraints: providing an onboard source of stored, pressurized air; controlling the mechanical motion of the articulated legs with pneumatic cylinders; distributing and controlling the air; controlling the pressure; reducing coordinated walking motion to sets of computer commands; interfacing a computer to the pneumatic control system; minimizing the weight and size of the machine; ensuring the safety of the system, including sensors on the machine for obstacle avoidance; and making the design changeable and adaptable during test trials on the competition floor.

Because the designers elected to power the walking machine pneumatically, an onboard fiber-wound pressure tank was selected as the energy source for the strategically placed pneumatic pistons that controlled the position and movements of the legs. As shown in Figure 11.29, the skeletal structure of the walking machine consisted of welded aluminum members with 16 double-acting pneumatic cylinders. The four corner legs have 3 degrees of freedom, and the front and rear-center legs have 2 degrees of freedom. The mechanical design of the six legs was such that control algorithms could easily produce forward, reverse, sidewise, and diagonal motion, as well as full-up and full-down motion of the main frame, by simple coordinated control of the 16 cylinders. Each leg included an axial cylinder that could be

(*a*) "Lurch"—Scotch yoke mechanism leg design (1989)

(*b*) "Airachnid"—First pneumatic design (1992)

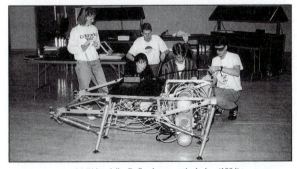

(*c*) "Airratic"—Refined pneumatic design (1994)

Figure 11.28 Student-designed walking machines from Colorado State University.

extended or retracted. Ten other cylinders were arranged to position the legs in different patterns in the horizontal plane.

For static stability the machine must be supported by a minimum of three legs at all times, requiring a total of six legs for locomotion. Most walking motions could be

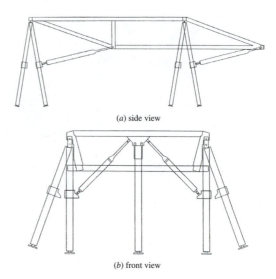

(*a*) side view

(*b*) front view

Figure 11.29 Aluminum frame and telescoping pneumatic legs.

divided into the coordinated movement of two sets of three legs each, called Group 1 Legs and Group 2 Legs. Figure 11.30 displays the flowchart used to develop the control code for one class of coordinated motion: forward motion of the machine.

The essence of the control system involves the coordination of 16 pneumatic cylinders using computer-controlled solenoid valves. Each solenoid valve switches air from the main cylinder to one side of the pneumatic cylinder and exhausts pressurized air from the other side, as shown in Figure 11.31. Needle valves were manually adjusted on each solenoid valve to govern cylinder actuating speed.

The complexity of the flowchart in Figure 11.30 warrants a microprocessor-based solution. This and similar flowcharts for each motion command were very helpful in designing the computer code necessary to control and coordinate the motion. The onboard computer controls the solenoid valves via 74373 tristate inverting buffer interface chips receiving signals from an 8-bit port on the computer. As shown in Figure 11.32, the 8-bit port is multiplexed to provide 18 bits of information for coordinated control of the cylinders.

The controller selected was a Motorola 68HC811E2 EVM 8-bit microprocessor-based single-board computer. It has 128 K of RAM, and 64 K acts as pseudo-ROM for downloading control programs from an external PC or laptop computer. Thirty I/O bits are provided and expanded via multiplexing to the 48 bits required for the control of all of the components of the machine. Computer code was created on a laptop computer in the high-level language C, then complied and downloaded to the Motorola EVM for testing. This provided a convenient development environment for testing and modifying the control strategies.

In addition to various walking motions, the machine included sensor feedback using two optical retroreflective sensors. The sensors were mounted on a rotating platform to allow scanning a portion of the region in front of the machine. A digital

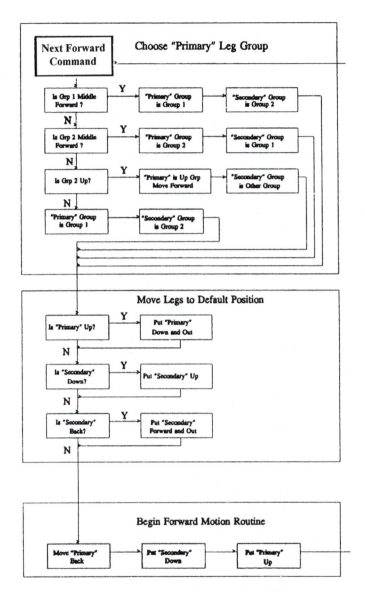

Figure 11.30 Flowchart for forward motion routine.

encoder provided position information corresponding to the sense state of the two sensors. The sensor data was read from the I/O bus of the microprocessor, and software control routines used the data to avoid obstacles.

In summary, we examined an example of a student-designed mechatronic system that received first place in the 1994 SAE Walking Machine Decathlon. It illustrated the use of pneumatic actuators for motion, optical sensors for feedback, and an onboard microcomputer for controlling the machine in both operator-controlled

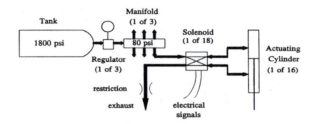

Figure 11.31 Pneumatic system.

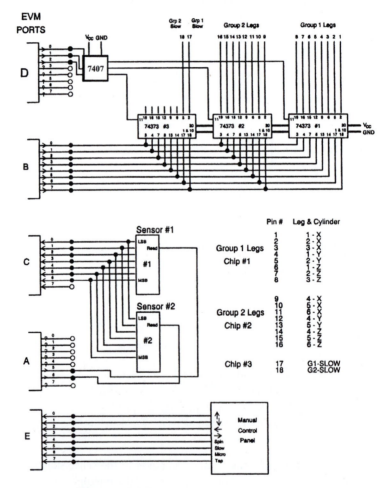

Figure 11.32 Computer ports and I/O board.

and autonomous modes. Projects such as this require a significant amount of time to complete. This project involved eight months of work. The experience gained in designing a complex mechatronic system is very valuable, particularly if the design is actually built, debugged, and challenged to achieve its design goals. In this project, the students were so excited about their success that they even programmed the machine to dance. This was so well received by the audience that they all joined in a country-style line dance at the finale.

11.7 LIST OF VARIOUS MECHATRONIC SYSTEMS

We conclude the book with a partial list of mechatronic systems that you can find in your everyday environment. A very large proportion of engineering systems designed today can be declared mechatronic systems.

- Airbag safety systems, antilock brake systems, remote automatic door locks, cruise control, stability and traction control, hybrid car power management, and other automobile systems
- NC mills, NC lathes, rapid prototyping systems, and other automated manufacturing equipment
- Copy machines, document scanners, and other semiautomatic office equipment
- MRI equipment, arthroscopic instruments, ultrasonic probes, and other medical diagnostic equipment
- Autofocus cameras, video and CD/DVD players, camcorders, and other sophisticated consumer electronic products
- Laser printers, hard-drive head-positioning systems, game controllers, and other computer peripherals
- Welding robots, automatic guided vehicles (AGVs), NASA Mars rover, and other robots
- Flight control actuators, landing gear systems, cockpit controls and instrumentation, and other subsystems on airplanes
- Garage door openers, security systems, heating, ventillation, and air conditioning (HVAC) controls, and other home support systems
- Washing machines, dishwashers, freezer automatic ice makers, and other home appliances
- Variable speed drills, digital torque wrenches, and other modern hand tools
- Materials testing machines, automobile test dummies, and other laboratory support equipment

- PLC-controlled bar-code-assisted conveyor systems and other factory automation systems
- Manual and semiautomatic controllers for hydraulic cranes and other construction equipment
- Automatic labeling systems and charge-coupled device (CCD) camera inspections in an IC manufacturing operation
- Video game and virtual reality input controllers

QUESTIONS AND EXERCISES

11.1. Document a complete and thorough answer to Class Discussion Item 11.1.

11.2. For selected mechatronic systems mentioned in the preceding list, specify the required electronics, sensors, and actuators.

11.3. For selected mechatronic systems mentioned in the preceding list, recommend a control architecture and support your choice.

11.4. Document a complete and thorough answer to Class Discussion Item 11.2.

11.5. Write a PicBasic Pro program to mimic the function of the sequential logic circuit shown in Class Discussion Item 11.2.

BIBLIOGRAPHY

Bolton, W., *Mechatronics,* 2nd ed., Addison Wesley, Harlow, Essex, England, 1999.

Ogata, K., *Modern Control Engineering,* 5th ed., Prentice-Hall, Englewood Cliffs, NJ, 2009.

Palm, W., *Modeling, Analysis, and Control of Dynamic Systems,* 2nd ed., John Wiley, New York, 1999.

Measurement Fundamentals

After you read, discuss, study, and apply ideas in this appendix, you will:

1. Be able to define SI units and use them in calculations

2. Know how to use statistics fundamentals to characterize measured data

3. Be able to compute the error associated with a measurement

A.1 SYSTEMS OF UNITS

Fundamental to the design, analysis, and use of any measurement system is a complete understanding of a consistent **system of units** used to quantify the physical parameters being measured. To define a system of units, we must select units of measure for fundamental quantities to serve as a basis for the definition of other physical parameters. Units for mass, length, time, temperature, electric current, amount of substance, and luminous intensity form one possible combination that serves this purpose. Other units used to measure physical quantities in mechatronic systems can be defined in terms of these seven **base units.**

■ **CLASS DISCUSSION ITEM A.1**
Definition of Base Units

Although everyone has intuitive knowledge of the physical quantities associated with the base units, it is difficult to define them in everyday terms. Try to define length. You will probably use some synonym for length in your definition. Also try to define the others. All are equally difficult to put into simple terms.

The seven base units we use to define mass, length, time, temperature, electric current, amount of substance, and luminous intensity are the kilogram, meter, second, Kelvin, ampere, mole, and candela. These units form the basis for the International System of Units, abbreviated **SI,** from the French *Le Systeme International d'Unites.*

The **kilogram** is the only unit defined in terms of a material standard. It is established by a platinum-iridium prototype in the laboratory of the Bureau des Poids et Mesures in Paris. Unfortunately, the name kilogram is confusing since it contains the prefix kilo, which conflicts with the SI prefixing conventions described in Section A.1.1.

The **meter** is defined as 1,650,763.73 wavelengths of the emission resulting from the transition between the $2p_{10}$ and $5d_5$ electron energy levels of the krypton 86 atom. This atomic standard for the meter was proposed long ago by Maxwell (1873) but not implemented until 1960. The earlier meter definition, the distance between two scribed lines on a platinum-iridium bar, like the kilogram, required a prototype for the definition. Now the practical measurement of the unit is deliberately separated from the definition, making the definition independent of a unique prototype. An alternative standard meter was defined in 1983 as the length of the path traveled by light in vacuum during a time interval of 1/299,792,458 sec.

The **second** is defined as the duration of 9,192,631,770 periods of the radiation corresponding to the transition between the two hyperfine levels of the ground state of the cesium 133 atom. The definition of the second was previously based on the mean solar second, which was defined as a fraction (1/86,400) of the earth's daily rotation. Irregularities in the earth's rotation amounting to 1 or 2 seconds per year limited accuracy.

The unit of absolute thermodynamic temperature is the **Kelvin.** The Kelvin scale has an absolute zero of 0 K, and no temperatures exist below this level. There is a misconception that all molecular motion ceases at this value. Actually, the molecular energy is at a minimum. A standard fixed calibration point on the temperature scale is the triple point of water, which is set at a value of 273.16 K to maintain consistency with the Celsius scale. Although the Kelvin scale is established using only the absolute zero and triple points, additional fixed points have been defined based on the boiling and melting points of other materials. These points are useful when calibrating temperature measurement devices. Temperature in Kelvin and temperature in degrees **Celsius** are related by the following equation:

$$T_C = T_K - 273.15 \tag{A.1}$$

where T_K is the Kelvin temperature and T_C is the Celsius temperature expressed in degrees Celsius (°C). Note that the triple point of water is 0.01°C, corresponding to 273.16 K. The Celsius temperature scale is sometimes referred to as the **centigrade scale** because it is calibrated to the 100°C temperature interval between the freezing point of water (0°C) and the boiling point of water (100°C). An interval or difference of temperature (ΔT) has the same value in both the Celsius and Kelvin scales ($\Delta T_C = \Delta T_K$).

The **ampere** is defined as the constant current that, if maintained in two straight parallel conductors of infinite length and negligible circular cross section and placed

1 meter apart in a vacuum, would produce a force between the conductors equal to 2×10^{-7} newtons per meter of length. Unfortunately, this creates a difficult measurement problem, since the definition is in terms of other base units. Therefore, any errors in the other base units compound the errors in the measurement of the ampere.

The **mole** is defined as the amount of substance that contains as many elementary entities as there are atoms in 0.012 kg of carbon 12 (^{12}C).

The **candela** is defined as the luminous intensity, in the perpendicular direction, of a surface area of 1/600,000 m^2 of a black body at the freezing point of platinum under a pressure of 101,325 N/m^2.

A.1.1 Three Classes of SI Units

SI units are divided into three classes: base units, derived units, and supplementary units. The complete set of SI **base units** and their symbols are listed in Table A.1.

Derived units are expressed as algebraic combinations of the base units. Any known physical parameter can be quantified using a derived unit. Some examples of derived units are listed in Table A.2. Several derived units have been given special names and symbols, which may be used themselves to express other derived units in a simpler way than in terms of base units. Some examples of these **supplemental units** are listed in Table A.3.

Often the base, derived, and supplemental units are modified with prefixes to enable convenient representation of large numerical ranges. The **prefixes** express orders of magnitude (powers of 10) of the unit, providing an alternative to scientific notation. The prefix names, symbols, and values are listed in Table A.4.

Table A.1 SI base units

Quantity	Name	Symbol
Length	Meter	m
Mass	Kilogram	kg
Time	Second	s
Electric current	Ampere	A
Thermodynamic temperature	Kelvin	K
Amount of matter	Mole	mol
Luminous intensity	Candela	cd

Table A.2 Examples of SI-derived units expressed in terms of base units

Quantity	Name	Expression
Area	Square meter	m^2
Volume	Cubic meter	m^3
Speed	Meter per second	m/s
Acceleration	Meter per second squared	(m/s)^2
Mass density	Kilogram per cubic meter	kg/m^3
Current density	Ampere per square meter	A/m^2

Table A.3 SI-derived units with special names (supplemental units)

Quantity	Name	Symbol	Expression
Frequency	Hertz	Hz	$1/s$
Force	Newton	N	$kg \cdot m/s^2$
Pressure, stress	Pascal	Pa	$N/m^2 = kg/m \cdot s^2$
Energy, work	Joule	J	$N \cdot m = kg \cdot m^2/s^2$
Power, radiant flux	Watt	W	$J/s = kg \cdot m^2/s^3$
Electric charge	Coulomb	C	$A \cdot s$
Voltage, electric potential	Volt	V	$W/A = kg \cdot m^2/A \cdot s^3$
Capacitance	Farad	F	$C/V = s^4 A^2/m^2 kg$
Electric resistance	Ohm	Ω	$V/A = m^2 kg/s^3 A^2$
Conductance	Siemens or mho	S or Ω	$1/\Omega = s^3 A^2/m^2 kg$
Magnetic field	Tesla	T	$N/A \cdot m = kg/s^2 A$
Magnetic flux	Weber	Wb	$T \cdot m^2 = m^2 kg/s^2 A$
Inductance	Henry	H	$V \cdot s/A = m^2 kg/s^2 A^2$

Table A.4 Unit prefixes

Name	Symbol	Quantity
yotta	Y	10^{24}
zetta	Z	10^{21}
exa	E	10^{18}
peta	P	10^{15}
tera	T	10^{12}
giga	G	10^{9}
mega	M	10^{6}
kilo	k	10^{3}
hecto	h	10^{2}
deca	da	10
deci	d	10^{-1}
centi	c	10^{-2}
milli	m	10^{-3}
micro	μ	10^{-6}
nano	n	10^{-9}
pico	p	10^{-12}
fempto	f	10^{-15}
atto	a	10^{-18}
zepto	z	10^{-21}
yocto	y	10^{-24}

EXAMPLE A.1	Unit Prefixes

The output of a 125 million watt power station can be expressed as

$$125,000,000 \text{ W} \quad \text{or} \quad 125 \text{ MW}$$

An example of a tiny interval of time, common in high-performance electronics, can be expressed as

$$5.27 \times 10^{-13} \text{ s} \quad \text{or} \quad 0.527 \text{ ps}$$

■ **CLASS DISCUSSION ITEM A.2**
Common Use of SI Prefixes

For each of the prefixes listed in Table A.4, think of or research an example of a measurable physical quantity for which the prefix is commonly used to express the value.

A.1.2 Conversion Factors

English units are still common in engineering practice in the United States. Table A.5 lists several factors that help when converting between English and SI units.

Table A.5 Useful English to SI conversion factors

Physical quantity	English unit	SI unit
Length	1 in.	2.540 cm
	1 ft	0.3048 m
	1 mi (mile)	1.609 km
Mass	1 lbm (pound mass)	0.4536 kg
Force	1 lbf (pound force)	4.448 N
Temperature	Fahrenheit temperature (T_F)	$T_k = 5/9 \cdot (T_F - 32) + 273.15$
Pressure	1 lb/in^2 (psi)	6.895×10^3 Pa
	1 atm	1.013×10^5 Pa
Power	1 Btu/h	0.2929 W
	1 hp	745.7 W
Magnetic field	1 gauss	1.000×10^{-4} tesla

■ **CLASS DISCUSSION ITEM A.3**
Physical Feel for SI Units

To help gain a physical feel for SI units, it is helpful to consider and remember concrete examples for each of the units. For each of the following common physical items, list the appropriate SI unit and approximate value:

■ Length of a typical human foot
■ Length of a city block
■ Mass of a 2-liter bottle of soda pop
■ Mass of an average adult human body
■ Force required to pick up a 2-liter bottle of soda pop
■ Force exerted by an average adult human body on a scale
■ Human body temperature
■ Comfortable room temperature
■ Atmospheric pressure
■ Typical air pressure in a building's pneumatic system
■ Power dissipated by a typical incandescent lightbulb
■ Typical maximum power generated by an automobile

A.2 SIGNIFICANT FIGURES

Whenever we deal with numerical data, we need to be aware of precision, accuracy, and different ways to present the data. Also, in establishing a rational approach to making numerical calculations with measured values, we must present decimal numbers with the appropriate number of digits.

The significant digits or **significant figures** in a number are those known with certainty. A measured value represented by N digits consists of $N - 1$ significant digits that are certain and 1 digit that is estimated. For example, when reading a dial on a pressure gage one might record 4.85 Pa. Here the 4 and 8 are certain, but the 5 may be an interpolated value. Hence, an observer is involved in the estimation of significance. If a number is reported with leading zeros, the leading zeros are not significant since they are only used to fix the decimal place (see Example A.2).

Today, with the common use of digital computers for data processing, one must be aware of the fact that a 12-digit number resulting from a computer calculation may have only 3 significant digits! The rest may be meaningless.

EXAMPLE A.2 | Significant Figures

Note the number of significant figures and the corresponding significant digits for each of the numbers that follow:

Number	Number of significant figures	List of significant digits
50.1	3	5, 0, 1
0.0501	3	5, 0, 1
5.010	4	5, 0, 1, 0

EXAMPLE A.3 | Scientific Notation

Scientific notation is useful in clearly representing the number of significant figures. Here are the representations of the numbers in Example A.2:

Number using scientific notation	Number of significant figures
5.01×10^1	3
5.01×10^{-2}	3
5.010×10^0	4

Mathematical computations present another difficulty when the argument values contain different numbers of significant figures. We must be careful in rounding off and retaining the appropriate number of significant figures in computed results.

When you are required to round a number to N significant figures, dispose of all digits to the right of the Nth place. If the discarded part exceeds one half of the Nth digit, increase the Nth digit by 1. If it is less than one half of the Nth digit, leave it as it is. If it is exactly one half of the Nth digit, a common rule is to leave the Nth digit unchanged if it is even or increase it by 1 if it is odd. It is important to be consistent when you apply these rules.

When adding quantities, determine the number of significant figures to the right of the decimal place in the least accurate number. Then retain only one more decimal place in the remaining numbers by truncation. Now add the numbers and round off to the same number of decimal places as in the least accurate number. This process is demonstrated in Example A.4.

Addition and Significant Figures EXAMPLE A.4

We wish to add the following numbers:

$$5.0365$$
$$+ 1.04$$
$$+ 6.093 14$$

Since the least number of significant figures to the right of the decimal place in the least accurate number (1.04) is two, we truncate the other numbers to three decimal places and add:

$$5.036$$
$$+ 1.04$$
$$+ 6.093$$

The result of this addition is

$$12.169$$

Since the number of decimal places in the least accurate number is two, we round off the result to two decimal places, giving

$$12.17$$

When subtracting two quantities, round off the more accurate number to the number of decimal places in the least accurate number. Subtract and provide a result with the same number of decimal places as in the least accurate number, as demonstrated in Example A.5.

Subtraction and Significant Figures EXAMPLE A.5

We wish to subtract the following numbers:

$$8.593 20$$
$$- 1.04$$

(continued)

(continued)   Since the least accurate number (1.04) has two decimal places, we round off the other number also to two decimal places and subtract

$$8.59$$
$$-1.04$$

The result of this subtraction is

$$7.55$$

When multiplying and dividing numbers, round off the more accurate numbers to one more significant figure than the least accurate number. Compute and round the result to the same number of significant figures as in the least accurate number as demonstrated in Example A.6.

EXAMPLE A.6   ## Multiplication and Division and Significant Figures

Given the multiplication and division problem:

$$(1.03)(51.7946)(3.01)(695.01)(7001.59)$$

We round off the more accurate numbers to four significant figures (one more than the three in 1.03 and 3.01):

$$(1.03)(51.79)(3.01)(695.0)(7002.)$$

The result of this multiplication after retaining three significant figures is

$$0.0000330 = 3.30 \times 10^{-5}$$

A.3 STATISTICS

When we process sets of data obtained from experimental measurements, we must handle the data in a rational, systematic, and organized fashion. The field of **statistics** provides models and rules for doing this properly.

Often, we seek a number or a small set of numbers to represent or characterize a large set of data. The first step is to assess the range of the data by noting the minimum and maximum values known as the **extreme values.** We then look at how the data points are distributed between these extreme values. This distribution can be graphically represented using a **histogram,** which results from sorting the values into subranges and displaying the number of data points in each subrange. The histogram for the set of experimental data given in Table A.6 is illustrated in Figure A.1.

The number of data points falling into each subrange of the histogram is called the **frequency.** The histogram may approximate a specific shape such as a normal, skewed, bimodal, or uniform distribution as illustrated in Figure A.2. The normal distribution represents a typical statistical spread of data around an average (mean) value. The skewed distribution represents some weighting in the statistics to one side of the mean. The bimodal distribution represents the case where there may be two

Table A.6 Set of experimental data

Index	Value
1	25.5
2	42.1
3	36.4
4	32.1
5	15.6
6	38.6
7	55.3
8	29.1
9	32.1
10	34.0
11	35.0

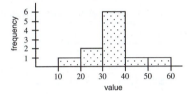

Figure A.1 Histogram of experimental data.

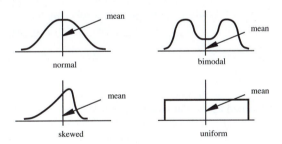

Figure A.2 Distributions of data.

merged populations with two different means. The uniform distribution represents completely random data.

Many data points are required to create a histogram. Statistics provides rules for distilling the histogram down to just a few numbers representing the characteristics of the data set. The most important statistical measure is the **arithmetic mean,** which is also called the **average** or simply the mean. Denoted by $\bar{x}$ it is the sum of each of the data values x_i divided by the number of data points N:

$$\bar{x} = \frac{\sum\limits_{i=1}^{} x_i}{N} \tag{A.2}$$

Other statistical measures that characterize a set of data are the **median,** which is the data point that has an equal number of data points on both sides; the **mode,** which is the value that occurs most frequently; and the **geometric mean,** defined as the Nth root of the product of the values:

$$GM = \sqrt[N]{x_1 x_2 \ldots x_N} \tag{A.3}$$

The geometric mean is more desirable than the arithmetic mean for averaging ratios since the reciprocal of the GM is equal to the GM of the reciprocals.

For the set of data in Table A.6, the mean is 34.2, the median is 34.0, the mode is 32.1, and the geometric mean is 32.7.

■ **CLASS DISCUSSION ITEM A.4**
Statistical Calculations

Verify the calculations for the mean, median, mode, and geometric mean for the data in Table A.6.

■ **CLASS DISCUSSION ITEM A.5**
Your Class Age Histogram

In one of your class lectures, form a single line in order of birth date with the youngest being first. Now form a birth year histogram by assembling into rows according to birth year. Store the histogram data (year, frequency) for use in Question A.2 at the end of the chapter.

The spread or dispersion of a data set over its range is characterized by another statistical measure, known as the **variance,** defined by

$$v = \sigma^2 = \sum_{i=1}^{N} \frac{(x_i - \bar{x})^2}{N - 1} \tag{A.4}$$

where x_i is an individual measurement and N is the total number of measurements, called the **sample size** of the experiment. The **standard deviation** σ also describes this distribution, but in the units of the individual measurements. It is defined as the square root of the variance:

$$\sigma = \sqrt{v} = \sqrt{\sum_{i=1}^{N} \frac{(x_i - \bar{x})^2}{N - 1}} \tag{A.5}$$

The standard deviation estimates the magnitude of the spread of the experimental data around the mean value. A small standard deviation indicates that the data set has a narrow spread.

A.4 ERROR ANALYSIS

The process of making measurements is imperfect, and uncertainty will always be associated with measured values. It is important to recognize sources of error and estimate the magnitude of error when one makes a measurement. Usually a manufacturer defines the accuracy of an instrument in published specifications, but other factors come into play.

There are three types of errors: systematic errors, random errors, and blunders. A **systematic error** is one that reoccurs in the same way each time a measurement is made. The method used to minimize the magnitude of systematic error is **calibration,** where the measurement instrument is used to record values from a standard input and is adjusted to compensate for any discrepancy. **Random errors** occur due to the stochastic variations in a measurement process. Some of the statistical tools presented in the previous section enable us to reduce the effects of these errors. **Blunders** occur when the engineer or scientist makes a mistake. Blunders can be avoided by careful design and review and through the use of methodical procedures.

Figure A.3 illustrates systematic and random errors. The center of the target represents the desired value, and the shot pattern represents measured data. The systematic error, called inaccuracy, is associated with the shift of the shot pattern from the center of the target and could be corrected by improved sighting, known as

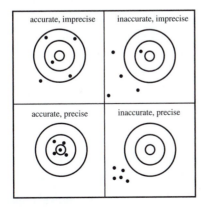

Figure A.3 Accuracy and precision.

calibration. The random error, called imprecision, is the size of the shot pattern and cannot be improved by adjusting the sighting. **Accuracy** is the closeness to the true value, and **precision** is the repeatability or consistency of the measurements.

Statistical calculations help us estimate a more precise value when a sample of imprecise measurements is taken in the presence of random errors. The average, or mean, provides this estimate.

A.4.1 Rules for Estimating Errors

When designing a measurement protocol to compute a parameter defined in terms of measured variables, it is necessary to estimate the error in the parameter due to the combined errors in the variables. A procedure for computing this overall error follows:

1. Prepare a table of data including the $\pm$ error estimate for each variable. Generally, the estimated error contains no more than two significant figures.

2. If the parameter to be computed is X, where X is a function of measured variables (v_i),

$$X = X(v_1, v_2, \ldots, v_n) \tag{A.6}$$

compute the partial derivatives $\partial X/\partial v_1, \partial X/\partial v_2, \ldots, \partial X/\partial v_n$ and evaluate each to three significant figures using the recorded data $(v_1, v_2, \ldots, v_n)$.

3. Compute the total absolute error E using

$$E = \Delta X = \left| \frac{\partial X}{\partial v_1} \Delta v_1 \right| + \left| \frac{\partial X}{\partial v_2} \Delta v_2 \right| + \cdots + \left| \frac{\partial X}{\partial v_n} \Delta v_n \right| \tag{A.7}$$

where Δv_i is the error in the recorded value v_i. Round E to two significant figures.

A more conventional error measure is the **root-mean-square (rms)** error given by the square root of the sum of the squares of the individual error terms:

$$E_{rms} = \sqrt{\left(\frac{\partial X}{\partial v_1} \Delta v_1 \right)^2 + \left(\frac{\partial X}{\partial v_2} \Delta v_2 \right)^2 + \cdots + \left(\frac{\partial X}{\partial v_n} \Delta v_n \right)^2} \tag{A.8}$$

The rms error measure yields a closer approximation to the actual error.

4. Compute X using Equation A.6 to one more decimal place than the rounded error E. For example, if $E = \pm 0.039$, and X is computed to be 8.9234, then this value is rounded to the same number of decimal places as E, giving 8.923. When computing X, treat $v_1, v_2, \ldots, v_n$ as exact numbers.

5. The result is

$$X = 8.923 \pm 0.039$$

The following list is a summary of important points to keep in mind when treating numbers and calculating errors in laboratory data analysis:

1. Note the number of significant figures that a particular instrument displays.

2. Record all values with the correct number of significant digits. If the instrument display does not provide a digital output, this may require that the observer estimate the accuracy of the measurement.

3. Different instruments in a system may have different numbers of significant digits.

4. When computing results using equations, retain the appropriate number of significant digits when performing calculations.

5. If events can be repeated, better estimates of values can be obtained by averaging. The standard deviation of these samples gives a measure of precision in the average.

QUESTIONS AND EXERCISES

A.1. Express each of the following quantities with a more appropriate SI prefix equivalent:

   a. 100,000,000 kg

   b. 0.000000025 m

   c. 16.9×10^{-10} s

A.2. Plot the histogram and calculate the mean, standard deviation, median, mode, and geometric mean for the data obtained from Class Discussion Item A.5.

A.3. What are the total absolute and rms errors in the calculated maximum stress of a rectangular cantilever beam with an end load of $12,520 \pm 10$ N? The beam geometry, as measured by a metric ruler accurate to 0.5 mm, is 0.95 m long by 11.8 cm wide by 12.1 cm high. The maximum stress occurs at the wall on the surface of the beam and is given by

$$\sigma_{max} = \frac{Mc}{I}$$

where M is the bending moment given by the product of the force and the beam length, c is the distance from the neutral axis given by half the height, and I is the area moment of inertia of the beam cross section given by

$$I = \frac{1}{12}wh^3$$

where w and h are the width and height of the beam cross section, respectively.

BIBLIOGRAPHY

Beckwith, T., Marangoni, R., and Lienhard, J., *Mechanical Measurements,* Addison-Wesley, Reading, MA, 1993.

Chapra, S. and Canale, R., *Introduction to Computing for Engineers,* McGraw-Hill, New York, 1994.

Chatfield, C., *Statistics for Technology,* Penguin Books, Middlesex, England, 1970.

Doeblin, E., *Measurement Systems Applications and Design,* 4th Edition, McGraw-Hill, New York, 1990.

B APPENDIX

Physical Principles

APPENDIX OBJECTIVES

After you read, discuss, study, and apply ideas in this appendix, you will be able to:

1. Identify possible relationships between various physical quantities

2. Identify approaches for measuring nearly all physical quantities

Sensor and transducer design always involves the application of some law or principle of physics or chemistry that relates the variable of interest to some measurable quantity. The following list summarizes many of the physical laws and principles that have potential application in sensor and transducer design. Some examples of applications are also provided. This list is extremely useful to a transducer designer who is searching for a method to measure a physical quantity. Practically every transducer applies one or more of these principles in its operation. The parameters related by the respective principles are highlighted.

- *Ampere's law:* The integral of the **magnetic field** around a closed loop is proportional to the **current** piercing the loop.

    A magnetic pickup sensor uses this effect as a nonintrusive method of measuring current in a conductor.

- *Archimedes' principle:* The buoyant **force** exerted on a submerged or floating object is equal to the weight of the fluid displaced. The **volume** displaced depends on the fluid **density.**

    A ball submersion hydrometer uses this effect to measure the density of a fluid (e.g., automotive coolant).

- *Bernoulli's equation:* Conservation of energy in a fluid predicts a relationship between **pressure** and **velocity** of the fluid.

    A pitot tube uses this effect to measure air speed of an aircraft. Video Demo B.1 shows an example of how to relate pressure readings to flow velocity.

Video Demo

B.1 Flow over a cylinder in a wind tunnel

- *Biot-Savart's law:* The contribution of a **current** element to a **magnetic field** at a point depends on the distance to the current element and the current direction.
- *Biot's law:* The rate of **heat conduction** through a medium is directly proportional to the **temperature** difference across the medium.

   This principle is basic to time constants associated with temperature transducers.
- *Blagdeno law:* The freezing **temperature** of a liquid drops and the boiling temperature rises with **concentration** of impurities in the liquid.
- *Boyle's law:* An ideal gas maintains a constant **pressure-volume** product with constant **temperature.**
- *Bragg's law:* The intensity of an X-ray beam diffracted by a **crystal lattice** is related to the crystal plane separation and the **wavelength** of the beam.

   An X-ray diffraction system uses this effect to measure the crystal lattice geometry of a crystalline specimen.
- *Brewster's law:* The **index of refraction** of a material is related to the angle of **polarized light** reflection or transmission.

   A Brewster's window on a laser tube is used to extract some of the power in the form of a laser beam. Lasers are used extensively in measurement systems.
- *Butterfly effect:* Chaotic nonlinear systems exhibit a sensitive dependence on initial conditions.
- *Centrifugal force:* A body moving along a curved path experiences an apparent outward **force** in line with the radius of curvature.
- *Charles' law:* An ideal gas maintains a constant **pressure-temperature** product with constant **volume.**
- *Christiansen effect:* Powders suspended in a liquid (i.e., a colloidal solution) result in altered fluid **refraction** properties.
- *Corbino effect:* **Current** flow is induced in a conducting disk rotating in a **magnetic field.**
- *Coriolis effect:* A body moving relative to a rotating frame of reference (e.g., the earth) experiences a **force** relative to the frame (see Internet Link B.1).

   A Coriolis flow meter uses this effect to measure mass flow rate in a u-tube in rotational vibration.
- *Coulomb's law:* **Electric charges** exert a **force** between each other.
- *Curie-Weiss law:* There is a transition **temperature** at which ferromagnetic materials exhibit **paramagnetic** behavior.
- *d'Alembert's principle:* **Acceleration** of a **mass** is equivalent to an equal and opposite applied **force.**
- *Debye frequency effect:* The **conductance** of an electrolyte increases (i.e., the **resistance** decreases) with **frequency.**
- *Doppler effect:* The **frequency** received from a wave source (e.g., sound or light) depends on the **speed** of the source.

Internet Link

B.1 Coriolis effect video demonstrations

A laser doppler velocimeter (LDV) uses the frequency shift of laser light reflected off of particles suspended in a fluid to measure fluid velocity.

■ *Edison effect:*   When metal is heated in a vacuum, it emits charged particles (i.e., **thermionic emission**) at a rate dependent on **temperature.**

A vacuum tube amplifier is based on this effect, where electrons are emitted and controlled to produce amplification of current.

■ *Faraday's law of electrolysis:*   The rate of **ion deposition** or depletion is proportional to the electrolytic **current.**

■ *Faraday's law of induction:*   A coil resists a change in **magnetic field** linkage with an **electromotive force.**

The induced voltages in the secondary coils of a linear variable differential transformer (LVDT) are a result of this effect.

■ *Gauss effect:*   The **resistance** of a conductor increases when **magnetized.**

■ *Gladstone-Dale law:*   The **index of refraction** of a substance is dependent on **density.**

■ *Gyroscopic effect:*   A body rotating about one axis resists rotation about other axes (see Internet Link B.2).

A navigation gyroscope uses this effect to track the orientation of a body with the aid of a gimbal-mounted flywheel that maintains constant orientation in space.

■ *Hall effect:*   A **voltage** is generated perpendicular to **current** flow in a **magnetic field.**

A Hall effect proximity sensor detects when a magnetic field changes due to the presence of a metallic object.

■ *Hertz effect:*   **Ultraviolet light** affects the discharge of a spark across a gap.

■ *Hooke's Law:*   Axial **stress** in a uniaxially loaded, linear elastic material is directly proportional to axial **strain.**

Resistance measurements from a strain gage can be converted to strain readings, which can be directly related to stresses in a loaded part.

■ *Johnsen-Rahbek effect:*   **Friction** at interfaces between a conductor, semiconductor, or insulator increases with **voltage** across the interfaces.

■ *Joule's law:*   **Heat** is produced by **current** flowing through a **resistor.**

The design of a hot-wire anemometer is based on this principle.

■ *Kerr effect:*   Applying a **voltage** across a substance can cause **optical polarization.**

Liquid crystal displays (LCDs) function as a result of this principle.

■ *Kohlrausch's law:*   An **electrolytic** substance has a limiting conductance (minimum **resistance**).

■ *Lambert's cosine law:*   The reflected **luminance** of a surface varies with the cosine of the **angle of incidence.**

■ *Lenz's law:*   Induced **current** flows in the direction to oppose the change in **magnetic field** that produces it.

Internet Link

B.2 Gyroscopic effect video demonstrations

- *Lorentz's force law:* A **current**-carrying coductor in a **magnetic field** experiences a **force.**

  Based on this law, a galvanometer measures current by measuring the deflection of a pivoted coil in a permanent magnetic field.

- *Magnetostrictive effect:* a ferromagnetic material **constricts** when surrounded by a **magnetic field.**

  This effect is used by magnetoresistive linear displacement sensors (see Video Demo B.2).

- *Magneto-rheological effect:* A magneto-rheological fluid's **viscosity** can increase dramatically in the presence of a **magnetic field.**

- *Magnus effect:* When fluid flows over a rotating body, the body experiences a **force** in a direction perpendicular to the flow.

- *Meissner effect:* A **superconducting** material within a **magnetic field** blocks this field and experiences no internal field.

- *Moore's law:* The density of transistors that can be manufactured on an integrated circuit doubles every 18 months.

- *Murphy's law:* Whatever can go wrong will go wrong and at the wrong time and in the wrong place.

  Your experiments in the laboratory will often demonstrate this law.

- *Nernst effect:* **Heat flow** across **magnetic field** lines produces a **voltage.**

- *Newton's law:* **Acceleration** of an object is proportional to **force** acting on the object.

- *Ohm's law:* **Current** through a **resistor** is proportional to the **voltage** drop across the resistor.

- *Parkinson's law:* Human work expands to fill the time allotted for it.

- *Peltier effect:* When **current** flows through the junction between two metals, **heat** is absorbed or liberated at the junction.

  Thermocouple measurements can be adversely affected by this principle.

- *Photoconductive effect:* When **light** strikes certain semiconductor materials, the **resistance** of the material decreases.

  A photodiode, which is used extensively in photodetector pairs, functions based on this effect.

- *Photoelectric effect:* When **light** strikes a metal cathode, electrons are emitted and attracted to an anode, resulting in **current** flow.

  The operation of a photomultiplier tube is based on this effect.

- *Photovoltaic effect:* When **light** strikes a semiconductor in contact with a metal base, a **voltage** is produced.

  The operation of a solar cell is based on this effect.

- *Piezoelectric effect:* **Charge** is displaced across a crystal when it is strained.

  A piezoelectric accelerometer measures charge polarization across a piezoelectric crystal subject to deformations due to the inertia of a mass.

Video Demo

B.2 Magneto-restrictive position sensor

A piezoelectric microphone's ability to convert sound pressure waves to a voltage signal is a result of this principle.

- *Piezoresistive effect:* **Resistance** is proportional to an applied **stress.**

   This effect is partially responsible for the response of a strain gage.

- *Pinch effect:* The cross section of a liquid conductor reduces with **current.**

- *Poisson effect:* A material deforms in a direction perpendicular to an applied **stress.**

   This effect is partially responsible for the response of a strain gage.

- *Pyroelectric effect:* A crystal becomes **polarized** when its **temperature** changes.

- *Raleigh criteria:* Relates the **acceleration** of a fluid to bubble formation.

- *Raoult's effect:* **Resistance** of a conductor changes when its length is changed.

   This effect is partially responsible for the response of a strain gage.

- *Seebeck effect:* Dissimilar metals in contact result in a **voltage** difference across the junction that depends on **temperature.**

   This is the primary effect that explains the function of a thermocouple.

- *Shape memory effect:* A deformed metal, when heated, returns to its original shape (see Video Demo B.3).

- *Snell's law:* Reflected and refracted rays of **light** at an optical interface are related to the angle of incidence.

- *Stark effect:* The **spectral lines** of an electromagnetic source split when the source is in a strong **electric field.**

- *Stefan-Boltzmann law:* The **heat** radiated from a black body is proportional to the fourth power of its **temperature.**

   The design of a pyrometer is based on this principle.

- *Stokes' law:* The **wavelength** of light emitted from a fluorescent material is always longer than that of the absorbed photons.

- *Tribo-electric effect:* Relative motion and **friction** between two dissimilar metals produces a **voltage** between the interface.

- *Wiedemann-Franz law:* The ratio of **thermal** to **electrical conductivity** of a material is proportional to its absolute **temperature.**

- *Wien effect:* The **conductance** of an electrolyte increases (i.e., the **resistance** decreases) with applied **voltage.**

- *Wien's displacement law:* As the **temperature** of an incandescent material increases, the spectrum of emitted **light** shifts toward blue.

Video Demo

B.3 Shape-memory alloy orthodontic wire

Mechanics of Materials

APPENDIX OBJECTIVES

After you read, discuss, study, and apply ideas in this appendix, you will:

1. Understand the basic relationships between stress and strain

2. Be able to determine the principal stress values and directions for a general state of planar stress

3. Be able to construct Mohr's circle for a state of planar stress

C.1 STRESS AND STRAIN RELATIONS

As shown in Figure C.1, when a cylindrical rod is loaded axially, it will lengthen by an amount ΔL and deform radially by an amount ΔD. The **axial strain** (ε_{axial}) is defined by the change in length per unit length:

$$\varepsilon_{axial} = \frac{\Delta L}{L} \tag{C.1}$$

Note that strain is a dimensionless quantity. The **axial stress** (σ_{axial}) is related to axial strain through **Hooke's law,** which states that for a uniaxially loaded linear elastic material the axial stress is directly proportional to the axial strain:

$$\sigma_{axial} = E\varepsilon_{axial} \tag{C.2}$$

where E is the constant of proportionality called the **modulus of elasticity** or **Young's modulus.** The axial stress in the rod is

$$\sigma_{axial} = F/A \tag{C.3}$$

where F is the axial force and A is the cross-sectional area of the rod. Therefore, the axial strain is related to the axial stress and load:

$$\varepsilon_{axial} = \frac{\sigma_{axial}}{E} = \frac{F/A}{E} \tag{C.4}$$

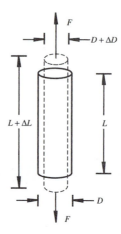

Figure C.1 Axial and transverse deformation of a cylindrical bar.

The **transverse strain** is defined as the change in width divided by the original width:

$$\varepsilon_{\text{transverse}} = \frac{\Delta D}{D} \qquad (\text{C.5})$$

The ratio of the transverse and axial strain is defined as **Poisson's ratio** (ν):

$$\nu = -\frac{\varepsilon_{\text{transverse}}}{\varepsilon_{\text{axial}}} \qquad (\text{C.6})$$

Note that for axial elongation ($\varepsilon_{\text{axial}} > 0$), $\varepsilon_{\text{transverse}}$ (from Equation C.6), and therefore ΔD (from Equation C.5) are negative, implying contraction in the transverse radial direction. Poisson's ratio for most metals is approximately 0.3, implying the transverse strain is -30% of the axial strain.

A general state of planar stress at a point, acting on an infinitesimal square element, is illustrated in Figure C.2a. It includes two normal stress components (σ_x and σ_y) and a shear stress component (τ_{xy}) whose values depend on the orientation of the element. At any point, there is always an orientation of the element that results in the maximum normal stress magnitude and zero shear stress ($\tau_{xy} = 0$). The two orthogonal normal stress directions corresponding to this orientation are called the **principal axes,** and the normal stress magnitudes are referred to as the **principal stresses** (σ_{max} and σ_{min}). Figure C.2b illustrates this orientation and its corresponding state of stress. The magnitude and direction of the principal stresses are related to the stresses in any other orientation by

$$\sigma_{\text{max}} = \left(\frac{\sigma_x + \sigma_y}{2}\right) + \sqrt{\left(\frac{\sigma_x - \sigma_y}{2}\right)^2 + \tau_{xy}^2} \qquad (\text{C.7})$$

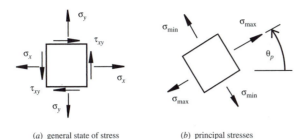

(a) general state of stress   (b) principal stresses

Figure C.2 General state of planar stress and principal stresses.

$$\sigma_{min} = \left(\frac{\sigma_x + \sigma_y}{2}\right) - \sqrt{\left(\frac{\sigma_x - \sigma_y}{2}\right)^2 + \tau_{xy}^2} \tag{C.8}$$

$$\tan(2\theta_p) = \frac{2\tau_{xy}}{\sigma_x - \sigma_y} \tag{C.9}$$

where θ_p is the angle from σ_x to σ_{max}, measured counterclockwise.

The principal stresses are important quantities when determining if a material will yield or fail when loaded because they determine the maximum values of stress, which can be compared to the yield strength of the material. The maximum shear stress is also important when assessing failure and is given by

$$\tau_{max} = \sqrt{\left(\frac{\sigma_x - \sigma_y}{2}\right)^2 + \tau_{xy}^2} = \frac{\sigma_{max} - \sigma_{min}}{2} \tag{C.10}$$

This relation can be used to rewrite Equations C.7 and C.8 as

$$\sigma_{max} = \sigma_{avg} + \tau_{max} \tag{C.11}$$

$$\sigma_{min} = \sigma_{avg} - \tau_{max} \tag{C.12}$$

where

$$\sigma_{avg} = \frac{\sigma_x + \sigma_y}{2} \tag{C.13}$$

The orientation of the element that results in τ_{max} is given by

$$\tan(2\theta_s) = -\frac{\sigma_x - \sigma_y}{2\tau_{xy}} \tag{C.14}$$

As with θ_p, θ_s is measured counterclockwise from the direction of σ_x. For the cylindrical bar in Figure C.1, with an element oriented in the axial (y) direction, $\sigma_{max} = \sigma_y = F/A$, $\sigma_x = 0$, and $\theta_p = 0$ because the element is aligned in the direction of the principal stress. Also, $\theta_s = 45°$ and $\tau_{max} = \sigma_y/2 = F/2A$.

The state of stress and its relation to the magnitude and direction of the principal stresses are often illustrated with **Mohr's circle,** which displays the relationship between the shear stress and the normal stresses in different directions (see Figure C.3).

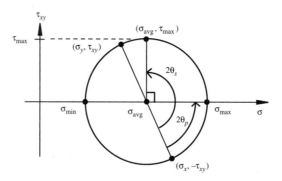

Figure C.3 Mohr's circle of plane stresses.

Internet Link

C.1 Mohr's circle equation derivation for uniaxial stress

Video Demo

C.1 Mohr's circle for uniaxial stress

C.2 Failure theories for brittle and ductile materials

Remember, tensile normal stresses are positive and compressive normal stresses are negative. For the example shown in Figure C.3, corresponding to the element shown in Figure C.2, both normal stresses are tensile. The sign of the shear stress is positive when it would cause the element to rotate clockwise about its center and negative when it would cause the element to rotate counterclockwise. For the element in Figure C.2, τ_{xy} is negative on the σ_x side of the element since it would cause the element to rotate counterclockwise, and τ_{xy} is positive on the σ_y side for the opposite reason. Note that the angle between the original stress directions and the principal stresses (θ_p) is measured in the same direction around the circle as with the actual element, but angles on the circle are twice the actual angles ($2\theta_p$). Since θ_p is measured counterclockwise from σ_x to σ_{max} in Figure C.2, the angle between the σ_x point and σ_{max} is $2\theta_p$ counterclockwise in Figure C.3. Also note that the orientation of the principal stresses and the orientation of the maximum shear stress are always 90° apart on Mohr's circle (45° apart on the actual element). This is confirmed by the fact that $\tan(2\theta_p)$ and $\tan(2\theta_s)$ are negative reciprocals of one another (see Equations C.9 and C.14). For more information, Internet Link C.1 points to a derivation of the equation for Mohr's circle for uniaxial stress, and Video Demo C.1 discusses and illustrates the results. Video Demo C.2 discusses how Mohr's circle can help one understand why brittle and ductile materials exhibit different fracture plans when they break.

■ **CLASS DISCUSSION ITEM C.1**
Fracture Plane Orientation in a Tensile Failure

When a metal bar fails under axial tension, the resulting fracture planes are oriented at 45° with respect to the bar's axis. Why?

BIBLIOGRAPHY

Beer, F. and Johnston, E., *Mechanics of Materials,* 5th Edition, McGraw-Hill, New York, 2008.

Dally, J. and Riley, W., *Experimental Stress Analysis,* 3rd Edition, McGraw-Hill, New York, 1991.

INDEX